CONT... BUSINESS STATISTICS

WITH CANADIAN APPLICATIONS

S. A. Hummelbrunner

John Gray

Len Rak

Prentice Hall Canada Inc.

Scarborough, Ontario

The authors wish to thank
Minitab Inc.
3081 Enterprise Drive,
State College, PA 16801 USA.
Telephone: (814) 238 3280 Fax (814) 238 4383
for their cooperation during the preparation of this text.
MINITAB is a registered trademark of Minitab Inc.
SPSS/PC+ is a registered trademark of SPSS Inc.
STATGRAPHICS is a registered trademark of Statistical
Graphics Corporation
SYSTAT and SYGRAPH are registered trademarks of SYSTAT, Inc.

Canadian Cataloguing in Publication Data

Hummelbrunner S. A. (Siegfried August)
Contemporary business statistics with Canadian applications

ISBN 0–13–184680–9

1. Statistics.   2. Commercial statistics.
3. Canada — Commerce — Statistics.   I. Gray, John, 1926- .
II. Rak, Len, 1957- .   III. Title.

HA29.H84 1993   519.5   C92–095649–1

Prentice Hall, Inc., Englewood Cliffs, New Jersey
Prentice-Hall International, Inc., London
Prentice-Hall of Australia, Pty., Ltd., Sydney
Prentice-Hall of India Pvt., Ltd., New Delhi
Prentice-Hall of Japan, Inc., Tokyo
Prentice-Hall of Southeast Asia (Pte.) Ltd., Singapore
Editora Prentice-Hall do Brasil Ltda., Rio de Janeiro
Prentice-Hall Hispanoamericana, S.A., Mexico

ISBN: 0–13–184680–9

Acquisitions Editor: Jacqueline Wood
Developmental Editor: Linda Gorman
Copy Editor: James Leahy
Production Editor: Elizabeth Long
Production Coordinator: Florence Rousseau
Cover Design: Pronk&Associates
Page Layout: Pronk&Associates

1 2 3 4 5 RRD 97 96 95 94 93

Printed and bound in USA by R.R. Donnelley & Sons,

Contents

Preface

Contemporary Business Statistics is intended for use in introductory statistics courses in business administration programs. In more general applications it also provides a comprehensive basis for those who wish to review and extend their understanding of business statistics.

The primary objective of the text is to introduce the student to statistical concepts through the solution of practical problems encountered in the business community. It also provides a supportive basis for topics in marketing, finance and accounting.

Contemporary Business Statistics is essentially a teaching text using the objectives approach. The systematic and sequential development of the material is supported by carefully selected and worked examples. These detailed, step-by-step solutions presented in a clear and uncluttered layout are particularly helpful in allowing students, in either the classroom setting or independent studies, to carefully monitor their own progress.

The first six chapters provide topics of descriptive statistics sequenced in an intuitive and appealing order for students. The remaining seven chapters deal with subject matter relating to inferential statistics. Each chapter is divided into seven parts: an Introduction, set of Objectives, Sections, Exercises, Review Exercises, Self-Test and Summary.

The Introduction gives a few simple and compelling reasons for the student to take the time and energy to learn this "stuff."

The Objectives act as a target and checklist of skills and concepts the student should possess after completing the chapter. As with any new body of knowledge, the student of statistics requires a complete understanding of the material before proceeding on to new topics.

There are from three to seven sections per chapter, with each section covering a major topic. The length of each section varies, depending on its complexity and the authors' depth of coverage.

An Exercise follows most sections, allowing the student to practise and develop the skills covered in the section. Odd- numbered exercises have solutions listed in the back of the text. Any difficulties with the exercises means the student should go back and review the applicable section before proceeding.

Exercises with a computer disk icon beside it can only be solved using the data enclosed with the text on computer disk and a statistical software package such as MYSTAT. These data disk exercises are large data sets using real data from or about Canadian corporations.

The Review Exercises are found at the end of each chapter and are designed to assist in the integration of all the material studied. Odd-numbered questions have solutions listed in the back of the text. Questions with a microcomputer icon beside it can be solved manually or using a statistical software package such as MYSTAT.

The Self-Test is a typical test used to evaluate the exit competencies after each chapter. The test usually takes one to two hours to complete, and students can check all their numerical solutions to the answers in the back of the text.

The Summary is found at the end of each chapter and gives a listing of all key terms and their location along with a summary of all the formulas used.

The use of computers has had a tremendous impact on Canadian Business especially in the area of statistical applications. Our goal is to expose the student to statistical software in easy-to- understand ways. Examples of graphical output from three major statistical packages have been used in Chapters 1, 2, 6, and 13 to familiarize students with such output as distinct from hand-prepared illustrations that are also used in the text. An Appendix provides some examples of how one such package (MINITAB) can be used to reduce the mechanical computational work involved in statistical procedures. A data disk has been enclosed containing files in two formats (MYSTAT and ASCII) so that any software package may be used in conjunction with the text.

SIEG HUMMELBRUNNER
JOHN GRAY
LEN RAK

September 1992

CHAPTER 1

Organizing and Presenting Data

Introduction

Statistical analysis is concerned with procedures by which data are collected, organized, presented, analyzed and interpreted. Raw data collected for the purpose of analysis must first be organized. Tables and graphs are widely used to display statistical data.

The purpose of a graph is to allow the reader to see the relationships in the data without intensive study. To obtain the desired impact, care must be taken to select the appropriate type of graph and to present the graph in a clear, readable manner.

Objectives

Upon completion of this chapter you will be able to
1. construct scatter diagrams from given sets of paired data;
2. construct single line graphs and multiple line graphs;
3. construct simple bar graphs and multiple bar graphs;
4. construct component line graphs and component bar graphs;
5. construct pie (circle) graphs;
6. construct bi- (two-) directional bar graphs;
7. construct high-low-close graphs.

SECTION 1.1 Looking at Data

Statistics is concerned with *collecting, organizing, summarizing, analyzing* and *presenting* data to assist in decision making.

Data collected in tabular form can be presented graphically in their entirety or in part by various methods to provide a pictorial indication of relationships without requiring intensive study of the data. In fact, a graphical illustration of data may be more revealing than the statistical measures that are commonly used to describe data.

To demonstrate this point let us look at a collection of four sets of data referred to as *Anscombe's Quartet* and presented in Table 1.1 below.

TABLE 1.1 Anscombe's Quartet

Set 1		Set 2		Set 3		Set 4	
x	y	x	y	x	y	x	y
12.0	10.84	12.0	9.13	12.0	8.15	8.0	5.56
4.0	4.26	4.0	3.10	4.0	5.39	19.0	12.50
9.0	8.81	9.0	8.77	9.0	7.11	8.0	8.84
7.0	4.82	7.0	7.26	7.0	6.42	8.0	7.91
14.0	9.96	14.0	8.10	14.0	8.84	8.0	7.04
6.0	7.24	6.0	6.13	6.0	6.08	8.0	5.25
10.0	8.04	10.0	9.14	10.0	7.46	8.0	6.58
11.0	8.33	11.0	9.26	11.0	7.81	8.0	8.47
8.0	6.95	8.0	8.14	8.0	6.77	8.0	5.76
13.0	7.58	13.0	8.74	13.0	12.74	8.0	7.71
5.0	5.68	5.0	4.74	5.0	5.73	8.0	6.89

Source: F.J. Anscombe, "Graphs in Statistical Analysis," *American Statistician*, 27 (February 1973).

○ **EXAMPLE 1.1a**

a) Review the four sets of data in Table 1.1 and identify similarities, oddities and relationships that you see in the data.

b) Plot the four sets of data in Table 1.1 on four graphs referred to as **scatter diagrams.**

● **SOLUTION**

a) When looking at the table you may have noticed that the X values for Sets 1, 2 and 3 are the same and take the whole number values 4, 5, ..., 14, while ten of the X values for Set 4 are the whole number 8. Otherwise, presentation of the data in the form of a table reveals very little about the data.

b) Representing the four sets of data in the form of scatter diagrams clearly indicates the differences and patterns in the data.

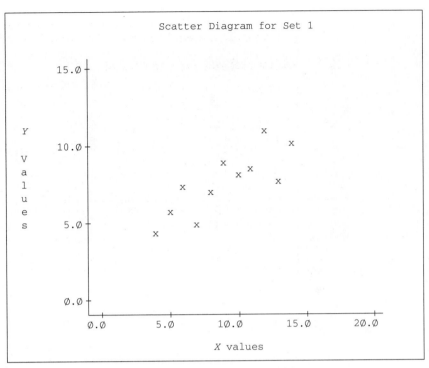

MINITAB®

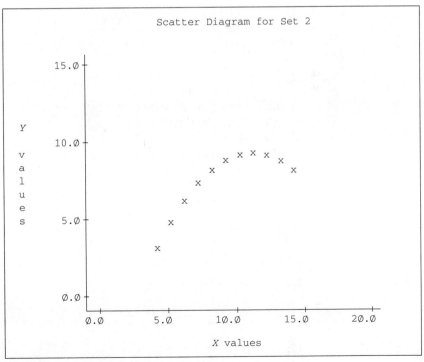

MINITAB®

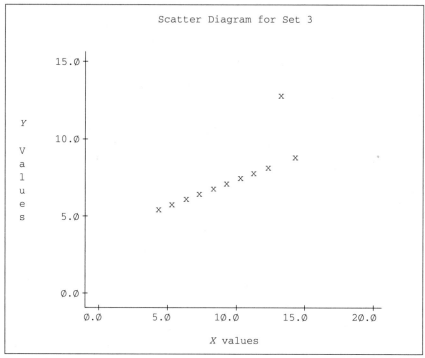

MINITAB®

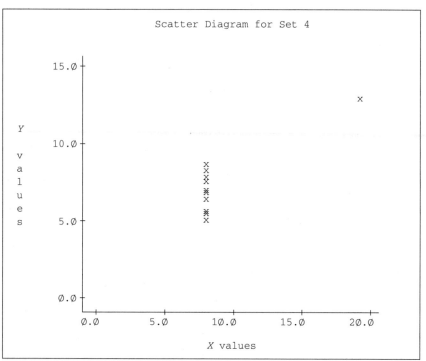

MINITAB®

EXERCISE 1.1

1. Plot the following set of numbers on a scatter diagram.

x	16	8	4	12	6	10	2	14	18
y	10	18	10	18	15	20	4	15	4

2. Use the horizontal axis of a scatter diagram for the years and the vertical axis for the associated values of the series of numbers.

Year	1983	1984	1985	1986	1987	1988	1989	1990
Value	15	13	11	9	7	5	3	1

3. *Data disk exercise* Using the enclosed data disk and the file name PUBLIC, plot a scatter diagram of company sales (the variable name is SALES) and interest expense (the variable name is INTEXP).

SECTION 1.2 ## Presenting Data—Graphs

Scatter diagrams are a particular method of presenting certain types of data. The most commonly used types of graphs are line graphs, bar graphs and specialty graphs such as component bar graphs, component line graphs, pie (circle) graphs, bi- (two-) directional graphs and high-low-close graphs.

Table 1.2 is a collection of data providing information about sales of North American cars for the period 1976 to 1986. Data from this table are used to demonstrate the construction of most of the common types of graphs referred to in the previous paragraph.

TABLE 1.2 Sales of North American Cars (% of Total Sales)

Year	Sub-compact Canada	Sub-compact U.S.	Compact Canada	Compact U.S.	Intermediate Canada	Intermediate U.S.	Full-size Canada	Full-size U.S.	Luxury Canada	Luxury U.S.
1976	8.9	12.1	30.9	28.3	31.4	33.1	27.2	22.1	1.6	4.4
1977	7.0	10.9	30.8	26.0	33.4	33.1	26.9	25.0	1.9	5.0
1978	11.8	13.2	30.4	24.3	32.3	32.8	23.4	23.3	2.0	6.4
1979	17.7	21.4	27.5	23.5	28.2	28.4	23.6	20.8	3.1	5.9
1980	18.9	25.4	30.9	25.5	27.8	27.9	20.0	16.4	2.5	4.9
1981	21.4	26.8	31.1	24.5	28.9	28.1	16.5	15.3	2.0	5.3
1982	32.4	30.2	25.8	19.2	30.0	28.1	10.4	16.1	1.4	6.4
1983	33.6	30.2	22.0	13.6	32.2	33.1	10.8	17.0	1.4	6.4
1984	33.0	29.0	25.0	16.5	29.0	30.9	11.1	15.5	1.9	8.1
1985	30.2	15.8	25.9	31.2	31.5	30.0	10.2	13.1	2.1	9.8
1986	n/a	16.1	n/a	30.0	n/a	31.0	n/a	13.6	n/a	9.3

Source: Canadian Motor Vehicle Manufacturers' Association, *Ward's Automotive Report*.

SECTION 1.3 Line Graphs

Line graphs are used primarily to present changes in data over a period of time. The values are represented by points that are joined by a line. The line starts at the first point and ends at the last. A single line graph results if one set of data is presented.

If two or more sets of related data are presented on the same chart, a **multiple line graph** can be constructed. In such cases, each line needs to be marked distinctly.

○ **EXAMPLE 1.3a**

Use the data in Table 1.2 to construct a line graph of North American luxury car sales in Canada and the U.S. for the period 1980 to 1985.

● **SOLUTION**

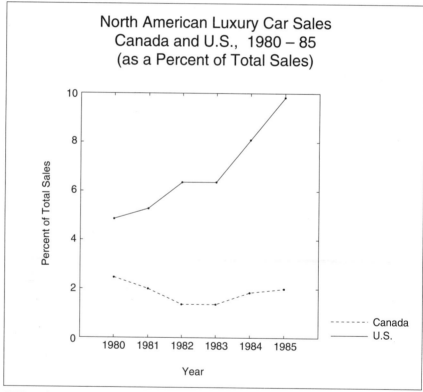

SYGRAPH® MODULE OF THE SYSTAT® PACKAGE

EXERCISE 1.3

1. The following data represent the average values of the U.S. dollar in Canadian cents for the period 1980 to 1991, reported on a yearly basis by the International Monetary Fund. Construct a single line graph to present the data.

Year	1980	1981	1982	1983	1984	1985
U.S. dollar in Cdn. cents	116.9	119.9	123.4	123.2	129.5	136.6

Year	1986	1987	1988	1989	1990	1991
U.S. dollar in Cdn. cents	139.0	132.6	123.1	118.4	116.7	113.6

2. Chrysler Corporation's car market share (in percents) in Canada for the period 1984 to 1990 was as follows:

Year	1984	1985	1986	1987	1988	1989	1990
Market share (%)	20.6	18.0	16.0	14.7	15.1	14.2	12.6

Construct a single line graph to represent the percents.

3. The operating results (in $000) for the A Company for the period 1983 to 1988 were as follows:

Year	1983	1984	1985	1986	1987	1988
Revenue	76	93	85	88	98	110
Cost of sales	35	45	40	45	55	60
Net income	20	22	18	15	17	23

Construct a multiple line graph to represent the data.

4. Promotional expenditures and net profit for Ace Inc. for the time period 1982 to 1987 (in $ millions) were as follows:

Year	1982	1983	1984	1985	1986	1987
Promotion	1.0	5.0	5.0	8.0	10.0	12.0
Net profit	7.2	8.2	10.0	15.6	18.0	15.0

Draw a multiple line graph to represent the data.

SECTION 1.4 Bar Graphs

A. Simple Bar Graphs

A **simple bar graph** shows the relationship among similar values by means of either *vertical* or *horizontal* bars. Vertical bar graphs are generally used when time is a factor, while horizontal bars are used for other sets of data.

In bar graphs the numerical scale should start at zero and should be devised to accommodate the magnitude of the numbers to be represented. Special consideration should be given to the width of the bars and any space separating the bars (usually about one-half the width of the bars).

○ **EXAMPLE 1.4a**

Use the data in Table 1.2 to construct a simple bar graph to portray Canadian car sales for 1980.

● **SOLUTION**

Because time is not a factor, a horizontal bar graph is appropriate. The vertical scale should be designed to accommodate the five types of cars, while the horizontal scale must allow for the maximum value of 30.9.

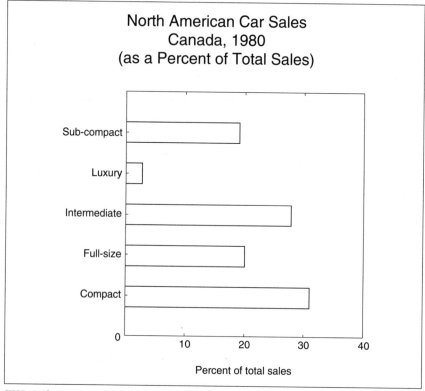

SYGRAPH® MODULE OF THE SYSTAT® PACKAGE

○ **EXAMPLE 1.4b**

Use the data in Table 1.2 to construct a bar graph of U.S. compact car sales for the time period 1980 to 1986.

● **SOLUTION**

Because the data to be portrayed cover a time period, a vertical bar graph should be used. The horizontal scale needs to accommodate the years 1980 to 1986, while the vertical scale must allow for a maximum value of 31.2.

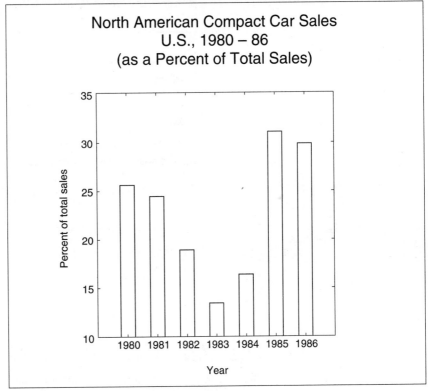

SYGRAPH® MODULE OF THE SYSTAT® PACKAGE

B. Multiple (Comparative) Bar Graphs

Multiple bar graphs are used to compare two or more sets of data.

○ **EXAMPLE 1.4c**

Use the data in Table 1.2 to construct a comparative bar graph of North American car sales in Canada and the U.S. in 1985.

● **SOLUTION**

Since time is not a factor, a horizontal bar graph is appropriate.

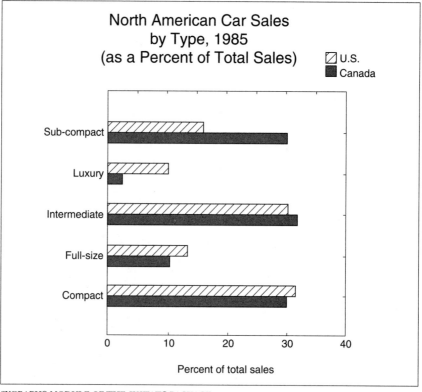

SYGRAPH® MODULE OF THE SYSTAT® PACKAGE

EXERCISE 1.4

1. The Quip Company had the following number of employees in its various departments at the end of last month: cutting, 30; machining, 45; folding, 24; shipping, 12. Represent the data on a bar graph.

2. Net sales (in $000) for Dreamers Inc. for the period 1980 to 1987 were as follows:

1980	1981	1982	1983	1984	1985	1986	1987
150	180	160	190	150	170	190	200

Represent the data on a bar graph.

3. Exports (in $ millions) for Company A and Company B from 1981 to 1987 were as follows:

Year	1981	1982	1983	1984	1985	1986	1987
Company A	20	16	12	15	10	7	8
Company B	2	5	8	10	13	17	19

Draw a multiple bar graph to represent the data.

4. Use the data in Table 1.2 to draw a multiple bar graph comparing North American car sales in Canada and the U.S., by type, for 1985.

SECTION 1.5 # Component Bar Graphs, Component Line Graphs, Pie Graphs

Component bar graphs, component line graphs and pie (circle) graphs are used to show the relationship among the parts of a total.

A. Component Bar Graphs

In a **component bar graph,** the area covered by the bar represents 100% of the total. Each component is assigned an area proportional to its percent of the total. If the data are not already in percent form, it is useful to compute what percent of the total each component represents and to obtain a running total of the percents.

○ **EXAMPLE 1.5a**

Use the data in Table 1.2 to construct a component bar graph of North American car sales in the U.S., by type, for 1982.

● **SOLUTION**

Type	%	Accumulated
Sub-compact	30.2	30.2
Compact	19.2	49.4
Intermediate	28.1	77.5
Full-size	16.1	93.6
Luxury	6.4	100.0

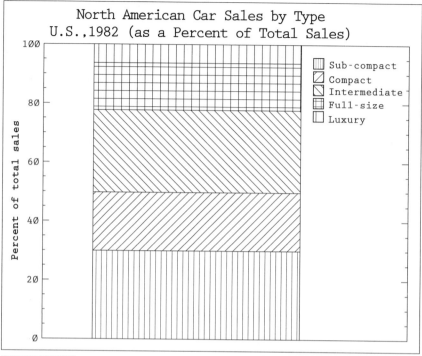

STATGRAPHICS®

B. Component Line Graphs

Component line graphs can be used to show the relative change in the components over a period of time. Each component is expressed as a percent of the total and assigned a vertical distance proportional to that percent. For each point in time, the combined vertical distances assigned to the components represent 100%.

○ **EXAMPLE 1.5b**

Use the data in Table 1.2 to construct a component line graph of North American car sales, by type, in Canada for the period 1980 to 1985.

● **SOLUTION**

Since the data are already in percent form, accumulate the percents for each year to facilitate the plotting of the points.

| | Accumulation of percents | | | | | |
Type	1980	1981	1982	1983	1984	1985
Sub-compact	18.9	21.4	32.4	33.6	33.0	30.2
Compact	49.8	52.5	58.2	55.6	58.0	56.1
Intermediate	77.6	81.4	88.2	87.8	87.0	87.6
Full-size	97.6	97.9	98.6	98.6	98.1	97.8
Luxury	100.0	100.0	100.0	100.0	100.0	100.0

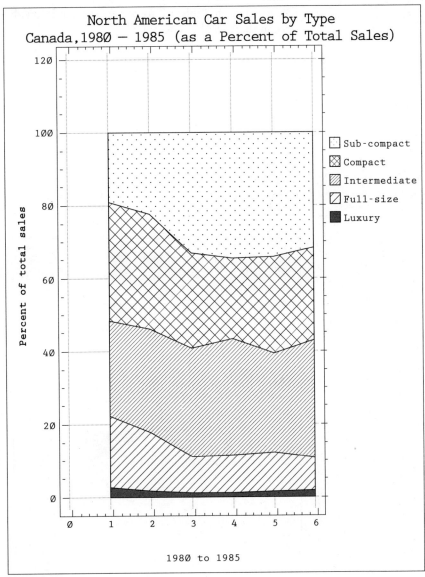

North American Car Sales by Type
Canada,1980 — 1985 (as a Percent of Total Sales)

1980 to 1985

STATGRAHICS®

C. Pie (or Circle) Graphs

Pie graphs serve the same purpose as component bar graphs. To construct a pie graph, change the data into percent form, multiply each percent by 360 to determine the number of degrees required for each component, and accumulate the degrees. A protractor is needed to construct an accurate pie graph.

○ **EXAMPLE 1.5c**

Use the data in Table 1.2 to construct a circle graph of North American car sales in the U.S., by type, for 1982.

● **SOLUTION**

Since the data are already in percent form, it is only necessary to convert the percents to degrees and accumulate.

Type	%	Degrees	Accumulated degrees
Sub-compact	30.2	108.7	108.7
Compact	19.2	69.1	177.8
Intermediate	28.1	101.2	279.0
Full-size	16.1	58.0	337.0
Luxury	6.4	23.0	360.0

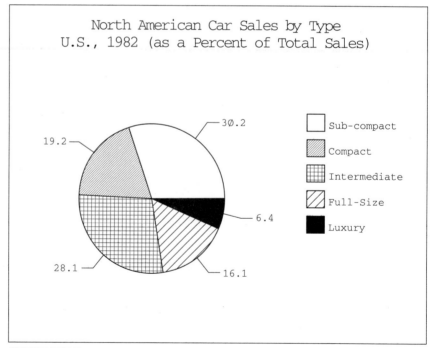

North American Car Sales by Type
U.S., 1982 (as a Percent of Total Sales)

STATGRAPHICS®

EXERCISE 1.5

1. A survey of inventory levels for manufacturing firms in three cities yielded the following information:

| | Percent of total respondents | | |
Inventory	Montreal	Toronto	Vancouver
High	24	10	18
Adequate	65	55	62
Low	11	35	20

Depict the inventory levels for the three cities by means of a component bar graph.

2. Revenues for the Sheridan *Sun* newspaper for 1985–1987 (in $ millions) were as follows:

Revenues	1985	1986	1987
Advertising	43	50	58
Subscriptions	32	35	40
Newsstand	12	10	8
Other	3	5	4

Represent the data by means of a component bar graph.

3. Operating expenses (in $000) for the Corner Boutique are as follows:

Operating expenses	1986	1987	1988
Salaries	24	28	36
Rent	6	6	8
Utilities	4	5	6
Other	5	8	10

Represent the data by means of a component line graph.

4. Component sales for HSI Inc. (in $ millions) are as follows:

Component	1982	1984	1986	1988
Radios	16	14	12	13
TV sets	18	20	24	27
Stereos	12	14	16	16
Parts	9	12	12	14

Construct a component line graph to represent the data.

5. Appliance World sold the following number of major appliances in 1988: ranges, 380; refrigerators, 300; freezers, 90; washers, 210; dryers, 120. Represent the data by means of

a) a component bar graph;

b) a circle graph.

6. Sales for a hardware store last month included the following items: tools, $96 000; chain saws, $76 800; bench saws, $51 200; radial saws, $32 000; grinders, $25 600; metal saws, $38 400. Represent the data by means of
a) a component bar graph;
b) a pie graph.

SECTION 1.6 Specialty Graphs

A. Bi- (Two-) Directional Bar Graphs

The **bi-directional bar graph** is used to portray percent changes from one time period to another. Increases are represented by bars constructed to the right of the vertical line drawn through the origin; decreases are represented by bars constructed to the left.

○ **EXAMPLE 1.6a**
Use the data in Table 1.2 to construct a two-directional bar graph of North American car sales, by type, in Canada for 1980 and 1985.

● **SOLUTION**
Determine the increase or decrease in car sales from 1980 to 1985 and express the change as a percent.

Type	1985	1980	Increase (decrease)	Percent change
Sub-compact	30.2	18.9	11.3	$\dfrac{11.3}{18.9} = 59.8\%$
Compact	25.9	30.9	(5.0)	$\dfrac{-5.0}{30.9} = -16.2\%$
Intermediate	31.5	27.8	3.7	$\dfrac{3.7}{27.8} = 13.3\%$
Full-size	10.2	20.0	(9.8)	$\dfrac{-9.8}{20.0} = -49.0\%$
Luxury	2.1	2.5	(0.4)	$\dfrac{-0.4}{2.5} = -16.0\%$

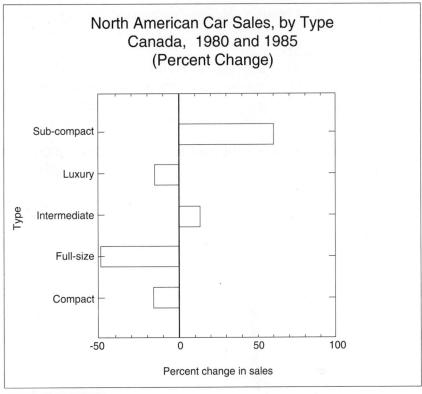

SYGRAPH® (MODIFIED)

B. High-Low-Close Graphs

The **high-low-close graph,** commonly found in the financial section of daily newspapers and in other financial publications, serves the special purpose of showing the periodic (daily, weekly, monthly, yearly) highs, lows and closing values of stock exchange indexes as well as the prices of shares traded on stock exchanges.

○ **EXAMPLE 1.6b**
The following are the daily highs, lows and closes (without decimals) of the TSE 300 Composite Index for the week covering Monday, April 1, 1991 to Friday, April 5, 1991.

Date	High	Low	Close
April 1	3496	3483	3486
April 2	3516	3490	3516
April 3	3535	3516	3516
April 4	3530	3513	3524
April 5	3537	3512	3518

Construct a high-low-close graph.

● SOLUTION

The construction of this type of graph is similar to that of multiple line graphs except that the three points for each date are joined by a vertical line, as shown below.

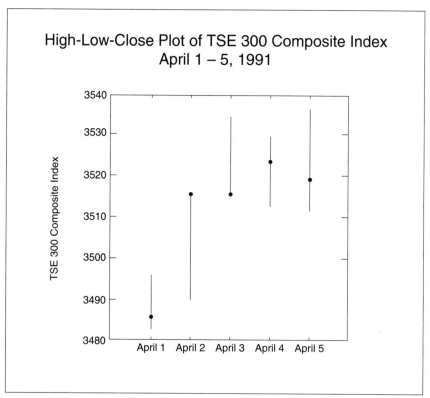

SYGRAPH® (MODIFIED)

EXERCISE 1.6

1. Draw a bi-directional bar graph to portray the percent change for the following information:

Department	A	B	C	D	E	F
Sales 1987	60	80	70	25	20	25
Sales 1988	66	72	56	35	18	20

2. Represent the percent change in the following sales figures by means of a two-directional bar graph.

Company	1	2	3	4	5	6
1984	40	50	20	45	20	25
1988	44	60	15	36	30	35

3. The monthly highs, lows and closes for the TSE 300 Composite Index for the period January to April 1991 are as follows:

Month	High	Low	Close
January	3272.90	3161.95	3272.90
February	3532.54	3293.48	3462.37
March	3571.71	3457.04	3495.67
April	3545.93	3468.81	3468.81

Construct a high-low-close graph.

4. Listed below are the monthly highs, lows and closes for Stelco Inc. A shares on the Toronto Stock Exchange for the period September 1990 to March 1991. Construct a high-low-close graph to represent the data.

Year	Month	High	Low	Close
1990	September	$15\frac{5}{8}$	$13\frac{1}{8}$	$13\frac{1}{4}$
	October	$13\frac{3}{4}$	$11\frac{3}{8}$	$11\frac{5}{8}$
	November	$12\frac{1}{4}$	$11\frac{1}{8}$	$11\frac{3}{4}$
	December	$12\frac{1}{4}$	11	$12\frac{1}{8}$
1991	January	$12\frac{1}{4}$	$9\frac{1}{4}$	$9\frac{3}{4}$
	February	$12\frac{1}{4}$	$9\frac{1}{2}$	$12\frac{1}{4}$
	March	$12\frac{3}{8}$	$6\frac{3}{4}$	$7\frac{1}{8}$

REVIEW EXERCISE

1. A person who is thinking of moving should consider the movers' vehicle volume used each month as a percent of available capacity.

Month	Jan	Feb	Mar	Apr	May	Jun
Volume	29	27	30	43	49	68
Month	Jul	Aug	Sep	Oct	Nov	Dec
Volume	90	98	60	30	30	28

a) Draw a horizontal line graph to represent the data.
b) When would you advise a person not to move? Explain.

2. Draw a multiple line graph to represent the following data:

	Unattached Ontario Residents Age 15 and over (000s)					
Sex	1985	1986	1987	1988	1989	1990
Male	437	441	463	507	498	528
Female	554	580	593	623	598	646

Source: Statistics Canada, *The Labour Force*, Catalogue 71–001.

3. Draw a multiple bar graph to describe the following data:

		Proportion of Households in Canada with:			
Year	VCR	Micro-wave	Two or more cars	Van or truck	Air conditioner
1988	52.0	53.0	25.1	24.3	20.8
1989	58.8	63.4	25.0	25.5	24.6

Source: Statistics Canada, *Perspectives on Labour and Income*, Catalogue 75–001E.

4. The following is a listing of the number of people employed in Ontario in 1988 in various industries:

Industry	Number of employees
Goods-producing industries:	
Forestry	10 400
Mining	31 100
Manufacturing	965 000
Construction	208 300
Service industries:	
Transportation and communication	295 000
Trade	733 500
Finance, insurance and real estate	292 600
Other	1 635 400

a) Draw a vertical bar graph to represent the goods-producing industries.
b) Draw a horizontal bar graph to represent the service industries.
c) What can be done to either graph to make it more readable?

5. The International Fabrics Institute has determined that the following are responsible for various dry-cleaning problems:

Responsibility	Problems	Percent of problems
Manufacturer	Colour loss, shrinkage	43.3%
Consumer	Stains, damage by bleach	34.7%
Dry cleaner	Colour loss, shrinkage	15.4%
Unknown	Mechanical and chemical damage	6.6%

Construct a component bar graph of dry-cleaning problems by responsibility.

6. Company size is sometimes measured by the number of persons a company employs. For 1988 and 1989 the distribution of the labour force in Ontario based on this method of measuring company size was as follows:

Company size	Number employed (000s)	
	1988	1989
Less than 20	785.8	846.7
20 to 49	391.8	403.9
50 to 99	313.4	329.6
100 or more	2653.4	2675.4

Source: Statistics Canada, *Perspectives on Labour and Income*, Catalogue 75–001E.

a) Draw a multiple bar graph to represent the data.
b) Represent the 1989 data by means of a circle graph.

7. The levels of education obtained by the Canadian labour force (000s) as of April 1990 are as follows:

Level of education	Male	Female
0 to 8 years	697	377
Some secondary	1595	1133
High-school diploma	1482	1491
Some post-secondary	684	624
College diploma	1880	1635
University degree	1085	787

Source: Statistics Canada, *The Labour Force*, Catalogue 71–001.

a) Use a circle (pie) graph to represent the male labour force.
b) Use a pie graph to represent the female labour force.

8. Average hourly earnings, including overtime, in January 1990 are as follows:

Province	Nfld	PEI	NS	NB	Que	Ont
Salaried employees	16.08	14.11	15.23	14.75	16.28	17.72
Hourly-paid employees	10.73	9.02	10.89	10.84	12.25	12.91

Province	Man	Sask	Alta	BC	Canada
Salaried employees	15.59	15.58	17.06	16.92	16.90
Hourly-paid employees	11.23	11.14	12.13	13.94	12.55

Source: Statistics Canada, *Employment Earnings and Hours*, Catalogue 72–002.

a) Compare salaried earnings and hourly wages by means of a multiple line graph.

b) Use a bi-directional bar graph to compare provincial averages with the national average for salaried employees.

9. A large retailer is analyzing departmental sales year over year and compiles the following data:

	Sales ($ millions)	
Department	May 1991	May 1992
Toys	1.2	1.14
Men's clothing	2.7	2.97
Women's clothing	8.8	10.12
Hardware	3.6	3.78
Furniture	13.5	11.88

Construct a two-directional bar graph showing the change in departmental sales from 1991 to 1992.

10. Shown below is a reproduction of a product information card for a new CD player and the data compiled from returned cards.

Model Number _____	Date Purchased _____	
Name _____	Address _____	
City _____	Province _____ Postal Code _____	
Name of Store _____		

1. Is this your first CD player? ☐ YES ☐ NO	4. Purchaser's Age ☐ UNDER 20 ☐ 31 – 40 ☐ 20 – 24 ☐ OVER 40 ☐ 25 – 30	6. What was the most important factor that influenced your purchase? ☐ PRICE ☐ SALESPERSON
2. Purchaser's sex ☐ FEMALE ☐ MALE	5. If a gift, what was the occasion? ☐ BIRTHDAY ☐ GRADUATION ☐ CHRISTMAS ☐ OTHER	☐ RECOMMENDED ☐ BRAND NAME ☐ ADVERTISING
3. Do you own a car? ☐ YES ☐ NO		☐ FEATURES

The following data were compiled from the returned cards:

Question no.	Survey results	
1	Yes (490)	No (50)
2	Male (308)	Female (182)
3	Yes (242)	No (198)
4	Under 20 (66)	20–24 (110)
	25–30 (92)	31–40 (42)
	Over 40 (130)	
5	Birthday (32)	Christmas (41)
	Graduation (28)	Other (12)
6	Price (102)	Salesperson (20)
	Recommended (80)	Advertising (32)
	Brand name (120)	Features (86)

a) What are the two main pieces of information the survey is designed to supply to the researcher?

b) Draw a vertical bar graph showing the factors influencing the decision to buy.

c) Draw a horizontal bar graph showing the reasons for purchasing a CD player.

d) Construct a pie graph summarizing the age distribution of the purchasers.

11. Draw a high-low-close graph to represent the following data:

	Campeau Corporation TSE Listing, 1989		
Month	High	Low	Close
January	$16\frac{3}{4}$	$15\frac{1}{4}$	$15\frac{3}{4}$
February	$18\frac{7}{8}$	$15\frac{1}{2}$	$17\frac{3}{8}$
March	$17\frac{1}{2}$	$15\frac{7}{8}$	$16\frac{1}{8}$
April	$18\frac{1}{4}$	$15\frac{3}{4}$	$17\frac{1}{2}$
May	$17\frac{1}{4}$	16.00	$16\frac{1}{8}$
June	$16\frac{3}{4}$	$15\frac{1}{2}$	$15\frac{7}{8}$
July	$16\frac{7}{8}$	$15\frac{3}{4}$	$16\frac{5}{8}$
August	$20\frac{1}{8}$	$16\frac{5}{8}$	$18\frac{3}{4}$
September	$22\frac{1}{4}$	$8\frac{7}{8}$	10.00
October	$10\frac{5}{8}$	$6\frac{1}{4}$	$7\frac{1}{4}$
November	$7\frac{5}{8}$	4.40	5.00
December	$5\frac{7}{8}$	3.25	3.65

12. The following table shows the TSE 300 Composite listing for the period March 1 to March 12, 1991. Construct a high-low-close graph to represent the data.

Date	High	Low	Close
March 1	3474.74	3445.46	3471.42
4	3519.06	3477.43	3519.06
5	3565.77	3520.88	3561.90
6	3598.05	3571.71	3571.71
7	3580.93	3563.31	3564.04
8	3572.18	3553.79	3571.02
11	3563.36	3554.47	3555.00
12	3558.25	3541.98	3544.41

SELF-TEST

1. Investors who put $10 000 into the New Horizon Fund or the Strong Investment Fund on January 1, 1987, came out about the same at the end of 1991. Use the given information to graph their rather different paths on a multiple line graph.

Date	87-01-01	87-12-31	88-12-31	89-12-31	90-12-31	91-12-31
New Horizon	10 000	11 800	13 920	12 980	14 430	16 100
Strong Inv't	10 000	12 380	12 000	11 650	13 160	16 120

2. After the introduction of the GST in Canada, prices for some goods and services went up, while others went down. Below are a few selected examples for analysis:

Item	Price before GST	Price after GST
Fast food	$ 3.30	$ 3.50
Luggage	80.00	78.27
Telephone	11.50	12.15
Cable TV	20.00	19.13
Taxi fare	8.50	8.91
Long-distance charges	55.00	52.61

a) Draw a multiple bar graph to represent the data.
b) Compare the prices of the items before and after GST by means of a horizontal bi-directional bar graph.

3. The Cousins family compiled a record of their yearly expenditures as follows:

Expenditure	1988	1989	1990	1991
Housing	$15 000	$15 000	$16 800	$15 500
Transportation	1 200	2 800	5 600	5 700
Food	6 200	6 500	5 900	7 800
Clothing	7 400	6 100	10 000	8 900
Entertainment	5 300	2 200	7 300	12 300

a) Construct a component line graph for the Cousins family expenditures for the period 1988 to 1991.

b) Construct a component bar graph for the 1990 expenditures.

c) Construct a circle graph for the 1991 expenditures.

4. The following table shows the high, low and closing values for Apple Computer (NASDAQ) shares for the period May 6 to May 10, 1991. Construct a high-low-close graph.

Date	High	Low	Close
Monday, May 6	$50\frac{1}{2}$	$48\frac{1}{4}$	$50\frac{1}{4}$
Tuesday, May 7	$51\frac{1}{4}$	$50\frac{1}{2}$	$50\frac{5}{8}$
Wednesday, May 8	$50\frac{3}{4}$	$49\frac{1}{4}$	$49\frac{3}{4}$
Thursday, May 9	$51\frac{1}{2}$	$49\frac{3}{4}$	$50\frac{3}{4}$
Friday, May 10	$53\frac{1}{4}$	$50\frac{3}{4}$	$51\frac{1}{4}$

Key Terms

CHAPTER 2

Frequency Distributions and Their Graphs

Introduction

Before we can analyze statistical data, they must first be organized. A simple way of doing this is to arrange the data in order of magnitude from lowest to highest value, or vice versa. Such an arrangement is called an array.

Sets of data often contain many values. In such cases arranging data in an array is very time-consuming. A more useful procedure is to group the data into classes. The resulting arrangement is called a frequency distribution.

For a visual picture of the characteristics of a frequency distribution, graphs of the distribution can be constructed. The three types of graphs used are the histogram, the frequency polygon and the cumulative frequency polygons (ogives).

The relative position of values in a frequency distribution is often of interest. These positional measures, called percentiles, can be determined graphically or by formula.

Objectives

Upon completion of this chapter you will be able to

1. arrange a small set of unorganized data into an array and use the array to determine the mean, the median and the mode of the data set;
2. arrange larger sets of unorganized data into frequency distributions;
3. convert a frequency distribution into a relative frequency distribution;
4. obtain, from a frequency distribution, the less-than and more-than cumulative frequency distributions and convert them into cumulative percent distributions;
5. construct the histogram and the frequency polygon of a frequency distribution;
6. construct the two types of cumulative frequency diagrams (ogives) for a frequency distribution;

7. graphically determine the values of specific percentiles from a less-than cumulative frequency diagram;
8. use formulas to determine the values of specific percentiles for grouped and ungrouped data;
9. compute the percentile rank of a given value in a frequency distribution.

SECTION 2.1 Frequency Distributions

A. Arranging Data in an Array

Table 2.1 is an example of unorganized data.

TABLE 2.1   Number of Units Produced

180	195	194	197	188	176	205	178
162	214	227	200	221	190	198	174
185	200	204	195	216	195	181	188
172	209	198	210	202	215	195	186

A simple way of organizing raw data is to arrange the values in *ascending* or *descending* order of magnitude. Such an arrangement is called an **array**.

○ **EXAMPLE 2.1a**
Arrange the data in Table 2.1 in the form of an array.

● **SOLUTION**
To facilitate the arrangement of the numbers in ascending order, rearrange groups of numbers in order:

162	195	194	195	188	176	181	174
172	200	198	197	202	190	195	178
180	209	204	200	216	195	198	186
185	214	227	210	221	215	205	188

Now list the numbers in their final order:

162	178	186	194	195	200	205	215
172	180	188	195	197	200	209	216
174	181	188	195	198	202	210	221
176	185	190	195	198	204	214	227

B. Advantages of an Array

Apart from the ease with which they can be set up, arrays have other advantages:

1. We can immediately read out the *lowest* value (162) and the *highest* value (227). By computing the *difference* between the two extreme (lowest and highest) values we obtain a statistical measure referred to as the **range** (65).
2. We can easily identify values that appear more than once and thus determine the *most frequently occurring* value. This value represents another statistical measure called the **mode**. Since the value 195 appears four times, the mode of the set of data is 195.
3. We can readily group the data into sections. Since the data set contains 32 values, the lower half ranges from 162 to 195, while the upper half ranges from 195 to 227. The *middle* value, called the **median**, is therefore 195.

C. *Arranging Data in a Frequency Distribution*

While an array of values is useful, arranging large data sets into arrays is tedious and time-consuming. In addition, it is difficult for most humans to recognize underlying relationships in a large, unorganized mass of data. A more useful procedure is to group the data into *classes*. Such an arrangement is called a **frequency distribution**.

○ **EXAMPLE 2.1b**
Construct a frequency distribution for the data set shown in Table 2.1.

● **SOLUTION**

Step 1   Determine the class width and the number of class intervals.
The first thing to consider in constructing a frequency distribution is the number of classes we wish to use and the width of the class interval. The number of **class intervals** is influenced by the number of observations that make up the data set. The larger the number of observations, the larger the number of classes that can be justified. However, it is recommended that the number of classes be no fewer than 6 and no more than 15.

To determine the number of classes and the width of the class intervals we first need to compute the range of values. As previously indicated, the range is the difference between the highest value and the lowest value, that is, $227 - 162 = 65$.

Because of the small number of observations (32 items), the number of classes should be kept low.

For 6 classes, the width of the class intervals would be $65 \div 6 = 10.8$; for 15 classes, the width of the class intervals would be $65 \div 15 = 4.3$.

The **class width** selected needs to be between 4.3 and 10.8. The width selected should be a convenient number such as 5 or 10. Because of the small number of observations in this example, a class width of 10 is a reasonable choice.

We can now obtain the number of class intervals by dividing the range by the class width ($65 \div 10 = 6.5$); this means 7 classes will have to be set up.

Step 2   Set up the classes.

To set up the classes it is now necessary to select the **lower limit** of the first class. While it is preferable to centre class intervals on or near frequently occurring values, it is usually acceptable to select a convenient value less than the smallest observed value as the lowest class limit.

Since the lowest observed value in the data set is 162, a convenient lowest class limit is 160. The lower limits of all classes can now be established by repeatedly adding the width of the class intervals until the highest value in the data set is reached. Accordingly, the lower class limits are 160, 170, 180, 190, 200, 210, 220, and the classes can now be set up as shown in Table 2.2.

Step 3   Determine the class frequencies.

The **class frequencies** are the number of observations that fall into the given classes. Each observation must be assigned to a class, but no observation can appear in more than one class.

The process of assigning the observations in the data set to the classes is referred to as *taking a tally*. This is done by checking off the observations and making a mark for each observation beside the class interval to which the observed value belongs.

For ease of counting, tally marks are usually arranged in groups of five (*////*), as shown in Table 2.2.

The class frequencies are now obtained by counting the tallies assigned to each class and then listing the count.

TABLE 2.2   Frequency Distribution

Class interval	Tally	Class frequencies
160 to under 170	/	1
170 to under 180	////	4
180 to under 190	//// /	6
190 to under 200	//// ////	9
200 to under 210	//// /	6
210 to under 220	////	4
220 to under 230	//	2

Note   The frequency distribution in Table 2.2 is an example of a *closed-ended* frequency distribution. If the first class (160 to under 170) read "under 170" and/or the last class (220 to under 230) read "220 and over," the distribution would be an *open-ended* frequency distribution.

D. *Relative Frequency Distributions*

The total number of observations in a data set influences the number of observations in each class. For purposes of comparison the class frequencies can be expressed as a fraction of the total number of observations. These fractions, presented in either fractional, decimal or percent form, constitute a *relative* frequency distribution.

○ **EXAMPLE 2.1c**

Construct a relative frequency distribution for the frequency distribution shown in Table 2.2.

● **SOLUTION**

The relative frequencies are obtained by dividing the class frequencies by the total number of observations and can be shown in the form of a common fraction, decimal or percent, as shown in Table 2.3.

TABLE 2.3 **Relative Frequency Distribution**

Class intervals	Class frequencies	Relative frequencies		
		Fraction	Decimal	Percent
160 to under 170	1	$\frac{1}{32}$	0.03125	3.125%
170 to under 180	4	$\frac{4}{32}$	0.12500	12.500%
180 to under 190	6	$\frac{6}{32}$	0.18750	18.750%
190 to under 200	9	$\frac{9}{32}$	0.28125	28.125%
200 to under 210	6	$\frac{6}{32}$	0.18750	18.750%
210 to under 220	4	$\frac{4}{32}$	0.12500	12.500%
220 to under 230	2	$\frac{2}{32}$	0.06250	6.250%
Totals	32	$\frac{32}{32}$	1.00000	100.000%

Note   The sum of the relative frequencies is always 1 (100%).

E. Cumulative Frequency Distributions

A **cumulative frequency distribution** shows a running total of the frequencies in the distribution. It can be used to determine how many observations are below or above certain values.

A cumulative *relative* frequency distribution is obtained by keeping a running total of the *relative* frequencies. The resulting values are fractions of the total observations in the data set and can be used to determine what percent of the observations are below or above specified values.

○ **EXAMPLE 2.1d**

Construct a cumulative frequency distribution for the class frequencies given in Table 2.3.

● **SOLUTION**

Compute and record the running totals for the class frequencies and obtain the cumulative relative frequencies by dividing the numbers in the running totals by the total class frequency as shown in Table 2.4.

TABLE 2.4   **Cumulative Frequency Distribution**

Class intervals	Class frequencies	Cumulative frequencies	Cumulative relative frequencies
160 to under 170	1	1	$\frac{1}{32}$ = 3.125%
170 to under 180	4	(1 + 4) = 5	$\frac{5}{32}$ = 15.625%
180 to under 190	6	(5 + 6) = 11	$\frac{11}{32}$ = 34.375%
190 to under 200	9	(11 + 9) = 20	$\frac{20}{32}$ = 62.500%
200 to under 210	6	(20 + 6) = 26	$\frac{26}{32}$ = 81.250%
210 to under 220	4	(26 + 4) = 30	$\frac{30}{32}$ = 93.750%
220 to under 230	2	(30 + 2) = 32	$\frac{32}{32}$ = 100.000%
Total	32		

Two types of cumulative frequency distribution are used:
1. the *less-than* cumulative frequency distribution, which shows the number and/or percent of the observations that have a value *less than* the **upper limit** of each class (Example 2.2d is this type);
2. the *more-than* cumulative frequency distribution, which shows the number and/or percent of the observations that have a value *more than* the *lower limit* of each class.

○ **EXAMPLE 2.1e**
For the frequency distribution given in Table 2.5, compute
a) the less-than cumulative frequencies and percents;
b) the more-than cumulative frequencies and percents.

TABLE 2.5   **Frequency Distribution**

Class interval	Frequency
0 to under 10	4
10 to under 20	8
20 to under 30	7
30 to under 40	3
40 to under 50	2
50 to under 60	1

● **SOLUTION**
a) To construct a less-than cumulative frequency distribution, accumulate the frequencies, starting with the lowest class, and continue through the highest class as shown in Table 2.6.

TABLE 2.6 **Less-Than Cumulative Frequency Distribution**

Class interval	Class frequency	Upper class limit	Cumulative frequencies	Cumulative percents
0 to under 10	4	10	4	$\frac{4}{25} = 16\%$
10 to under 20	8	20	$4 + 8 = 12$	$\frac{12}{25} = 48\%$
20 to under 30	7	30	$12 + 7 = 19$	$\frac{19}{25} = 76\%$
30 to under 40	3	40	$19 + 3 = 22$	$\frac{22}{25} = 88\%$
40 to under 50	2	50	$22 + 2 = 24$	$\frac{24}{25} = 96\%$
50 to under 60	1	60	$24 + 1 = 25$	$\frac{25}{25} = 100\%$

Notes

1. In a less-than cumulative frequency table, the cumulative frequencies column indicates the number of observations that have a value less than the upper limit of each class; that is,

   4 observations have a value less than 10;
   12 observations have a value less than 20;
   19 observations have a value less than 30;
   22 observations have a value less than 40;
   24 observations have a value less than 50;
   25 (all) observations have a value less than 60.

2. In a similar way the cumulative percents column shows the percent of observations that have a value less than the upper limit of each class; that is,

   16% of the observations have a value less than 10;
   48% of the observations have a value less than 20;
   76% of the observations have a value less than 30;
   88% of the observations have a value less than 40;
   96% of the observations have a value less than 50;
   100% of the observations have a value less than 60.

b) To construct a more-than cumulative frequency distribution, accumulate the frequencies, starting with the highest class, and continue through the lowest class as shown in Table 2.7.

TABLE 2.7   More-Than Cumulative Frequency Distribution

Class interval	Class frequency	Upper class limit	Cumulative frequencies	Cumulative percents
0 to under 10	4	0	$21 + 4 = 25$	$\dfrac{25}{25} = 100\%$
10 to under 20	8	10	$13 + 8 = 21$	$\dfrac{21}{25} = 84\%$
20 to under 30	7	20	$6 + 7 = 13$	$\dfrac{13}{25} = 52\%$
30 to under 40	3	30	$3 + 3 = 6$	$\dfrac{6}{25} = 24\%$
40 to under 50	2	40	$1 + 2 = 3$	$\dfrac{3}{25} = 12\%$
50 to under 60	1	50	1	$\dfrac{1}{25} = 4\%$

Notes

3. In a more-than cumulative frequency distribution the cumulative frequencies column indicates the number of observations that have a value equal to or more than the lower limit of each class; that is,

> 25 out of 25 observations have a value of 0 or more;
> 21 out of 25 observations have a value of 10 or more;
> 13 out of 25 observations have a value of 20 or more;
> 6 out of 25 observations have a value of 30 or more;
> 3 out of 25 observations have a value of 40 or more;
> 1 out of 25 observations has a value of 50 or more.

4. In a similar way, the cumulative percents column shows the percent of observations that have a value equal to or more than the lower limit of each class; that is,

> 100% of the observations have a value of 0 or more;
> 84% of the observations have a value of 10 or more;
> 52% of the observations have a value of 20 or more;
> 24% of the observations have a value of 30 or more;
> 12% of the observations have a value of 40 or more;
>   4% of the observations have a value of 50 or more.

EXERCISE 2.1

1. Last week's production for a plant of 24 employees was as follows:

213	240	260	203	234	245	245	247
267	256	219	227	266	251	245	239
238	245	250	225	253	242	248	271

a) Arrange the data in the form of an array.

b) From the array in (a) determine
   i) the range;
   ii) the mode;
   iii) the median.
c) Arrange the data in the form of a frequency distribution, using equal class intervals of size 10 starting with 200.
d) Compute the less-than cumulative frequencies and percents.
e) Compute the more-than cumulative frequencies and percents.

2. The following data represent the daily earnings of the employees in an assembly plant:

120	60	132	96	116	92	116	116
92	100	160	166	196	176	108	126
68	192	144	128	152	128	144	130
100	116	140	96	168	126	98	144
116	72	180	66	176	88	164	120

a) Arrange the data in the form of an array.
b) From the array in (a) determine
   i) the range;
   ii) the mode;
   iii) the median.
c) Arrange the data in the form of a frequency distribution using equal class intervals of size 20 with 60 as the lowest class limit.
d) Compute the less-than cumulative frequencies and percents.
e) Compute the more-than cumulative frequencies and percents.

3. Consider the following frequency distribution:

Class	0 to under 10	10 to under 20	20 to under 30	30 to under 40	40 to under 50	50 to under 60	60 to under 70
Frequency	4	6	10	8	6	4	2

a) Construct the less-than and more-than cumulative frequency and percent distributions.
b) How many of the observations in the given frequency distribution have
   i) a value less than 40?
   ii) a value of 20 or more?
c) What percent of the observations in the given frequency distribution have
   i) a value less than 50?
   ii) a value of 40 or more?

4. Consider the following information about the wages of a group of employees:

Weekly wages	$360 to under $400	$400 to under $440	$440 to under $480	$480 to under $520	$520 to under $560	$560 to under $600	$600 to under $640
Number of employees	24	40	55	36	25	12	3

a) Construct the less-than and more-than cumulative frequency and percent distributions.
b) How many employees earn weekly wages of
   i) $440 or more?
   ii) less than $560?
c) What percent of the employees earn weekly wages of
   i) $520 or more?
   ii) less than $600?

5. ***Data disk exercise***   Using the enclosed data disk and the file name PUBLIC, determine **(i)** the range and **(ii)** the median for company sales (the variable name is SALES).

Graphs of Frequency Distributions

A. The Histogram of a Frequency Distribution

The **histogram** of a frequency distribution is a vertical bar graph of the class frequencies. The horizontal axis is used for the class limits and the vertical axis for the class frequencies.

○ **EXAMPLE 2.2a**
Construct a histogram for the frequency distribution in the following table.

TABLE 2.8

Test score	Number of students
40 to under 50	1
50 to under 60	3
60 to under 70	9
70 to under 80	14
80 to under 90	11
90 to under 100	2

● **SOLUTION**
The lowest class limit (40) in the distribution is arbitrarily marked as the starting point for the horizontal scale. The remaining class limits are then marked at equal intervals.

The vertical scale starts at zero and must accommodate the largest class frequency (14).

Vertical bars are then drawn over each class interval to represent the number of observed values in that class. In our example a bar of height "1" is drawn over the line segment "40 to 50"; a bar of height "3" is drawn over the line segment "50 to 60"; and so on.

For ease of construction it is best to draw first the bar over the interval showing the highest frequency. In our case, a bar of height "14" over the line segment "70 to 80" should be the first bar drawn. The remaining bars can then be drawn easily to the left and right of the highest bar.

Note that, unlike simple bar graphs, there is no space between the bars in a histogram.

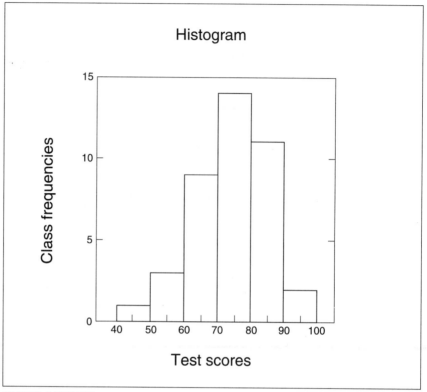

SYSTST® (MODIFIED)

B. The Frequency Polygon of a Frequency Distribution

The **frequency polygon** of a frequency distribution is a simple line graph joining the *midpoints* of the bars of a histogram.

○ **EXAMPLE 2.2b**

Construct the frequency polygon for the frequency distribution given in Example 2.2a (Table 2.8).

● **SOLUTION**

Mark the class midpoint (45) of the lowest class (40 to under 50) on the horizontal scale. Mark the remaining class midpoints (55, 65, 75, 85, 95) at equal intervals. Mark one additional point at each end of the scale (35 and 105).

Use the vertical axis for the class frequencies. The scale used starts at zero and must accommodate the greatest class frequency. Now plot the frequency in each class as a dot above the class midpoint. Join the dots by straight line segments. Complete the polygon by connecting the points representing the frequencies in the lowest and highest class respectively to the extra midpoints on the horizontal axis.

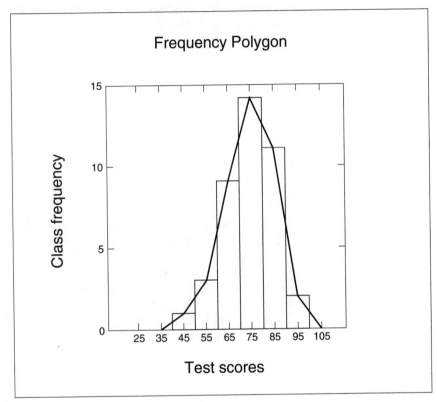

SYSTAT® (MODIFIED)

C. *Cumulative Frequency Diagrams (Ogives)*

Cumulative frequency diagrams, also referred to as **ogives,** are constructed by plotting the cumulative frequencies. Depending on the type of cumulative frequency distribution (as introduced in Section 2.1), we distinguish between two cumulative frequency diagrams:

1. the *less-than ogive*, obtained by plotting the less-than cumulative frequencies against the upper class limits;
2. the *more-than ogive*, obtained by plotting the more-than cumulative frequencies against the lower class limits.

○ **EXAMPLE 2.2c**

For the frequency distribution used in Example 2.2a (Table 2.8),
a) construct the less-than frequency diagram;
b) determine how many students scored less than 65 on the test.

● **SOLUTION**

First obtain the less-than cumulative frequencies and the less-than cumulative percents as shown below:

Test score	Number of students	Less-than cumulative frequency	Less-than cumulative percents
40 to under 50	1	1	2.50
50 to under 60	3	4	10.00
60 to under 70	9	13	32.50
70 to under 80	14	27	67.50
80 to under 90	11	38	95.00
90 to under 100	2	40	100.00
Total	40		

The horizontal axis of the diagram is used to mark the class limits, while the vertical axis is used for the cumulative frequencies or the cumulative relative frequencies (percents).

Since the cumulative relative frequencies range from 0% to 100% for any frequency distribution, the cumulative percents are preferred for the vertical scale.

Now plot the less-than cumulative percents against the upper limits of the corresponding class. For example, plot

2.5% against the upper limit (test score) 50;
10.0% against the upper limit (test score) 60;
32.5% against the upper limit (test score) 70;
67.5% against the upper limit (test score) 80;
95.0% against the upper limit (test score) 90;
100.0% against the upper limit (test score) 100.

By now, a point is plotted against every class limit except the lowest limit of 40.

Since no observation in the frequency distribution has a value less than the lowest limit, zero is plotted against the lowest limit of 40. The points can now be joined to obtain the less-than ogive.

The less-than frequency diagram indicates the number of observations in the frequency distribution that have a value less than a chosen value.

To determine how many students scored less than 65, draw a perpendicular line from the point marked 65 on the horizontal axis to the less-than ogive.

The point of intersection indicates the percent of students who scored less than 65 on the test.

To read the percent on the vertical scale, draw a line at right angles to the vertical scale from the point of intersection. The indication is that approximately 20% of the number of students in the class, that is 8 students, scored less than 65 on the test.

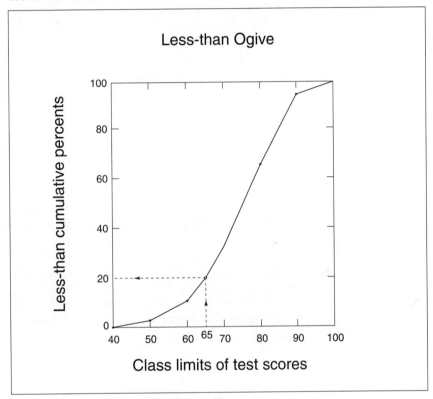

SYGRAPH® MODULE OF THE SYSTAT® PACKAGE

○ EXAMPLE 2.2d

For the frequency distribution used in Example 2.2a (Table 2.8),
a) construct the more-than frequency diagram;
b) determine how many students scored 85 or more on the test.

● SOLUTION

First determine the more-than cumulative frequencies and the more-than cumulative percents as shown:

Test score	Number of students	More-than cumulative frequency	More-than cumulative percents
40 to under 50	1	40	100.00
50 to under 60	3	39	97.50
60 to under 70	9	36	90.00
70 to under 80	14	27	67.50
80 to under 90	11	13	32.50
90 to under 100	2	2	5.00
Total	40		

The axes are used in the same way as for the less-than ogive, and the scales are identical. However, the more-than cumulative percents are plotted against the lower limits of the corresponding class. For example, plot

100.0% against the lower limit (test score) 40;
97.5% against the lower limit (test score) 50;
90.0% against the lower limit (test score) 60;
67.5% against the lower limit (test score) 70;
32.5% against the lower limit (test score) 80;
5.0% against the lower limit (test score) 90.

By now, a point is plotted against every class limit except the highest limit 100.

Since no observation in a frequency distribution has a value more than the highest limit, zero is plotted against the highest class limit 100. The points can now be joined to graph the more-than ogive.

The more-than frequency diagram indicates the number of observations in the frequency distribution that have a value more than or equal to a chosen value.

To determine how many students scored 85 or more, draw a perpendicular from the point marked 85 on the horizontal axis to the more-than ogive. The point of intersection represents the number of students who scored 85 or more on the test.

Draw a line from that point at right angles to the vertical axis. The line indicates that approximately 20% of the number of students, that is 8 students, scored 85 or more on the test.

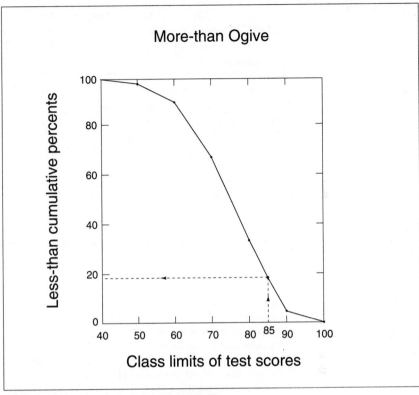

SYGRAPH® MODULE OF THE SYSTAT® PACKAGE

EXERCISE 2.2

1. For Question **3** in Exercise 2.1,
    a) construct the histogram;
    b) construct the frequency polygon;
    c) construct the less-than ogive;
    d) construct the more-than ogive;
    e) graphically determine the number of observations that have
        i) a value less than 45, and
        ii) a value of 25 or more.

2. For Question **4** in Exercise 2.1,
    a) construct the histogram;
    b) construct the frequency polygon;
    c) construct the less-than frequency diagram;
    d) construct the more-than frequency diagram;
    e) graphically determine the number of employees who earned weekly wages
        i) of $250 or more, and
        ii) less than $215.

3. For the following frequency distribution,

Class	100 to under 120	120 to under 140	140 to under 160	160 to under 180	180 to under 200	200 to under 220	220 to under 240
Frequency	4	18	44	92	54	26	2

a) construct the histogram;
b) construct the frequency polygon;
c) construct the less-than ogive;
d) construct the more-than ogive;
e) determine the number of observations having
   i) a value less than 190, and
   ii) a value of 150 or more.

4. For the following frequency distribution,

Class	0 to under 4	4 to under 8	8 to under 12	12 to under 16	16 to under 20	20 to under 24	24 to under 28
Frequency	42	76	135	140	127	56	24

a) construct the histogram;
b) construct the frequency polygon;
c) construct the less-than ogive;
d) construct the more-than ogive;
e) determine the number of observations that have
   i) a value of 10 or more, and
   ii) a value less than 16.

5. *Data disk exercise*   Using the enclosed data disk and the file name PUBLIC, construct a histogram consisting of nine bars for the number of employees (the variable name is NOEMPL).

SECTION 2.3 Percentiles

A. Meaning of the Term Percentile

Percentiles divide a set of data into 100 equal parts. They are used as *positional* measures to indicate what percent of the observations in the data set have a value less than a specified value. Percentiles are numbered from 1 to 100. The position number or rank of a desired percentile is selected from the set of whole numbers defined by $x = \{1, 2, 3, 4, \ldots, 98, 99, 100\}$.

The value of a particular percentile is represented by the symbol P_x. Thus the symbol P_{20} refers to the value of an observation that lies at the 20th percentile. In Example 2.2c we determined that 20% of the number of students in the class scored less than 65, that is $P_{20} = 65$.

B. Graphical Determination of Percentiles

The value of a desired percentile as well as the rank number of a given value in a data set can be obtained graphically from the less-than ogive.

○ **EXAMPLE 2.3a**

The frequency distribution below represents the hourly earnings of a group of employees.
a) Determine the hourly earnings of an employee at the 90th percentile.
b) Determine the percentile rank of an employee who earns $7.20 per hour.

Hourly earnings (in dollars)	Number of employees
6.00 to under 6.50	3
6.50 to under 7.00	28
7.00 to under 7.50	64
7.50 to under 8.00	56
8.00 to under 8.50	42
8.50 to under 9.00	22
9.00 to under 9.50	2
9.50 to under 10.00	3

● **SOLUTION**

First compute the less-than cumulative frequencies and percents as shown below.

Hourly earnings (in dollars)	Number of employees	Cumulative frequencies	Cumulative percents
6.00 to under 6.50	3	3	1.4
6.50 to under 7.00	28	31	14.1
7.00 to under 7.50	64	95	43.2
7.50 to under 8.00	56	151	68.6
8.00 to under 8.50	42	193	87.7
8.50 to under 9.00	22	215	97.7
9.00 to under 9.50	2	217	98.6
9.50 to under 10.00	3	220	100.0

Now construct the less-than ogive (see Figure 2.5).

a) To determine the hourly earnings at the 90th percentile, draw a line parallel to the horizontal axis from the 90% mark to the less-than ogive. From the point of intersection draw a line parallel to the vertical axis to the horizontal axis and read the value at the horizontal scale. This indicates that hourly earnings of $8.60 lie at the 90th percentile.

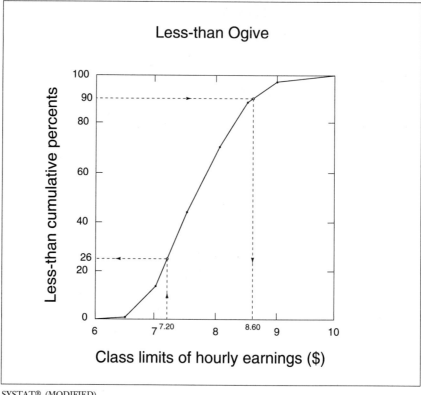

SYSTAT® (MODIFIED)

b) To determine the percentile rank for hourly earnings of $7.20, reverse the process used in (a). Locate the position of $7.20 on the horizontal scale. Draw a line parallel to the vertical axis to the less-than ogive. From the point of intersection draw a line parallel to the horizontal axis to the vertical axis and read off the percent on the vertical scale. This indicates the hourly earnings of $7.20 lie at the 26th percentile.

C. Special Cases

Some of the more frequently used percentiles have specific names, as follows:
1. The **median** is the value that occupies the *middle* position:

$$\text{MEDIAN} = P_{50}$$

2. The **quartiles** are the values that occupy the *quarter* positions:

$$\text{FIRST QUARTILE,} \quad Q_1 = P_{25}$$
$$\text{THIRD QUARTILE,} \quad Q_3 = P_{75}$$

3. The **deciles** are the values that occupy the positions evenly divisible by 10:

$$D_1 = P_{10}; \qquad D_2 = P_{20}; \qquad D_3 = P_{30}; \qquad \text{etc.}$$

○ **EXAMPLE 2.3b**

For the frequency distribution in Example 2.3a, graphically determine the value of
a) the median;
b) Q_1 and Q_3;
c) D_6.

● **SOLUTION**

Use the less-than ogive constructed in Example 2.3a and reproduced below. From the percentile positions marked on the vertical scale, draw lines parallel to the horizontal axis to the less-than ogive. From the points of intersection, draw lines parallel to the vertical axis to intersect the horizontal axis.

Read off the hourly earnings on the horizontal scale.
a) The value of the median (P_{50}) = 7.64 .
b) The value of Q_1 (P_{25}) = 7.19 ;
   the value of Q_3 (P_{75}) = 8.16 .
c) The value of D_6 (P_{60}) = 7.83 .

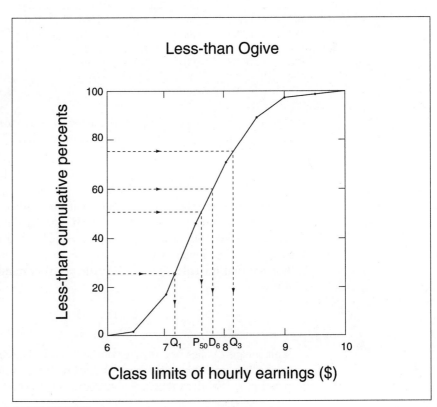

STATGRAPHICS®

D. Computing Percentiles by Formula — Ungrouped Data

Provided the data are first arranged in an *ascending* array, the position number of the percentiles within the array can be computed by using the formula

$$\text{Position number of } P_x \text{ within the array } = \frac{N + 1}{100}(x)$$

where P_x = the value at the xth percentile;
    x = the number of the desired percentile;
    N = the total number of observations in the data set.

    This formula can also be used to determine the position number of percentiles for arrays that have been grouped by value. In this case N is the sum of the frequencies in the groups of values.

○ **EXAMPLE 2.3c**

For the following data (arranged into an array in Example 2.1a), determine
a) the median;
b) the first quartile;
c) the third quartile.

162	178	186	194	195	200	205	215
172	180	188	195	197	200	209	216
174	181	188	195	198	202	210	221
176	185	190	195	198	204	214	227

● **SOLUTION**

a) The number of observations in the array, $N = 32$; for the median, $x = 50$; the position number of the median within the array

$$= \frac{(N + 1)}{100}(x) = \frac{(32 + 1)}{100}(50) = \frac{33}{2} = 16.50.$$

The position number 16.5 indicates that the median lies halfway between the values located at the 16th and 17th positions in the array. The value occupying the 16th position is 195; the value located at the 17th position is also 195. The median is 195.

b) For the first quartile, Q_1, $x = 25$; the position number of Q_1 within the array

$$= \frac{(32 + 1)}{100} = \frac{33}{4} = 8.25.$$

This means Q_1 lies one-quarter of the distance between the values located at the 8th and 9th position in the array. The value occupying the 8th position is 185; the value at the 9th position is 186.

$$Q_1 = 185 + 0.25(186 - 185) = 185 + 0.25 = 186.25$$

c) For the third quartile, Q_3, $x = 75$; the position number of Q_3 within the array

$$= \frac{(32 + 1)}{100}(75) = \frac{33}{4}(3) = 24.75.$$

Q_3 lies three-quarters of the distance between the values located at the 24th and 25th position in the array. The value located at the 24th position is 204; the value at the 25th position is 205.

$$Q_3 = 204 + 0.75(205 - 204) = 204 + 0.75 = 204.75$$

E. *Computing Percentiles by Formula — Grouped Data*

For frequency distributions the *value* at a given percentile can be obtained by using the formula

$$P_x = L_x + (i)\left(\frac{\dfrac{Nx}{100} - \Sigma f_c}{f_x}\right)$$

where P_x = the value at the xth percentile;
x = the number of the desired percentile;
N = total number of observations;
i = the size of the class interval;
L_x = the *lower* limit of the class containing the xth percentile;
f_x = the number of observations in the class containing the xth percentile;
Σf_c = the *cumulative* frequencies below the class containing the xth percentile.

Note The symbol Σ, read "sigma," is the Greek capital letter "S" and denotes the summation of values.

○ **EXAMPLE 2.3d**
For the frequency distribution given below (see Example 2.3a), compute
(a)P_{85}; (b) D_6; (c) Q_1; (d) Q_3; (e) the median.

Hourly earnings (in dollars)	Number of employees
6.00 to under 6.50	3
6.50 to under 7.00	28
7.00 to under 7.50	64
7.50 to under 8.00	56
8.00 to under 8.50	42
8.50 to under 9.00	22
9.00 to under 9.50	2
9.50 to under 10.00	3

● **SOLUTION**

First compute the less-than cumulative frequencies and percents as shown below.

Hourly earnings (in dollars)	Number of employees	Cumulative frequencies	Cumulative percents
6.00 to under 6.50	3	3	1.4
6.50 to under 7.00	28	31	14.1
7.00 to under 7.50	64	95	43.2
7.50 to under 8.00	56	151	68.6
8.00 to under 8.50	42	193	87.7
8.50 to under 9.00	22	215	97.7
9.00 to under 9.50	2	217	98.6
9.50 to under 10.00	3	220	100.0

a) The values of the variables appearing in the formula are now determined as follows:

$x = 85$ This is the rank number of the desired percentile.

$N = 220$ This is the total number of observations in the frequency distribution; it is the final number in the cumulative frequencies column.

$i = 0.50$ This is the width of the class intervals; it is determined by computing the difference between two successive class limits (e.g., $6.50 - 6.00 = 0.50$).

$L_x = 8.00$ This is the *lower* limit of the class containing the desired percentile. To determine this value, locate the cumulative percent that is closest to but greater than the rank number of the desired percentile. For the 85th percentile, the cumulative percent closest to 85 but greater than 85 is 87.7. Now read off the lower limit of the class in the column listing the class limits.

$f_x = 42$ This is the frequency in the class containing the desired percentile. This number is located in the frequency column next to the class.

$\sum f_c = 151$ This is the cumulative frequency below the class containing the desired percentile. The number is located in the cumulative frequencies column just before the class containing the desired percentile.

Now substitute in the formula and compute P_{85}:

$$P_{85} = 8.00 + (0.50) \left(\frac{\frac{(220)(85)}{100} - 151}{42} \right)$$

$$= 8.00 + (0.50) \left(\frac{187 - 151}{42} \right)$$

$$= 8.00 + (0.50) \frac{36}{42}$$

$$= 8.00 + 0.43$$

$$= 8.43$$

The value at the 85th percentile is \$8.43; that is, 85 percent of the employees earn hourly wages of less than \$8.43.

b) $D_6 = P_{60}$; $x = 60$; $N = 220$; $i = 0.50$; $L_x = 7.50$; $f_x = 56$; $\sum f_c = 95$
Use the following pattern to locate L_x, f_x and $\sum f_c$ in the frequency distribution table:

Use the following pattern to locate L_x, f_x, and $\sum f_c$ in the frequency distribution table:

$$D_6 = P_{60} = 7.50 + (0.50) \left(\frac{\frac{(220)(60)}{100} - 95}{56} \right)$$

$$= 7.50 + (0.50) \left(\frac{132 - 95}{56} \right)$$

$$= 7.50 + (0.50) \frac{37}{56}$$

$$= 7.50 + 0.33$$

$$= 7.83$$

The 6th decile is \$7.83; that is, 60 percent of the employees earn less than \$7.83 per hour.

c) $Q_1 = P_{25}; x = 25; N = 220; i = 0.50; L_x = 7.00; f_x = 64; \Sigma f_c = 31$

$$Q_1 = 7.00 + (0.50) \left(\frac{\frac{(220)(25)}{100} - 31}{64} \right)$$

$$= 7.00 + (0.50) \frac{24}{64}$$

$$= 7.00 + 0.19$$

$$= 7.19$$

The first quartile is $7.19; that is, 25 percent of the employees earn less than $7.19.

d) $Q_3 = P_{75}; x = 75; N = 220; i = 0.50; L_x = 8.00; f_x = 42; \Sigma f_c = 51$

$$Q_3 = 8.00 + (0.50) \left(\frac{\frac{(220)(75)}{100} - 51}{42} \right)$$

$$= 8.00 + (0.50) \frac{14}{42}$$

$$= 8.00 + 0.17$$

$$= 8.17$$

The third quartile is 8.17; that is, 75 percent of the employees earn less than $8.17.

e) Median = $P_{50}; x = 50; N = 220; i = 0.50; L_x = 7.50; f_x = 56; \Sigma f_c = 95$

$$\text{Median} = 7.50 + (0.50) \left(\frac{\frac{(220)(50)}{100} - 95}{56} \right)$$

$$= 7.50 + (0.50) \frac{15}{56}$$

$$= 7.50 + 0.13$$

$$= 7.63$$

The median is 7.63; that is, 50 percent of the employees earn less than $7.63.

F. Computing the Percentile Rank of a Given Value

The percentile rank of any given value in a frequency distribution can be found using the percentile formula. The difference in this type of problem is that the value of P_x is known, while x has to be determined.

This can be done by determining n, i, L_x, f_x and Σf_c, and then substituting in the formula and solving the resulting equation for x.

○ **EXAMPLE 2.3e**

For the frequency distribution given in Example 2.3a, determine the percentile rank of hourly earnings of (a) \$8.90;   (b) \$6.45.

● **SOLUTION**

a) $P_x = 8.90$; $N = 220$; $i = 0.50$; $L_x = 8.50$ (since 8.90 is located between 8.50 and 9.00); $f_x = 22$; $\sum f_c = 193$

$$8.90 = 8.50 + (0.50)\left(\frac{\dfrac{220\,x}{100} - 193}{22}\right)$$

$$8.90 = 8.50 + (0.50)\frac{2.20\,x - 193}{22}$$

$$8.90 - 8.50 = (0.50)\frac{2.20\,x - 193}{22}$$

$$(0.40)(22) = (0.50)(2.20\,x - 193)$$

$$8.80 = 1.10\,x - 96.50$$

$$105.30 = 1.10\,x$$

$$x = 95.7$$

Hourly earnings of \$8.90 lie at the 96th percentile.

b) $P_x = 6.45$; $N = 220$; $i = 0.50$; $L_x = 6.00$; $f_x = 3$; $\sum f_c = 0$ (the cumulative frequency below 6.00 is zero)

$$6.45 = 6.00 + (0.50)\left(\frac{\dfrac{220\,x}{100} - 0}{3}\right)$$

$$0.45 = (0.50)\frac{2.20\,x - 0}{3}$$

$$3(0.45) = (0.50)(2.20\,x - 0)$$

$$1.35 = 1.10\,x$$

$$x = 1.23$$

Hourly earnings of \$6.45 lie at the 2nd percentile.

EXERCISE 2.3

1. For Question **3** in Exercise 2.1, determine graphically the value of
   a) the 65th percentile;
   b) the 4th decile;
   c) the first quartile;
   d) the median;
   e) the percentile rank of a score of 15.

2. For Question **4** in Exercise 2.1, determine graphically the value of
   a) the 35th percentile;
   b) the 9th decile;
   c) the third quartile;
   d) the median;
   e) the percentile rank of weekly wages of $550.

3. Use the percentile formula to verify your solutions to Question **1**.

4. Use the percentile formula to verify your solutions to Question **2**.

5. For the frequency distribution given in Question **3** of Exercise 2.2,
   a) determine graphically the value of
      i) Q_1,
      ii) Q_3,
      iii) the median,
      iv) the percentile rank of the value 190;
   b) verify your answers in **(a)** by using the percentile formula.

6. For the frequency distribution given in Question **4** of Exercise 2.2,
   a) determine graphically the value of
      i) Q_1,
      ii) Q_3,
      iii) the median,
      iv) the percentile rank of the value 16;
   b) verify your answers in **(a)** by using the percentile formula.

 7. ***Data disk exercise***   Using the enclosed data and file name PUBLIC, determine
   (i) Q_1 and **(ii)** Q_3
   for the outstanding common shares (the variable name is COMMON).

REVIEW EXERCISE

1. The Government of Canada has determined that, on average, 100 high-school students will achieve the following levels of education:

Grade	9	10	11	12
Number of students	8	27	26	39

   a) Construct the less-than and more-than cumulative frequency and percent distributions.
   b) What percent of students
      i) do not graduate from high school?
      ii) have less than Grade 11 education?

2. The following open-ended distribution of share prices was recorded by the Toronto Stock Exchange at the close of trading on Friday.

Share price	Number of stocks
Penny stocks (under $5.00)	492
$5.00 to under $10.00	264
$10.00 to under $15.00	168
$15.00 to under $20.00	96
$20.00 to under $25.00	60
$25.00 and over	120

a) Construct the less-than and more-than cumulative frequencies and the percent distributions.

b) How many of the stocks were not penny stocks?

c) What percent of the stocks were
    i) $15.00 or more?
    ii) $25.00 or more?
    iii) less than $20.00?
    iv) under $10.00?

d) Compute the share price for a stock at
    i) the 60th percentile;
    ii) the first quartile.

e) Compute the percentile rank for a stock trading at $12.00.

3. The average hourly earnings, including overtime, for salaried employees in each province are as follows:

BC	Alta	Sask	Man	Ont
$16.92	$17.06	$15.58	$15.58	$17.72

Que	NB	NS	PEI	Nfld
$16.28	$14.75	$15.23	$14.11	$16.08

a) Organize the data in an ascending array.

b) Which province appears to be
    i) the poorest?
    ii) the richest?

c) Determine
    i) the range;
    ii) the mode;
    iii) the median.

4. The manager of a college pub recorded the number of beverages each student consumed on a Friday night and organized the data as follows:

Number of beverages	1	2	3	4	5	6 or more
Number of students	28	48	43	11	8	2

a) Construct the less-than and the more-than cumulative frequency and percent distributions.

b) How many students consumed
    i) 3 beverages or more?
    ii) less than 5 beverages?

c) What percent of the students had
   i) more than 5 beverages?
   ii) 2 beverages or less?

5. The data below list the test marks, out of 100, for a night-school statistics class:

74	45	53	84	63	72	88	92	56	74	79	63
97	49	67	73	78	80	51	66	61	80	68	64
58	66	70	78	64	69	42	85	73	65	68	62
55	63	71	79	87	62	67	72	77	71	61	51

a) Arrange the data in the form of an array.
b) Determine
   i) the range;
   ii) the mode;
   iii) the median.
c) Organize the data into a frequency distribution with a lowest class limit of 40 and class intervals of 10.
d) Compute the less-than cumulative frequencies and percent.
e) Compute the more-than cumulative frequencies and percent.
f) Compute the test score that lies at
   i) the 4th decile;
   ii) the third quartile.
g) Compute the percentile rank for a test score of 81.
h) Construct the histogram and the frequency polygon.
i) Construct the less-than and more-than frequency diagrams.

6. A survey of Canada's "Top 500" corporations produced the following average starting salaries by academic area:

Medicine	38 742	Marketing	28 684
Engineering	35 849	Accounting	27 551
Computer science	35 849	Business admin.	26 650
Physics	32 852	Hotel management	25 447
Mathematics	29 538	Education	24 779
Chemistry	28 814	Liberal arts	23 719

a) Determine the range for average starting salaries.
b) What is the median starting salary?
c) List the academic areas whose average starting salaries fall within the middle 50% of the data.

7. For the following set of data,
   a) draw the histogram and the frequency polygon;
   b) construct the less-than and more-than frequency diagrams.

Driver's age	Probability of having an accident
Under 25	42%
25 to under 35	30%
35 to under 45	18%
45 to under 55	11%
55 to under 65	26%
65 and over	41%

8. For the given distribution of hourly earnings,
   a) construct the histogram and the frequency polygon;
   b) construct the less-than and more-than ogives;
   c) compute
      i) the median,
      ii) the 9th decile,
      iii) $Q_3 - Q_1$.

Hourly earnings (in dollars)	Number of employees
5.00 to under 10.00	6
10.00 to under 15.00	16
15.00 to under 20.00	9
20.00 to under 25.00	7
25.00 to under 30.00	4

9. A recent survey of expenditures per consumer for vacations showed the following results:

Amount spent for vacations	Number of respondents
$0 to under $500	179
$500 to under $1000	187
$1000 to under $1500	82
$1500 to under $2000	49
$2000 to under $2500	53

   a) Construct the less-than and more-than frequency diagrams for the given data.
   b) What is the name of the point of intersection of the two curves?
   c) Compute the value of the median.
   d) Compute $Q_3 - Q_1$.
   e) Determine the percent of the respondents who spent
      i) less than $1500;
      ii) more than $500.
   f) Given that a particular respondent spent $2200,
      i) compute the percentile rank for that respondent;
      ii) determine how many respondents spent more than $2200.
   g) Compute how many respondents spent less than $400.

10. Last year the distribution of drivers by age group in Alberta was as follows:

Driver's age	16 to under 25	25 to under 35	35 to under 45	45 to under 55	55 to under 65	65 and over
Frequency	6%	25%	27%	18%	17%	7%

a) Construct the following
  i) the histogram;
  ii) the frequency polygon;
  iii) the less-than ogive;
  iv) the more-than ogive.
b) Use the cumulative relative frequencies to determine the percent of drivers in the following age groups:
  i) under 55;
  ii) between 25 and 45;
  iii) 35 and over;
  iv) under 25 and 65 or over.
c) Compute the following:
  i) the first quartile;
  ii) the median;
  iii) the 85th percentile;
  iv) the percentile rank of a 37-year-old driver.

SELF-TEST

1. The owner of a small business has kept track of the number of customers for each of the last 24 business days:

| 67 | 57 | 43 | 59 | 80 | 25 | 48 | 51 | 62 | 53 | 65 | 44 |
| 78 | 23 | 34 | 84 | 60 | 74 | 38 | 75 | 26 | 88 | 33 | 41 |

a) Organize the data in the form of a descending array.
b) Determine
  i) the range;
  ii) the median.
c) Organize the data into a frequency distribution with equal class intervals of size 10.

2. The co-ordinator of a business administration program gave a short test to a group of first-year students during the first week of classes. The results of the test were as follows:

Correct answers	6	5	4	3	2	1	0
Number of students	42	72	65	18	12	5	2

a) Construct a histogram for the data.
b) Construct the more-than and less-than frequency diagrams (ogives).
c) What percent of the students had three or more correct answers?
d) What percent of the students had fewer than four correct answers?

3. The hourly wages paid by Heavy Metal Inc. to its labour force are summarized below:

Hourly rate of pay	$10.00 to under $15.00	$15.00 to under $20.00	$20.00 to under $25.00	$25.00 to under $30.00	$30.00 to under $35.00
Number of employees	9	22	14	10	5

Compute
a) the first quartile;
b) the median;
c) the fourth decile;
d) the 80th percentile;
e) the percentile rank of an employee earning $28.00;
f) the percentile rank of an employee earning $12.00.

Key Terms

Summary of Formulas

1. Percentiles

a) Ungrouped Data:

$$\text{Position number of } P_x \text{ within array} = \frac{N+1}{100}(x)$$

b) Grouped Data:

$$P_x = L_x + (i)\left(\frac{\frac{Nx}{100} - \Sigma f_c}{f_x}\right)$$

CHAPTER 3

Measures of Central Tendency

Introduction

Percentiles provide useful information about the relative position of observations in an array or in frequency distributions. However, much additional information can be obtained by computing *central (average) values*. These values, referred to as *measures of central tendency*, are the values that are most representative of the array or the frequency distribution, since they are the central values around which the data tend to cluster.

Objectives

Upon completion of this chapter you will be able to
1. differentiate between the mean, the median and the mode of data;
2. determine the value of the mean, the median and the mode of ungrouped data;
3. determine the value of the mean, the median and the mode of grouped data;
4. recognize the relationships among the three measures of central tendency for symmetrical and skewed distributions;
5. be familiar with the advantages and disadvantages of the three measures.

SECTION 3.1 Central Values (Averages)

Presenting statistical data by means of tables and graphs is a first step toward analyzing data. Additional insights can be obtained by using values that are representative of a set of data.

The most common types of central values are the *mean*, the *median* and the *mode*. These three measures are referred to as **measures of central tendency** since they are indicators of typical middle values in the data set.

The *mean* is the arithmetic average of all values in the data set; the *median* indicates the middle value; the *mode* is the most frequently occurring value.

The Mean—Arithmetic Average

A. The Mean of Ungrouped Data

The mean (arithmetic average) of a set of observations is determined by adding the values and dividing by the number of values.

○ **EXAMPLE 3.2a**
The test scores for a group of students on a test were 73, 86, 52, 6, 93, 74, 81, 70, 68. Determine the mean test score.

● **SOLUTION**
To determine the mean, add the values and divide by the number of values:
The sum of the values = 73 + 86 + 52 + 6 + 93 + 74 + 81 + 70 + 68 = 603;
the number of values = 9;

$$\text{the mean} = \frac{\text{the sum of the values}}{\text{the number of values}} = \frac{603}{9} = 67.$$

B. Formula for Computing the Mean

The computations performed in Example 3.2a can be described by the following notation:

μ (pronounced "mu") = the mean of the data set;
N = the number of observations in the data set;
x = the individual values of the N observations;
$\sum x$ = the sum of the individual values.

Note The symbol $\sum$, pronounced "sigma," is the summation symbol indicating that values are to be added.

$$\mu = \frac{\sum x}{N} = \frac{73 + 86 + 52 + 6 + 93 + 74 + 81 + 70 + 68}{9} = \frac{603}{9} = 67$$

C. The Weighted Mean

Weighted mean calculations occur when specific values appear more than once in the data set. The frequency with which each value occurs is used as a weighing factor. The resulting products are added up and divided by the total frequency.

○ **EXAMPLE 3.2b**
The following data summarize the test scores received by a group of students:

Test score	4	5	6	7	8	9	10
Number of students	3	17	24	45	21	9	1

Determine the average test score for the group of students.

● **SOLUTION**

Test score x	Frequency f	Weighted test score fx
4	3	(4)(3) = 12
5	17	(5)(17) = 85
6	24	(6)(24) = 144
7	45	(7)(45) = 315
8	21	(8)(21) = 168
9	9	(9)(9) = 81
10	1	(10)(1) = 10
Total	$N = \sum f = 120$	$\sum fx = 815$

The test scores are represented by x; the number of students (frequency) achieving a particular test score is represented by f; and the resulting products are represented by fx.

The mean value μ is now obtained by dividing the sum of the products fx by the number of observations N:

$$\mu = \frac{\sum fx}{N} = \frac{815}{120} = 6.79$$

From the above, the weighted average formula is

$$\mu = \frac{\sum fx}{N}$$

where μ = the mean of the data set;
x = the observed values;
f = the number of times each value occurs;
$N = \sum f$ = the number of observations in the data set;
fx = the weighted values of the observed values;
$\sum fx$ = the sum of the weighted values.

D. The Mean of Grouped Data

Computing the mean of grouped data involves a weighted average calculation. The formula used is the same as that used for calculating a weighted average. However, there is an important difference in the meaning of the symbol x.

For grouped data the individual values included in a class are no longer known. *The class midpoint is assumed to be the central value* for the individual values in the class, and the symbol x is used to represent the class midpoints. As a consequence, the calculation results in an *estimate* of the mean value for the frequency distribution.

The formula for finding the mean of data grouped into classes is

$$\mu = \frac{\sum fx}{N}$$

where μ = the mean of the data set;

    x = the individual class midpoints;

    f = the number of observations in each class;

    $N = \sum f$ = the total number of observations in the data set;

    $\sum fx$ = the sum of the weighted values of the class midpoints.

○ EXAMPLE 3.2c

The following is a summary of the weekly wages of a group of employees.

Weekly wages	Number of employees
$360 to under $400	24
$400 to under $440	40
$440 to under $480	55
$480 to under $520	36
$520 to under $560	25
$560 to under $600	12
$600 to under $640	3

Determine the average weekly wages.

● SOLUTION

The given data and the required calculations are shown in Table 3.1:

TABLE 3.1 Calculation for Grouped Mean

Column 1	Column 2	Column 3	Column 4
Weekly wages $	Number of employees f	Class midpoint x	Weighted class midpoint fx
$360 to under $400	24	380	(24)(380) = 9 120
400 to under 440	40	420	(40)(420) = 16 800
440 to under 480	55	460	(55)(460) = 25 300
480 to under 520	36	500	(36)(500) = 18 000
520 to under 560	25	540	(25)(540) = 13 500
560 to under 600	12	580	(12)(580) = 6 960
600 to under 640	3	620	(3)(620) = 1 860
Totals	$N = \sum f = 195$		$\sum fx = 91\ 540$

Explanation of numbers in Table 3.1:

1. The given class intervals of weekly wages are listed in Column 1.
2. The number of employees in each class (represented by f) are listed in Column 2.
3. The numbers in Column 2 are added to obtain $N = \sum f = 195$.
4. The class midpoints, represented by x, are listed in Column 3. The class midpoints are found by adding successive class 360 + 400 limits and dividing by 2, e.g., $\dfrac{360 + 400}{2} = 380$. This repetitive calculation can

be simplified by computing the class midpoint of the lowest class by the method described above and obtaining the remaining class midpoints by adding the width of the class interval. In Table 3.1 the midpoint of the lowest class is 380, and the width of the class intervals is 40. The successive remaining class midpoints are obtained by adding 40 — for example, 380 + 40 = 420; 420 + 40 = 460; and so on.

5. The weighted class midpoints fx, listed in Column 4, are obtained by multiplying the numbers in Column 2 and Column 3 in order.

6. The numbers in Column 4 are added to obtain $\sum fx$ = 91 540. Now use the formula and substitute:

$$\mu = \frac{\sum fx}{N} = \frac{91\ 540}{195} = 469.44$$

EXERCISE 3.2

1. The results of a test for a statistics class were as follows:
   35, 63, 83, 74, 47, 63, 63, 88, 97, 78, 82, 55, 71
   Determine the mean test score.

2. The number of units produced by a group of workers were
   84, 76, 68, 64, 71, 77, 62, 87, 64, 64, 75, 66
   Determine the average production per worker.

3. The weekly hours for a group of employees were as follows

Weekly hours	36	37	38	39	40	41	42	43	44
No. of employees	4	9	12	12	14	22	26	19	2

Compute the mean weekly hours worked per employee.

4. Determine the mean test score for a group of job applicants from the following set of data:

Test score	45	50	55	60	65	70	75	80	85
Frequency	7	12	36	73	66	47	35	16	8

5. Compute the mean of the following frequency distribution:

Class	0 to under 10	10 to under 20	20 to under 30	30 to under 40	40 to under 50	50 to under 60	60 to under 70	70 to under 80	80 to under 90
Frequency	7	20	30	27	15	10	5	5	3

6. Determine the arithmetic mean of the given set of data:

Class	5.00 to under 5.50	5.50 to under 6.00	6.00 to under 6.50	6.50 to under 7.00	7.00 to under 7.50	7.50 to under 8.00	8.00 to under 8.50	8.50 to under 9.00	9.00 to under 9.50
Frequency	5	22	49	63	45	25	5	4	2

7. **Data disk exercise**   Using the enclosed data disk and the file name MARKET, determine the mean score for Sample One, Question One (the variable name is S1Q1).

The Median

A.   The Median of Ungrouped Data

The median is the value that occupies the middle position in a set of data.

To determine the median of ungrouped data, arrange the values in the form of an array. The position of the median can then be determined by the formula $\frac{N+1}{100}(x)$, introduced in Section 2.3.

Since, for the median, $x = 50$, the formula for finding the position number of the median in an array simplifies to $\frac{N+1}{2}$.

This means that the position of the median in an array can be quickly determined by adding "1" to the number of observations in the data set and dividing by "2."

○ **EXAMPLE 3.3a**

The test scores for the group of students in Example 3.2a were 73, 86, 52, 6, 93, 74, 81, 70, 68. Determine the median.

● **SOLUTION**

First arrange the test scores in order of magnitude

Position number	1	2	3	4	5	6	7	8	9
Test score	6	52	68	70	73	74	81	86	93

The number of values (test scores), $N = 9$;

the position number of the median $= \frac{N+1}{2} = \frac{9+1}{2} = 5$;

the fifth position in the array is occupied by the test score 73;
the median test score is 73.

○ **EXAMPLE 3.3b**

The hourly wages for a group of employees are

$9.00, \$6.00, \$7.00, \$20.00, \$8.40, \$4.40, \$9.60, \$7.80.$

Determine the median hourly pay.

● **SOLUTION**

First arrange the hourly rates of pay in order of magnitude:

Position number	1	2	3	4	5	6	7	8
Hourly rate	4.40	6.60	7.00	7.80	8.40	9.00	9.60	20.00

The number of values, $N = 8$;

The position number of the median $= \dfrac{N + 1}{2} = \dfrac{8 + 1}{2} = 4.5$.

The fractional value 4.5 indicates that the median is located between the fourth and the fifth position.

The fourth position shows a value of $7.80;

the fifth position shows a value of $8.40.

The median is assumed to be the average value of the two middle positions

$= \dfrac{7.80 + 8.40}{2} = 8.10$.

Notes

1. If the number of observations in an array is an *odd* number, the computed middle position is always a *whole* number. The array has a single middle position, and the median is the value occupying that position.
2. If the number of observations in an array is an *even* number, the computed middle position is always a *fraction*. The array has two middle positions and the median is the average of the two values occupying the two middle positions.

B. *The Median of Data Grouped by Value*

○ **EXAMPLE 3.3c**

Determine the median for the test scores in Example 3.2b (see data below).

● **SOLUTION**

Test scores	4	5	6	7	8	9	10
No. of students	3	17	24	45	21	9	1
Cumulative no.	3	20	44	89	110	119	120

The test scores are arranged in order of magnitude and cumulative numbers are obtained by listing the running totals.

The total number of test scores, $N = 120$;

the median position $= \dfrac{120 + 1}{2} = 60.5$.

The middle positions are the 60th and 61st positions.

The cumulative totals indicate the positions occupied by the various test scores. Test score "4" occurs three times; the lowest three positions are occupied by a test score of "4." The next seventeen positions (positions 4 to 20) are occupied by a test score of "5"; positions 21 to 44 are occupied by a test score of "6"; positions 45 to 89 are occupied by a test score of "7." Therefore, positions 60 and 61 are occupied by a test score of "7." The median is 7.

C. The Median of Grouped Data

For data arranged in class intervals, the median is the 50th percentile and can be found by using the percentile formula introduced in Section 2.3:

$$P_x = L_x + (i) \left(\frac{\frac{Nx}{100} - \Sigma f_c}{f_x} \right)$$

○ **EXAMPLE 3.3d**

Determine the median weekly wages for the group of employees in Example 3.2c (see data in the solution below).

● **SOLUTION**

To determine a percentile, proceed as explained in Section 2.3.

Compute the cumulative frequencies and cumulative percents, determine the values needed and substitute in the formula.

Weekly wages	Number of employees	Cumulative frequencies	Cumulative percents
$360 to under $400	24	24	12.3
400 to under 440	40	64	32.8
440 to under 480	55	119	61.0
480 to under 520	36	155	79.5
520 to under 560	25	180	92.3
560 to under 600	12	192	98.5
600 to under 640	3	195	100.0

$x = 50$; $N = 195$; $i = 40$; $L_x = 440$; $f_x = 55$; $\Sigma f_c = 64$

$$P_{50} = 440 + (40) \left(\frac{\frac{(195)(50)}{100} - 64}{55} \right)$$

$$= 440 + (40) \left(\frac{97.50 - 64}{55} \right)$$

$$= 440 + (40) \frac{33.50}{55}$$

$$= 440 + 24.40$$

$$= 464.40$$

The median weekly wage is $464.40.

EXERCISE 3.3

1. Compute the median for the data in Exercise 3.2, questions **1**, **3** and **5**.

2. Compute the median for the data in Exercise 3.2, questions **2**, **4** and **6**.

 3. *Data disk exercise* Using the enclosed data disk and the file name MARKET, determine the median score for Sample One, Question Two (the variable name is S1Q2).

SECTION 3.4 The Mode

The mode is the most frequently occurring value in a set of data.

A. The Mode of Ungrouped Data

The mode of ungrouped data is readily determined by arranging the data in the form of an array and locating the most frequently occurring value. If no value is repeated, the data have no mode.

○ **EXAMPLE 3.4a**
Test scores for a group of students were 12, 9, 16, 9, 14, 17, 12, 17, 14, 14, 19, 14. Determine the mode.

● **SOLUTION**
First arrange the test scores in an array:

$$9, 9, 12, 12, 14, 14, 14, 14, 16, 17, 17, 19$$

The most frequent test score of 14 is the mode.

B. The Mode of Grouped Data

For data arranged in a frequency distribution, the mode is assumed to be located in the class having the highest frequency. The mode can be determined either graphically from the histogram of the distribution or by the formula

$$\text{MODE} = L_{MO} + (i) \left(\frac{d_1}{d_1 + d_2} \right)$$

where L_{MO} = the lower limit of the modal class (the class having the highest frequency);
i = the width of the modal class;
d_1 = the frequency in the modal class minus the frequency in the class below the modal class;
d_2 = the frequency in the modal class minus the frequency in the class above the modal class.

○ **EXAMPLE 3.4b**

Determine the mode for the data used in Example 3.2c (reproduced below) (a) graphically;   (b) by formula.

● **SOLUTION**

Weekly wages	Number of employees	
$360 to under $400	24	
400 to under 440	40	$d_1 = 55 - 40 = 15$
440 to under 480	55	← Modal class
480 to under 520	36	$d_2 = 55 - 36 = 19$
520 to under 560	25	
560 to under 600	12	
600 to under 640	3	

a) *Graphical solution.* First construct the histogram and locate the *modal class* represented by the highest bar. Mark the endpoints of the bar A and B as shown. Mark the top right corner of the bar representing the class below the modal class as C. Mark the top left corner of the bar representing the class above the modal class as D.

   Join BC and AD and mark the point of intersection of the two line segments E. Draw a line from E perpendicular to the horizontal axis at F. Read the value for F from the horizontal scale as the value of the mode = 458 (approximately).

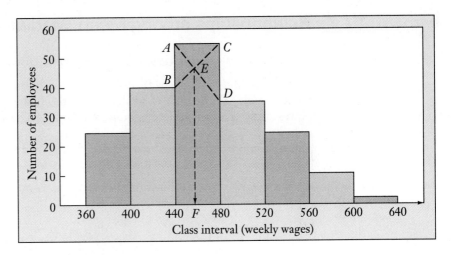

b) *Solution by formula.*

L_{MO} = 440 (lower limit of modal class)
i = 40 (width of modal class)
the frequency in the modal class = 55
the frequency in the class below = 40
the frequency in the class above = 36
d_1 = 55 − 40 = 15; d_2 = 55 − 36 = 19

$$\text{MODE} = 440 + (40)\,\frac{15}{15 + 19} = 440 + (40)\,\frac{15}{34} = 440 + 17.65 = 457.65$$

EXERCISE 3.4

1. Determine the mode for the data in Exercise 3.2, Question **1**.

2. Determine the mode for the data in Exercise 3.2, Question **2**.

3. Graphically determine the mode for the frequency distribution in Exercise 3.2, Question **5**, and verify your answer using the formula.

4. Graphically determine the mode for the frequency distribution in Exercise 3.2, Question **6**, and verify your answer using the formula.

5. *Data disk exercise* Using the enclosed data disk and the file name MARKET, determine the most frequent response (mode) for Sample One, Question Three (the variable name is S1Q3).

SECTION 3.5 Relationships among Mean, Median and Mode

A. Location of Central Values for Symmetrical Distributions

In a *symmetrical* (bell-shaped) distribution, the mean, the median and the mode *coincide*. Their value is located under the highest point on the graph representing the distribution, as indicated on the diagram.

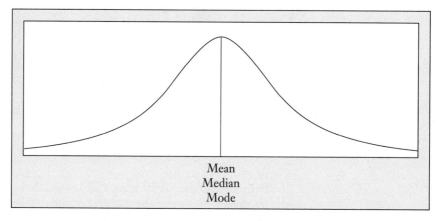

Mean
Median
Mode

B. Location of Central Values for Skewed Distributions

Distributions that are not symmetrical are referred to as **skewed distributions**. For skewed distributions, only the mode is located under the highest point.

A distribution is *positively skewed* if more of the extreme values lie to the *right* of the highest point. The mode is least influenced by extreme values and lies under the highest point on the graph. The median is somewhat affected by extreme values and is located to the right of the mode. The mean is most affected and lies farther to the right.

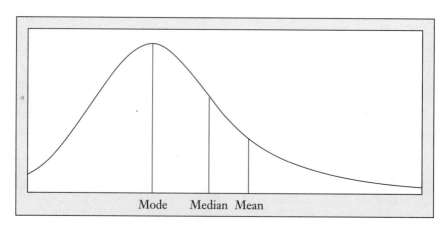

Mode   Median Mean

A distribution is *negatively skewed* if more of the extreme values lie to the *left* of the highest point on the graph. The mode is located under the highest point and the median and mean are located to the left of the mode, as indicated on the diagram.

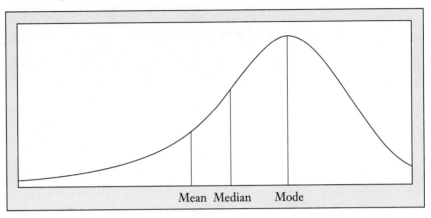

Note that for skewed distributions the median is located between the mean and the mode. If any two of the three central values are known, an estimate of the third central value can be obtained by using the relationship

$$\text{MEAN} - \text{MODE} = 3(\text{MEAN} - \text{MEDIAN})$$

This indicates that the difference between the mean and the mode is approximately three times the difference between the mean and the median.

C. Advantages and Disadvantages of the Three Measures of Central Tendency

1. **The Mean**   *–Advantages*
   –the mean reflects all values;
   –the basic calculation is readily understood;
   –the mean has mathematical properties that are useful in many statistical procedures.

   –Disadvantages
   –the mean is unduly influenced by extreme values;
   –it cannot be computed for open-ended distributions.

2. **The Median** *–Advantages*
   –the concept is easy to understand;
   –the median can be determined for any distribution, including open-ended distributions;
   –it is not unduly influenced by extreme values.

–*Disadvantages*
–arranging data in an array is time-consuming;
–the median lacks the useful mathematical properties that make the mean the preferred statistical measure of central tendency.

3. **The Mode**   –*Advantages*
–the mode can be obtained for any distribution;
–it is not affected by extreme values;
–it can be obtained for qualitative data.

–*Disadvantages*
–not all sets of data have a modal value;
–some sets of data have more than one modal value;
–multiple modal values are usually difficult to interpret;
–the mode lacks useful mathematical properties.

○ **EXAMPLE 3.5a**

The following were the scores on a mathematics test:

14	11	13	18	18	16	10
18	17	20	20	11	14	18
15	8	17	8	13	18	9
15	16	13	11			

a) Determine the median and the mode.
b) Estimate the mean using the formula

$$\text{MEAN} - \text{MODE} = 3(\text{MEAN} - \text{MEDIAN})$$

c) Compute the actual mean.
d) Comment on the nature of the distribution.

● **SOLUTION**

a) First arrange the test scores in an array:

8	8	9	10	11	11	11
13	13	13	14	14	15	15
16	16	17	17	18	18	18
18	18	19	20			

The data contain 25 test scores, $N = 25$;

the median position $= \dfrac{N+1}{2} = \dfrac{25+1}{2} = 13$;

the 13th position is occupied by a test score of 15;
the median is 15.
The most frequent score is 18, occurring 5 times;
the mode is 18.

b)
$$\text{MEAN} - \text{MODE} = 3(\text{MEAN} - \text{MEDIAN})$$
$$\mu - 18 = 3(\mu - 15)$$
$$\mu - 18 = 3\mu - 45$$
$$27 = 2\mu$$
$$\mu = 13.5$$

The estimated mean is 13.5.

c)
$$\mu = \frac{\sum x}{N} = \frac{360}{25} = 14.4$$

The actual mean is 14.4.

d) Since the mode is the highest value (18), followed by the median (15) and the mean (14.4), the distribution is negatively skewed.

EXERCISE 3.5

1. Test scores for a statistics class were as follows:

5	11	4	18	8	15	17
20	14	7	16	13	8	16
20	11	8	8	12	8	14
10	15	5	8	15	9	14

a) Determine the median and the mode.
b) Compute the true mean.
c) Use the formula MEAN − MODE = 3(MEAN − MEDIAN) to estimate the mean.
d) Comment on the nature of the distribution.

2. For a particular set of observations the mode was determined to be 840 and the median was 810. Assuming a moderately skewed distribution,
   a) sketch the distribution and label the positions of the mean, median and mode;
   b) calculate the approximate value of the mean.

3. In a frequency distribution the mean is 38 and the mode is 33. Determine whether the distribution is symmetrical, positively skewed or negatively skewed.

4. Given that the difference between the mean and the median of a frequency distribution is 8, estimate the difference between the median and the mode.

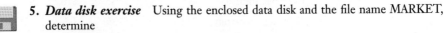 5. *Data disk exercise* Using the enclosed data disk and the file name MARKET, determine
   i) the mean;
   ii) the median
   for Sample One, Question Four (the variable name is S1Q4). Estimate the mode and comment on the distribution.

REVIEW EXERCISE

1. Sales (in $ billions) for General Motors Corporation from 1985 to 1990 were as follows:

Year	1985	1986	1987	1988	1989	1990
Sales	96.3	102.8	101.7	120.4	123.2	124.7

For the time period, determine
a) the company's average sales;
b) the median sales.

2. The number of employees working for Canada's 10 largest banks in 1990 were as follows:

Royal Bank of Canada	42 910
National Bank of Canada	13 873
Canadian Imperial Bank of Commerce	35 811
Hongkong Bank of Canada	3 000
Laurentian Bank of Canada	3 000
Toronto-Dominion Bank	24 549
Bank of Nova Scotia	30 114
Bank of Montreal	33 500
Citibank Canada	725
Barclays Bank of Canada	1 500

a) Compute the average number and median number of employees for the group of banks.
b) Do your answers to (a) reasonably describe the group of banks by size? Comment on the factors influencing the measures calculated in (a).

3. Sales (in $ billions) of Canada's largest companies have been summarized below:

Sales	4 to under 8	8 to under 12	12 to under 16	16 to under 20
Number of companies	19	7	1	2

a) Calculate the average sales for the group of companies.
b) Determine the median value of their sales.

4. The trustee for a bankrupt hardware store compiled the following inventory list by unit price. Compute the average price per item.

Cost per item	$2	$3	$5	$10	$20	$30	$50
Number of items	251	182	87	93	41	6	18

5. An ergonomic study found that a customer-service representative spent an average of 48 seconds on the phone helping a customer. After checking the accuracy of the results, management discovered that one of the 10 phone calls used to calculate the mean should have read 22 seconds instead of 72 seconds. Compute the average duration of the phone calls, using the corrected data.

6. If the arithmetic mean of 22 observations is 553, what will be the mean if one of the values is changed from 959 to 453?

7. If 12 students weigh an average 67 kg each, what will be the mean weight per student if another student whose weight is 106 kg is included in the calculation?

8. Preliminary records showed that Bruce averaged 26 points per game in the college intramural basketball league. On rechecking, it was discovered that he had played a tenth game in which he had scored 36 points. What was Bruce's point average per game?

9. Anne and Randy calculated their average hydro bill for the first six months of the year was $98.23. Randy wanted to know how much they had paid each month but could only locate the following hydro bills:

January	$135.29	February	$128.73	March	$?
April	$67.22	May	$50.98	June	$77.21

What was the amount of the hydro bill for March?

10. The mean of 10 numbers is 177. Nine of the numbers are as follows:
   58   82   47   23   15   79   36   61   25
   a) Determine the size of the missing number.
   b) Considering the magnitude of the missing number, is the mean a representative measure for the group of numbers? Explain.

11. Joan Condie likes to "play the stock market." She recently purchased the following shares:

Company	Number of shares	Price per share
IBM	100	$110
Canadian Tire	500	24
BCE Inc.	200	42

Calculate the average price per share paid by Joan.

12. Joe owns some homes, which he rents to students at the following monthly rates per room:

Monthly rent	$250	$300	$350	$400
Number of rooms	6	5	3	2

Determine the average rent per room received by Joe.

13. The ages of video renters at Jumbo Video last Saturday were as follows:

21	15	19	29	16	19	23	23	12	39
22	15	19	37	49	46	19	22	15	10

   a) Determine the median and the mode.
   b) Estimate the value of the mean, using the formula given in Section 3.5.
   c) Compute the actual mean.
   d) Identify the type of skewness of the data.

14. The height, in metres, of a group of stores located consecutively along Yonge

Street is as follows:

5	28	30	28	13	21	26	28
12	13	21	26	28	11	8	30

a) Determine the mean and the median.
b) Estimate the value of the mode using the values obtained in **a)** .
c) Determine the value of the actual mode.
d) Comment on the nature of the distribution.

15. For the following frequency distribution,
   a) calculate the mean, the median and the mode;
   b) identify the type of distribution by skewness.;
   c) graphically determine the value of the mode.

Profit ($ millions)	Number of firms
0 to under 10	8
10 to under 20	11
20 to under 30	15
30 to under 40	5
40 to under 50	1

16. Given the following data,
   a) graphically determine the value of the mode;
   b) identify the nature of the skewness in the distribution;
   c) compute the mean, the median and the mode.

Class	0 to under 20	20 to under 40	40 to under 60	60 to under 80	80 to under 100
Frequency	2	4	24	15	5

17. The median unit cost and the mode unit cost of a company's inventory are $5.50 and $5.10 respectively.
   a) Estimate the mean unit cost of the inventory by formula.
   b) Identify the nature of skewness of the data.

18. Given that the mean and the mode for a set of test scores are 65 and 71 respectively,
   a) estimate the median test score by formula;
   b) comment on the nature of the distribution.

19. Given the three measures of central tendency, state whether each of the following distributions is symmetrical, positively skewed, or negatively skewed.

	Mean	Median	Mode
a)	92	93	95
b)	0.65	0.52	0.49
c)	10.0	10.0	10.0

20. Utilize the value of the three given measures to sketch each of the following

distributions and indicate the nature of symmetry or skewness.

	Mean	Median	Mode
a)	111	1111	11111
b)	0.009	0.008	0.001
c)	349	359	364
d)	15 990	15 990	15 990

SELF-TEST

A. Questions **1** through **4** are based on the following data:

Linden Life Insurance Company
Seniority of Employees (in Months)
with Less Than Five Years' Service with the Company

51	2	24	20	11	43	42	22	24	59	12	38
33	25	24	8	18	31	47	24	3	35	17	45
28	55	51	13	8	30	7	34	42	17	27	24

1. Determine the median and the mode.
2. Estimate the mean, using the values obtained for the median and the mode in Question **1**.
3. Compute the true mean.
4. Identify the nature of skewness in the data.

B. Questions **5** through **8** are based on the following information:

Linden Life Insurance Company
Seniority of Employees (in Years)

Seniority	Number of employees
0 to under 5	35
5 to under 10	39
10 to under 15	28
15 to under 20	13
20 to under 25	15
25 to under 30	11
30 to under 35	7
35 to under 40	2

5. Compute Q_1 and Q_3.
6. Compute the mean, the median and the mode.
7. Comment on the skewness of the distribution.
8. Graphically determine the value of the mode.

C. Answer the following question.

9. Marty was talking to Lise about his flight from Vancouver to Toronto, a distance of 4537 km. The plane left at 9.00 a.m. and arrived in Toronto at 4:00 p.m. Based on that information, Marty calculated the plane's speed to be 648.1 km/h. Lise pointed out to him that his calculation was incorrect because of the three-hour difference in time zones. What was the plane's true average speed per hour?

Key Terms

Measures of central tendency   59
Skewed distributions   70

Summary of Formulas

1. The Arithmetic Mean:

a) Ungrouped data:

$$\mu = \frac{\sum x}{N}$$

b) Grouped data:

$$\mu = \frac{\sum fx}{N}$$

2. The Median:

a) Ungrouped data:

$$\text{Position number of median in array} = \frac{N + 1}{2}$$

b) Grouped data:

$$P_{50} = L_x + (i) \left(\frac{\frac{Nx}{100} - \sum f_c}{fx} \right)$$

3. The Mode:

Grouped data:

$$\text{MODE} = L_{MO} + (i) \left(\frac{d_1}{d_1 + d_2} \right)$$

4. Empirical Relationship:

$$\text{MEAN} - \text{MODE} = 3(\text{MEAN} - \text{MEDIAN})$$

CHAPTER 4

Measures of the Variability of Data

Introduction

The measures of central tendency considered in Chapter 3 are useful because data tend to cluster around central values. However, as the individual values in a distribution of data differ from each other, central values provide only an incomplete picture of the features of the distribution.

For example, a mean of 6.79, a median of 7 and a mode of 7 (the three central values in Example 3.2b) indicate a fairly symmetrical distribution but provide no clue as to the spread of the data. Similarly, a mean of \$469.44, a median of \$464.40, and a mode of \$457.65 (the three central values computed for the grouped data in Example 3.2c) indicate a positively skewed distribution but again provide no information about the dispersion of the data.

To obtain a more complete picture of the nature of the distribution, the *variability* (or *dispersion* or *spread*) of the data needs to be measured.

The **measures of variability** considered in this chapter include the *range*, the *interquartile range*, the *average deviation from the mean*, the *variance*, the *standard deviation* and the *coefficient of variation*.

Objectives

Upon completion of this chapter you will be able to
1. determine the range and interquartile range of grouped and ungrouped data;
2. determine the average deviation from the mean for grouped and ungrouped data;
3. compute the variance and the standard deviation for grouped and ungrouped data;
4. compute the coefficient of variation of different data sets and interpret the results.

SECTION 4.1   The Range and the Interquartile Range

A. The Range

The **range** is a measure of the distance between the *highest* and the *lowest* value in a set of data. For grouped frequency distributions, the highest value in the data set is assumed to be the upper limit of the class containing the greatest values. The lowest value is the lower limit of the class containing the smallest values.

○ **EXAMPLE 4.1a**
Determine the range in test scores for the following data (see Example 3.2b).

Test score	4	5	6	7	8	9	10
Number of students	3	17	24	45	21	9	1

● **SOLUTION**
The highest test score is 10;
the lowest test score is 4.
The range = HIGHEST VALUE − LOWEST VALUE = $10 - 4 = 6$.

○ **EXAMPLE 4.1b**
Determine the range in weekly wages for the following data (see Example 3.2c).

Weekly wages	Number of employees
$360 to under $400	24
$400 to under $440	40
$440 to under $480	55
$480 to under $520	36
$520 to under $560	25
$560 to under $600	12
$600 to under $640	3

● **SOLUTION**
The upper limit of the highest class is $640; the lower limit of the lowest class is $360. The range = $640 - 360 = \$280$.

Note   The range is easily computed and its meaning is readily understood. However, it is at best a very rough indicator of dispersion since the calculation does not allow for the distribution of the data between the two extreme values.

B. The Interquartile Range

The **interquartile range** is a measure of the distance between the *first* and *third quartiles* in a set of data. This means it measures the spread of the *middle*

50% of the observed values. Measuring the spread of the middle 50% of the values is useful for distributions that have extremely high or low values.

○ EXAMPLE 4.1c
Determine the interquartile range for the data used in Example 4.1a.

● SOLUTION

Test score	4	5	6	7	8	9	10
Number of students	3	17	24	45	21	9	1
Cumulative number	3	20	44	89	110	119	120

$N = 120$.

The position of the first quartile Q_1 in the array

$$= \frac{N+1}{100}(x) = \frac{120+1}{100}(25) = \frac{121}{4} = 30.25.$$

This means Q_1 lies one-quarter the distance between the test scores obtained by the two students occupying positions 30 and 31. Since both students had a test score of 6, $Q_1 = 6$.

The position number of the third quartile (Q_3)

$$= \frac{120+1}{100}(75) = \frac{121}{4}(3) = 90.75.$$

This means Q_3 lies three-quarters the distance between the test scores obtained by the two students occupying positions 90 and 91. Since both students had a test score of 8, $Q_3 = 8$.

The interquartile range $= Q_3 - Q_1 = 8 - 6 = 2$.

○ EXAMPLE 4.1d
Determine the interquartile range in weekly wages for Example 4.1b.

● SOLUTION
For grouped data, the percentile formula

$$P_x = L_x + (i)\frac{\dfrac{Nx}{100} - \Sigma f_c}{f_x}$$

can be used to compute the values of Q_3 and Q_1.

Weekly wages	Number of employees	Cumulative total	Cumulative percents
$360 to under $400	24	24	12.3
$400 to under $440	40	64	32.8
$440 to under $480	55	119	61.0
$480 to under $520	36	155	79.5
$520 to under $560	25	180	92.3
$560 to under $600	12	192	98.5
$600 to under $640	3	195	100.0

For Q_1,

$$x = 25; \quad N = 195; \quad i = 40; \quad L_x = 400; \quad \Sigma f_c = 24; \quad f_x = 40.$$

$$Q_1 = P_{25} = 400 + (40) \left(\frac{\frac{(195)(25)}{100} - 24}{40} \right)$$

$$= 400 + (40) \left(\frac{48.75 - 24}{40} \right)$$

$$= 400 + (40) \frac{24.75}{40}$$

$$= 400 + 24.75 = 424.75$$

For Q_3,

$$x = 75; \quad N = 195; \quad i = 40; \quad L_x = 480; \quad \Sigma f_c = 119; \quad f_x = 36.$$

$$Q_3 = P_{75} = 480 + (40) \left(\frac{\frac{(195)(75)}{100} - 119}{36} \right)$$

$$= 480 + (40) \left(\frac{146.25 - 119}{36} \right)$$

$$= 480 + (40) \frac{27.25}{36}$$

$$= 480 + 30.28 = 510.28$$

The interquartile range = $510.28 - 424.75 = \$85.53$.

EXERCISE 4.1

1. A survey of Canada's "Top 500" corporations produced the following average starting salaries by academic area:

Physics	$32 852	Chemistry	$28 814
Liberal arts	$23 719	Accounting	$27 551
Mathematics	$29 538	Hotel management	$25 447
Engineering	$36 614	Medicine	$38 742
Computer science	$35 849	Education	$24 779
Marketing	$28 684	Business admin.	$26 650

a) Determine the range of starting salaries.
b) What academic areas make up the interquartile range?
c) What minimum starting salary is required to put you into the top quartile?
d) Below what amount will your starting salary put you into the bottom quartile?

2. The numbers in the chart below were compiled from a survey.

66	50	59	26	46	49	51	62
48	68	20	37	40	23	67	20
18	54	5	49	44	71	31	29
50	64	35	44	53	50	37	15
59	37	13	36	53	64	55	44

a) Determine the range and the interquartile range for the ungrouped data.
b) Group the data in a frequency distribution using equal intervals of size 10 with a lowest class limit of 0.
c) Determine the range and the interquartile range for the frequency distribution.
d) Compare the answers to parts **(a)** and **(c)**. What is gained and what is lost by grouping data?

3. The data in the frequency distribution below represent the hourly wages paid to a group of employees.

Class interval	Frequency
$14.00 to under $15.00	11
$15.00 to under $16.00	39
$16.00 to under $17.00	53
$17.00 to under $18.00	35
$18.00 to under $19.00	15
$19.00 to under $20.00	5
$20.00 to under $21.00	2

Determine   **(a)** the range;   **(b)** the interquartile range.

4. For the following frequency distribution determine
   (a) the range;   **(b)** the interquartile range.

Class interval	Frequency
10 to under 30	14
30 to under 50	28
50 to under 70	40
70 to under 90	28
90 to under 110	16
110 to under 130	9
130 to under 150	2

5. *Data disk exercise* Using the enclosed data disk and the file name PUBLIC, determine **(i)** range; and **(ii)** interquartile range for sales (the variable name is SALES) of metal mining companies. Metal mining companies fall under the standard industrial classification (the variable name is SIC) code 58.

SECTION 4.2 The Average Deviation from the Mean

A. *Deviation from the Mean Defined*

The term **deviation** refers to the *difference* between the value of an individual observation in a data set and the mean of the data set. Since the mean is a central value, some of the deviations will be positive and some will be negative.

○ **EXAMPLE 4.2a**
For the set of test scores 17, 11, 13, 15, 9, 16, 12, 11, compute the deviations of the test scores and the sum of the deviations from the mean.

● **SOLUTION**
First determine the mean of the test scores

$$N = 8;$$

$$\sum x = 17 + 11 + 13 + 15 + 9 + 16 + 12 + 11 = 104;$$

$$\mu = \frac{\sum x}{N} = \frac{104}{8} = 13.$$

Now compute the deviations $(x - \mu)$ as shown below and add the deviations.

Test scores	Deviation from mean
x	$(x - \mu)$
17	$17 - 13 = 4$
11	$11 - 13 = -2$
13	$13 - 13 = 0$
15	$15 - 13 = 2$
19	$9 - 13 = -4$
16	$16 - 13 = 3$
12	$12 - 13 = -1$
11	$11 - 13 = -2$
$\sum x = 104$	$\sum(x - \mu) = 0$

Note   The sum of the deviations from the mean is always zero.

B. *Average Deviation from the Mean — Ungrouped Data*

The **average deviation from the mean** is a measure of the average magnitude (absolute value — numerical value without + or − sign) of the deviations from the mean.

The average deviation from the mean can be determined by using the formula

$$\text{AVERAGE DEVIATION FROM THE MEAN} = \frac{\sum |x - \mu|}{N}$$

where x is the value of an observation in the data set;

μ is the mean of the data set $= \dfrac{\sum x}{N}$;

N is the number of observations in the data set;

$|x - \mu|$ is the absolute value of the deviation of an observation from the mean;

$\sum |x - \mu|$ is the sum of the absolute values of the deviations from the mean.

○ **EXAMPLE 4.2b**

For the set of test scores used in Example 4.2a, compute the average deviation from the mean.

● **SOLUTION**

Test scores x	Deviations from mean $(x - \mu)$	Absolute deviations $\lvert x - \mu \rvert$
17	$17 - 13 = \ \ 4$	4
11	$11 - 13 = -2$	2
13	$13 - 13 = \ \ 0$	0
15	$15 - 13 = \ \ 2$	2
9	$9 - 13 = -4$	4
16	$16 - 13 = \ \ 3$	3
12	$12 - 13 = -1$	1
11	$11 - 13 = -2$	2
$\sum x = 104$	$\sum(x - \mu) = \ \ 0$	$\sum \lvert x - \mu \rvert = 18$

$$\text{AVERAGE DEVIATION FROM THE MEAN} = \frac{\sum \lvert x - \mu \rvert}{N} = \frac{18}{8} = 2.25$$

This means that, on average, the test scores deviate from the mean by 2.25.

Note   The meaning of the average deviation is easily understood and is a better measure of dispersion than the range or interquartile range since it takes into account all observations in the data set.

EXERCISE 4.2

1. The weekly incomes for a group of eight employees are $581.00, $570.00, $625.00, $660.00, $630.00 and $492.00. Compute the average deviation from the mean.

2. The following are the index numbers of production in an industry for the last 10-year period:

   100   122   120   112   114   90   89   101   109   113

   Determine the average deviation from the mean.

3. After a company's annual golf tournament, the best seven rounds were published as follows:

   71   73   78   81   82   82   86

   Compute the average deviation from the mean of the rounds.

4. A random selection of homeowners produced the following annual gas bills:

   $785 $1218 $826 $988 $803

   Calculate the average deviation from the mean of the gas bills.

 5. **Data disk exercise**   Using the enclosed data disk and the file name MARKET, determine the average deviation from the mean for Sample Three, Question One (the variable name is S3Q1).

SECTION 4.3 The Variance and Standard Deviation

A. Computation for Ungrouped Data

The variance and the standard deviation are the most widely used measures of variability. The **variance** is the *average squared deviation* from the mean. It is similar to the average deviation from the mean in that it is based on the deviation of the individual values from the mean. It differs from the average deviation because the deviations are squared before summing.

The **standard deviation** is readily obtained from the variance as the *square root of the variance*. The symbol used for the standard deviation is the small Greek letter σ (pronounced "sigma"). Because of the relationship between the variance and the standard deviation, the variance is usually designated by the symbol σ^2.

The basic computational process is similar to that used for finding the average deviation from the mean and is summarized in the basic formulas for computing the variance and the standard deviation.

$$\text{VARIANCE,} \quad \sigma^2 = \frac{\Sigma(x - \mu)^2}{N}$$

$$\text{STANDARD DEVIATION,} \quad \sigma = \sqrt{\sigma^2}$$

where N is the number of observations in the data set;
x is the value of an observation;
μ is the mean $= \dfrac{\Sigma x}{N}$;
$x - \mu$ is the deviation of an observation from the mean;
$(x - \mu)^2$ is the square of a deviation from the mean;
$\Sigma(x - \mu)^2$ is the sum of the squared deviations.

○ **EXAMPLE 4.3a**

Compute the variance and the standard deviation for the data set used in Example 4.2a.

● **SOLUTION**

Test scores x	Deviations from mean $(x - \mu)$	Squared deviation $(x - \mu)^2$
17	4	16
11	−2	4
13	0	0
15	2	4
9	−4	16
16	3	9
12	−1	1
11	−2	4
$\sum x = 104$	$\sum(x - \mu) = 0$	$\sum(x - \mu)^2 = 54$

$$\text{MEAN,} \quad \mu = \frac{\sum x}{N} = \frac{104}{8} = 13$$

$$\text{VARIANCE,} \quad \sigma^2 = \frac{\sum(x - \mu)^2}{N} = \frac{54}{8} = 6.75$$

$$\text{STANDARD DEVIATION,} \quad \sigma = \sqrt{\sigma^2} = \sqrt{6.75} = 2.5981$$

B. Short-Cut Formula for Computing the Variance for Ungrouped Data

The basic formula for computing the variance, $\sigma^2 = \dfrac{\sum(x - \mu)^2}{N}$, is normally used only for a set of data that contains few observations. For larger sets of data, the calculation of the deviation and the squared deviation for each observation is avoided by using an alternate formula derived by mathematical manipulation of the basic formula.

This alternate formula is

$$\sigma^2 = \frac{\sum x^2}{N} - \left(\frac{\sum x}{N}\right)^2 \quad \text{or} \quad \sigma^2 = \frac{\sum x^2}{N} - \mu^2$$

where N is the number of observations;
   x is the value of an observation;
   x^2 is the squared value of an observation;
   $\sum x$ is the sum of the values of the N observations;
   $\sum x^2$ is the sum of the squared values,

○ **EXAMPLE 4.3b**
Use the short-cut formula to compute the variance and the standard deviation for the data set used in Example 4.2a.

● **SOLUTION**

Test scores x	Squared scores x^2
17	$17^2 = 289$
11	$11^2 = 121$
13	$13^2 = 169$
15	$15^2 = 225$
9	$9^2 = 81$
16	$16^2 = 256$
12	$12^2 = 144$
11	$11^2 = 121$
$\sum x = 104$	$\sum x^2 = 1406$

$$\text{VARIANCE,} \quad \sigma^2 = \frac{\sum x^2}{N} - \left(\frac{\sum x}{N}\right)^2 = \frac{1406}{8} - \left(\frac{104}{8}\right)^2$$

$$= 175.75 - 13^2 = 175.75 - 169 = 6.75$$

$$\text{STANDARD DEVIATION,} \quad \sigma = \sqrt{\sigma^2} = \sqrt{6.75} = 2.5981$$

Notes

1. The results in examples 4.3a and 4.3b are, of course, the same.

2. The value $\dfrac{\sum x}{N}$ in the short-cut formula equals μ and leads to the alternate version of the short-cut formula, $\sigma^2 = \dfrac{\sum x^2}{N} - \mu^2$. This form of the short-cut formula can be used to advantage when the mean has to be determined.

○ **EXAMPLE 4.3c**

Compute the variance and the standard deviation for the following hourly wages of a group of workers, using (a) the basic formula; (b) the short-cut formula.

$14.60	$15.40	$14.00	$14.40	$15.80
$15.60	$16.40	$18.60	$19.20	$21.00

● **SOLUTION**

(a)   **(b)**

Hourly wages x	Deviation from mean $(x - \mu)$	Squared deviation $(x - \mu)^2$	Hourly wages x	Squared hourly wages x^2
14.60	14.60 − 16.50 = −1.90	3.61	14.60	213.1600
15.40	15.40 − 16.50 = −1.10	1.21	15.40	237.1600
14.00	14.00 − 16.50 = −2.50	6.25	14.00	196.0000
14.40	14.40 − 16.50 = −2.10	4.41	14.40	207.3600
15.80	15.80 − 16.50 = −0.70	0.49	15.80	249.6400
15.60	15.60 − 16.50 = −0.90	0.81	15.60	243.6400
16.40	16.40 − 16.50 = −0.10	0.01	16.40	268.9600
18.60	18.60 − 16.50 = 2.10	4.41	18.60	345.9600
19.20	19.20 − 16.50 = 2.70	7.29	19.20	368.6400
21.00	21.00 − 16.50 = 4.50	20.25	21.00	441.0000
165.00		0.00 48.74	165.00	2771.24

(a) $\sum x = 165.00$

$$\mu = \frac{\sum x}{N} = \frac{165.00}{10} = 16.50$$

$$\sigma^2 = \frac{\sum(x - \mu)^2}{N} = \frac{48.74}{10} = 4.874$$

$$\sigma = \sqrt{\sigma^2} = \sqrt{4.874} = 2.207714$$

(b) $\sum x = 165.00$

$$\sigma^2 = \frac{\sum x^2}{N} - \mu^2$$

$$= \frac{2771.24}{10} - 16.50$$

$$= 277.124 - 272.25$$

$$= 4.874$$

$$\sigma = \sqrt{4.874} = 2.207714$$

C. *Variance and Standard Deviation for Grouped Data*

For grouped data the short-cut formula is modified to allow for the weighing of the class midpoints by the frequencies:

$$\sigma^2 = \frac{\sum fx^2}{N} - \left(\frac{\sum fx}{N}\right)^2 \quad \text{or} \quad \sigma^2 = \frac{\sum fx^2}{N} - \mu^2$$

where N is the number of observations;
  x is the midpoint of a class;
  x^2 is the square of a class midpoint;
  f is the frequency in a class;
  fx is the weighted value of a class midpoint;
  $\sum fx$ is the sum of the weighted values of the class midpoints;
  fx^2 is the weighted value of a squared class midpoint;
  $\sum fx^2$ is the sum of the weighted values of the squared class midpoints.

○ **EXAMPLE 4.3d**

The frequency distribution of the test scores obtained by a group of students is shown in the table below in columns 1 and 2. Compute

a) the mean;
b) the variance;
c) the standard deviation.

● **SOLUTION**

Column 1	Column 2	Column 3	Column 4
Class (test scores)	Class frequency f	Class midpoint x	Squared midpoint x^2
0 to under 10	2	5	25
10 to under 20	4	15	225
20 to under 30	12	25	625
30 to under 40	28	35	1225
40 to under 50	44	45	2025
50 to under 60	43	55	3025
60 to under 70	23	65	4225
70 to under 80	7	75	5625
80 to under 90	5	85	7225
90 to under 100	2	95	9025
Totals	$N = \sum f = 170$		

Column 1	Column 5	Column 6
Class (test scores)	Weighted midpoint fx	Weighted squared midpoint fx^2
0 to under 10	(2)(5) = 10	(2)(25) = 50
10 to under 20	(4)(15) = 60	(4)(225) = 900
20 to under 30	(12)(25) = 300	(12)(625) = 7 500
30 to under 40	(28)(35) = 980	(28)(1225) = 34 300
40 to under 50	(44)(45) = 1980	(44)(2025) = 89 100
50 to under 60	(43)(55) = 2365	(43)(3025) = 130 075
60 to under 70	(23)(65) = 1495	(23)(4225) = 97 175
70 to under 80	(7)(75) = 525	(7)(5625) = 39 375
80 to under 90	(5)(85) = 425	(5)(7225) = 36 125
90 to under 100	(2)(95) = 190	(2)(9025) = 18 050
Totals	$\sum fx = 8330$	$\sum fx^2 = 452\,650$

Step 1   Determine N by adding the frequencies in Column 1, $N = \sum f = 170$.

Step 2   Determine the class midpoints x (see Column 3).

Step 3   Determine the value of the squared midpoints x^2 (see Column 4).

Step 4   Determine the value of the weighted midpoints fx by multiplying the values in Column 2 and Column 3, and add the values to obtain $\sum fx$ (see Column 5).

Step 5   Determine the values of the weighted squared midpoints fx^2 by multiplying the values in Column 2 and Column 4, and add the values to obtain $\sum fx^2$ (see Column 6). (The values in Column 6 can also be obtained by multiplying the values in Column 3 and Column 5: $(fx)x = fx^2$.)

a) The mean, $\mu = \dfrac{\sum fx}{N} = \dfrac{8330}{170} = 49$

b) The variance, $\sigma^2 = \dfrac{\sum fx^2}{N} - \mu^2 = \dfrac{452\ 650}{170} - 49^2$

$$= 2662.6471 - 2401 = 261.6471$$

c) The standard deviation, $\sigma = \sqrt{\sigma^2} = \sqrt{261.6471} = 16.1755$

D.   The Coefficient of Variation

The **coefficient of variation** is a measure of the *relative* magnitudes of the standard deviation and mean of a data set. It is used to compare the variability of two or more data sets.

$$\text{COEFFICIENT OF VARIATION, } CV = \frac{\sigma}{\mu}$$

where σ is the standard deviation of the data set;
$\quad\quad \mu$ is the mean of the data set.

○ **EXAMPLE 4.3e**
The daily closing prices of the common shares of two companies for a two-week trading period are listed below:

Company A	$8.00	$8.40	$7.80	$8.30	$8.60
	$9.00	$8.70	$8.30	$7.70	$7.40
Company B	$150.00	$154.00	$148.00	$151.00	$157.00
	$157.00	$160.00	$152.00	$148.00	$153.00

Compare the variation in the prices of the two stocks over the two-week period.

● **SOLUTION**
To compare the variability in the two sets of data, determine the mean, the standard deviation and the coefficient of variation for both sets.

Company A		Company B	
Price		Price	
x	x^2	x	x^2
8.00	64.00	150	22 500
8.40	70.56	154	23 716
7.80	60.84	148	21 904
8.30	68.89	151	22 801
8.60	73.96	157	24 649
9.00	81.00	160	25 600
8.70	75.69	157	24 649
8.30	68.89	152	23 104
7.70	59.29	148	21 904
7.40	54.76	153	23 409
$\sum x = 82.20$	$\sum x^2 = 677.88$	$\sum x = 1530$	$\sum x^2 = 234236$

For Company A,

$$\mu = \frac{82.20}{10} = 8.22$$

$$\sigma^2 = \frac{677.88}{10} - 8.22^2$$

$$= 67.788 - 67.5684$$

$$= 0.2196$$

$$\sigma = \sqrt{0.2196} = 0.468615$$

$$CV = \frac{\sigma}{\mu} = \frac{0.46815}{8.22}$$

$$= 0.0570 = 5.70\%$$

For Company B,

$$\mu = \frac{1530}{10} = 153.00$$

$$\sigma^2 = \frac{23\ 423}{10} - 153^2$$

$$= 23\ 423.60 - 23\ 409$$

$$= 14.60$$

$$\sigma = \sqrt{14.60} = 3.8210$$

$$CV = \frac{\sigma}{\mu} = \frac{3.8210}{153.00}$$

$$= 0.024974 = 2.50\%$$

Conclusion Based on absolute magnitudes, the variability of the share price of Company A with a standard deviation of $0.47 is decidedly smaller than that of the share price for Company B showing a standard deviation of $3.82. However, in terms of the price level, the price of the shares of Company A is more than twice as variable as the price of the shares of Company B.

EXERCISE 4.3

1. During the first quarter, the number of rejected units produced per week by a group of workers were as follows:

Worker	A	B	C	D	E	F
No. rejected	84	76	68	64	71	77

Worker	G	H	I	J	K	L
No. rejected	62	80	64	64	75	66

For the given data, compute
a) the mean;
b) the variance;
c) the standard deviation.

2. For the following set of marks, compute
   a) the mean;
   b) the variance;
   c) the standard deviation.

91	79	87	87	63	69	62	69	41	60	84	87
89	45	48	94	92	87	79	71	84	62	80	67

3. For the set of test scores given below, determine
   a) the mean;
   b) the variance;
   c) the standard deviation.

Test score	1	2	3	4	5	6	7	8	9	10
No. of students	2	4	12	28	44	48	28	7	5	2

4. The following information was compiled about the number of days absent by a group of employees:

No. of days absent	2	4	6	8	10	12	14	16	18
No. of employees	4	13	21	36	38	27	15	4	2

Compute
a) the average number of days absent;
b) the variance;
c) the standard deviation.

5. For the frequency distribution below compute
   a) the mean;
   b) the variance;
   c) the standard deviation.

Class	Frequency
0 to under 10	44
10 to under 20	62
20 to under 30	77
30 to under 40	52
40 to under 50	32
50 to under 60	20
60 to under 70	12
70 to under 80	6
80 to under 90	4
90 to under 100	1

6. Compute **(a)** the mean; **(b)** the variance; **(c)** the standard deviation for the frequency distribution below.

Class interval	Number in class
100 to under 150	28
150 to under 200	56
200 to under 250	85
250 to under 300	65
300 to under 350	32
350 to under 400	18
400 to under 450	10
450 to under 500	6

7. Determine the coefficient of variation of the following two sets of class marks and comment on their variability.

Class A	12	4	18	15	9	11	3	7
Class B	84	56	73	62	78	48	93	80

Class A	17	20	13	8	14	12	6
Class B	66	75	73	88	56	64	83

8. Compare the variability in the data contained in the following two frequency distributions:

Hourly wages	Number
$10.00 to under $12.00	12
$12.00 to under $14.00	25
$14.00 to under $16.00	49
$16.00 to under $18.00	46
$18.00 to under $20.00	38
$20.00 to under $22.00	18
$22.00 to under $24.00	9
$24.00 to under $26.00	3

Weekly salary	Number
$300.00 to under $400.00	10
$400.00 to under $500.00	19
$500.00 to under $600.00	20
$600.00 to under $700.00	18
$700.00 to under $800.00	10
$800.00 to under $900.00	6
$900.00 to under $1000.00	4
$1000.00 to under $1100.00	2
$1100.00 to under $1200.00	1

9. ***Data disk exercise***   Using the enclosed data disk and the file name MARKET, determine **(a)** the variance; **(b)** the standard deviation for Sample One, Question One (the variable name is S1Q1).

10. ***Data disk exercise***   Using the enclosed data disk and the file name MARKET, determine the coefficient of variation for Sample One, Question One (the variable name is S1Q1) for subjects living in **(a)** Toronto (code is T) and **(b)** Montreal (code is M). (The variable name is CITY.)

REVIEW EXERCISE

1. For the following test scores, calculate
   a) the mean;   **b)** the median;
   c) the range;   **d)** the interquartile range;
   e) the average deviation from the median;
   f) the variance;   **g)** the standard deviation.

55	59	87	59	91	77	68	86
58	87	73	83	84	94	89	82

2. For the years of seniority listed below, calculate
   a) the mean;   **b)** the median;
   c) the range;   **d)** the interquartile range;
   e) the average deviation from the mean;
   f) the variance;   **g)** the standard deviation.

15	12	6	15	12	7	16	13	7	16
13	8	17	13	10	17	14	10	19	20

3. In a recent Canada-wide marketing research study of 1200 homes, the following data were collected:

Family income	Number of families
$10 000 to under $20 000	46
$20 000 to under $30 000	152
$30 000 to under $40 000	271
$40 000 to under $50 000	282
$50 000 to under $60 000	254
$60 000 to under $70 000	101
$70 000 to under $80 000	71
$80 000 to under $90 000	14
$90 000 to under $100 000	5
$100 000 to under $110 000	3
$110 000 to under $120 000	1

Calculate
a) the mean;   **b)** the median;
c) the range;   **d)** the interquartile range;
e) the 6th decile;   **f)** the 90th percentile;
g) the variance;   **h)** the standard deviation.

4. A recent review of a compact disc distributor's product line is summarized as follows:

Selling price	Number of titles
$0 to under $5	51
$5 to under $10	184
$10 to under $15	241
$15 to under $20	1468
$20 to under $25	1339
$25 to under $30	255
$30 to under $35	98
$35 to under $40	14

Determine
a) the mean;   b) the median;
c) the range;   d) the interquartile range;
e) the third decile;   f) the 85th percentile;
g) the variance;   h) the standard deviation.

5. One number is missing from the data below.

55	95	78	95	19	77	86	78
68	85	?	37	38	48	49	28

a) Given that the mean is 66.5, determine the missing number.
b) Determine
   i) the median;
   ii) the interquartile range;
   iii) the average deviation from the mean.

6. The following 20 observations have been written in ascending order. The value of three data points is missing, but their position in the array is known and indicated by "xx."

xx	50	xx	56	57	63	68	72	73	73
75	79	81	82	85	87	87	88	xx	99

a) Determine
   (i) the median;   (ii) the interquartile range.
b) Why can the interquartile range be determined but not the range or the mean?

7. Corvette convertibles have become classic show cars commanding high selling prices. In 1991, cars in excellent condition were selling for the following amounts:

Model year	1953	1954	1955	1956	1957	1958	1959
Amount	91 437	51 939	61 045	45 530	52 026	40 890	39 585

For the given selling prices, determine the average deviation from the mean.

8. The number of urban housing starts in Canada between 1983 and 1990 were as follows:

Year	1983	1984	1985	1986	1987	1988	1989	1990
Starts	134 207	110 874	139 408	170 863	215 340	189 635	183 323	155 824

For the given data, determine the average deviation from the mean.

9. For the given data, determine
   a) the average deviation from the mean;
   b) the variance and the standard deviation.

Class	1 to 3	4 to 6	7 to 9	10 to 12	13 to 15	16 to 18
Frequency	2	9	29	58	37	12

10. Frequency distributions are sometimes reported by class midpoint instead of class intervals, as shown for the data below.

Class midpoint	0.5	1.5	2.5	3.5	4.5
Frequency	28	42	47	36	15

Compute
   a) the average deviation from the mean;
   b) the variance and the standard deviation.

11. *Performance Car* magazine's tire test scores for 1990 were as follows:

Make of tire	Test score
Dunlop – D40M2	8836
Yokohama – AUS	8725
Goodyear Eagle – VR	8523
Michelin – MXX2	8147
Bridgestone – ER90	8079
Firestone Firehawk	7637
B.F. Goodrich – COMP T/A	6907

For the test scores, compute
   (a) the mean;   (b) the variance;   (c) the standard deviation.

12. The following are the weights of cows (in kilograms) shipped from Tim Rak's cattle farm to the slaughterhouse:

$$841 \quad 820 \quad 910 \quad 885 \quad 945 \quad 905 \quad 875$$

For the shipment, compute
   (a) the mean;   (b) the variance;   (c) the standard deviation.

13. Statistics Canada keeps track of the country of origin for all new Canadians. Between 1980 and 1989 the greatest number of immigrants came to Canada from the following eight countries:

Country of origin	Number of immigrants
Vietnam	92 873
United Kingdom	91 025
India	87 117
Hong Kong	77 752
China	69 093
Poland	66 164
United States	61 259
Philippines	34 853

Compute

(a) the mean; (b) the average deviation from the mean; (c) the variance; (d) the standard deviation.

14. The 10 highest-paid chief executive officers of North American corporations were given compensation packages in 1991 as follows (in thousands of U.S. dollars).

18 301	16 730	14 822	12 290	11 676
11 233	8 483	7 568	7 463	7 143

Compute
(a) the mean; (b) the average deviation from the mean;
(c) the variance; (d) the standard deviation.

15. A typical college switchboard operator handles more calls in September than at any other time of the year. To monitor any overloading of the switchboard, a tracking system is used to keep a record of the elapsed time before the calls are answered. The system produced the following data for a typical day in September:

Elapsed time (seconds)	0 to 3	4 to 7	8 to 11	12 to 15	16 to 19
Number of phone calls	48	923	1721	1387	69

Calculate (a) the mean; (b) the median; (c) the variance; (d) the standard deviation.

16. For the following data, compute
(a) the mean; (b) the median; (c) the variance;
(d) the standard deviation.

Size (in metres)	0 to under 2	2 to under 4	4 to under 6	6 to under 8	8 to under 10
Frequency	1426	2292	2202	1332	465

17. An alderman living near Abbotsford school requested a population analysis of the neighbourhood served by the school to determine if a request for a playground should be granted. The following data were provided:

Number of children per household	0	1	2	3	4	5	6
Number of households	15	89	71	46	28	7	4

Determine the mean and the standard deviation for the number of children per household.

18. Calculate the mean and the standard deviation for the following values of orders received by Ace Distributors:

Dollar value of order	2	4	6	8	10	12	14	16	18
Frequency	4	13	21	36	38	27	15	4	2

19. Because of the large registration for a charity golf tournament, the field was split into a morning group and an afternoon group. The morning group had 60 participants, with an average score of 82 and a variance of 6.25. The afternoon group had 80 participants, with an average score of 76 and a variance of 9.0.
a) Compare the variability of the scores for the two groups.
b) Determine the mean score for all participants.

20. Truck A delivered 84 boxes having an average weight of 4.2 kg with a variance of 0.81 kg. Truck B delivered 180 boxes with an average weight of 4.0 kg and a variance of 0.64 kg.
a) Comment on the variability in the weight of the boxes delivered by the two trucks.
b) Compute the mean for the combined shipments.

21. A collection of 35 blood samples had an average protein proportion of 6.3% and a standard deviation of 0.4%. If one of the blood samples with a protein proportion of 13.1% was found to be contaminated, determine the mean protein proportion for the 34 good blood samples.

22. A teacher with a class size of 40 students computes the class average for a test to be 61 with a standard deviation of 12. Shortly afterwards, John Doe, whose test score was 22, withdrew from the class. Recalculate the class mean on the test for the remaining students.

23. Three companies produced the following sales data:

	Company A	Company B	Company C
Average yearly sales	$2 500 000 000	$125 000 000	$5 750 000
Standard deviation	$25 000 000	$1 875 000	$287 500

a) Determine the coefficient of variation for each company's sales.
b) If you were looking at long-term job prospects based on stability, which company should you choose to work for? Explain.

24. Ashley and Riley both sell for an electronics equipment distributor. Their monthly sales for the last six months are summarized below.

	Jul	Aug	Sept	Oct	Nov	Dec
Ashley	11 500	9 000	16 500	11 500	11 000	12 500
Riley	32 500	25 000	40 000	32 500	29 000	33 000

a) Compute the average monthly sales for each salesperson.
b) Compute the standard deviation for each.
c) Determine the coefficient of variation for each.
d) Compare the variation in monthly sales between the two.

SELF-TEST

A. Questions **1** through **5** are based on the following frequency distribution of stocks in a pension fund:

Stock price	Number of stocks
$0.00 to under $5.00	4
5.00 to under 10.00	65
10.00 to under 15.00	80
15.00 to under 20.00	120
20.00 to under 25.00	135
25.00 to under 30.00	40
30.00 to under 35.00	11
35.00 to under 40.00	5

1. What is the range of prices in the portfolio?

2. Compute the interquartile range.

3. Compute the median stock price.

4. Compute the mean stock price.

5. Calculate the variance and the standard deviation.

B. Questions **6** through **12** are based on the following house prices for a standard two-storey home in various parts of Toronto for April 1990 and April 1991 (in $000).

April 1990	350	360	240	265	315	380	250	415
April 1991	270	300	205	210	300	295	225	320

6. What was the range of prices in 1991?

7. Determine the interquartile range for 1991.

8. Determine the mean and the median for both 1990 and 1991.

9. Calculate the average deviation from the mean for 1990.

10. Determine the standard deviation for both 1990 and 1991.

11. Determine the coefficient of variation for each year.

12. Comment on the variation in house prices caused by the recession of 1991.

Key Terms

Average deviation from the mean   85
Coefficient of variation   92
Deviation   84
Interquartile range   80
Measures of variability   79
Range   80
Standard deviation   87
Variance   87

Summary of Formulas

1. RANGE = HIGHEST VALUE − LOWEST VALUE

2. INTERQUARTILE RANGE = $Q_3 - Q_1$

3. AVERAGE DEVIATION FROM THE MEAN = $\dfrac{\sum |x - \mu|}{N}$

4. Variance:

a) Ungrouped data:

$$\sigma^2 = \frac{\sum (x - \mu)^2}{N}$$

$$\text{or} \quad \sigma^2 = \frac{\sum x^2}{N} - \left(\frac{\sum x}{N}\right)^2 = \frac{\sum x^2}{N} - \mu^2$$

b) Grouped data:

$$\sigma^2 = \frac{\sum f x^2}{N} - \left(\frac{\sum f x}{N}\right)^2 = \frac{\sum f x^2}{N} - \mu^2$$

5. STANDARD DEVIATION, $\sigma = \sqrt{\text{VARIANCE}}$

6. COEFFICIENT OF VARIATION, $CV = \dfrac{\sigma}{\mu}$

CHAPTER 5

Index Numbers

Introduction

The use of index numbers in business has grown steadily since they were first introduced at the turn of the century. Indexes provide an easy way of expressing changes that occur in daily business. Converting data to indexes makes working with very large or small numbers easier and provides a basis for many types of analysis.

Objectives

Upon completion of this chapter you will be able to
1. construct simple price, quantity and value indexes;
2. construct weighted price, quantity and value indexes;
3. construct special-purpose (composite) indexes;
4. interpret indexes to identify trends in a data set;
5. use the consumer price index to determine the purchasing power of the dollar and to compute real income;
6. shift the base to make two series comparable;
7. splice an old series with a new series of index numbers.

SECTION 5.1 **Nature of Index Numbers**

An **index number** results from the comparison of two values measured at different points in time. The comparison of the two values is stated in the form of a ratio, changed into the form of a percent; when the percent symbol is dropped, the result is called an index number.

○ **EXAMPLE 5.1a**
The price of an article was $12.00 in 1981 and $18.00 in 1991. Compare the two prices to create an index number.

103

● **SOLUTION**

The change in price over the time period 1981 to 1991 can be measured in relative terms by writing the ratio

$$\frac{\text{Price in 1991}}{\text{Price in 1981}} = \frac{18.00}{12.00} = 1.50 = 150\%$$

The desired price index number = 150.

Note 1   The construction of the index number requires the selection of one of the two numbers as the denominator of the ratio. The point in time at which the selected number was measured is referred to as the **base period** — in our example 1981 has been selected as the base period. Usually the *chronologically earliest* year is chosen as base period.

Note 2   The index for the base period is always 100. The difference between an index number and 100 indicates the change that has taken place. The index number 150 indicates that the price in 1991 was 50% higher than the price in 1981.

A wide variety of index numbers can be constructed. They are used in comparing and analyzing economic data and have become a widely accepted tool for measuring changes in business activity.

The most important index numbers are
1. *price indexes*, measuring relative change in price;
2. *quantity indexes*, measuring relative change in quantity;
3. *value indexes*, measuring relative change in value.

Price indexes, quantity indexes and value indexes can be **simple index numbers** or **aggregate index numbers**. Indexes that compare individual items over time are called simple indexes, while indexes that involve a group of commodities are referred to as aggregate indexes.

SECTION 5.2   Constructing and Interpreting Simple Indexes

A.   Simple Price Indexes

A **simple price index** can be constructed by means of the formula

$$P = \frac{P_n}{P_0}(100)$$

where P_0 is the price in the chosen base period;
P_n is the price in any other given period.

○ **EXAMPLE 5.2a**

Construct index numbers for the prices listed below for the period 1980 to 1992 and interpret their meaning using 1980 as the base year.

Year	1980	1983	1986	1989	1992
Price	$20	$22	$25	$20	$18

● **SOLUTION**

The price in the chosen base year 1980, $P_0 = 20$.

a) For 1983, $P_n = 22$.

$$P = \frac{P_n}{P_0}(100) = \frac{22}{20}(100) = 1.1(100) = 110$$

Interpretation   The price of the article in 1983 was 10% higher than the price in 1980.

b) For 1986, $P_n = 25$.

$$P = \frac{P_n}{P_0}(100) = \frac{25}{20}(100) = 1.25(100) = 125$$

The price in 1986 was 25% higher than the price in 1980.

c) For 1989, $P_n = 20$.

$$P = \frac{P_n}{P_0}(100) = \frac{20}{20}(100) = 1.00(100) = 100$$

The price in 1989 was the same as in 1980.

d) For 1992, $P_n = 18$.

$$P = \frac{P_n}{P_0}(100) = \frac{18}{20}(100) = 0.90(100) = 90$$

The price in 1992 was 10% lower than the price in 1980.

B. *Simple Quantity Indexes*

The formula for constructing a **simple quantity index** is

$$Q = \frac{Q_n}{Q_0}(100)$$

where Q_0 is the quantity in the base period;
        Q_n is the quantity in any other given period.

○ **EXAMPLE 5.2b**

Assume that the number of articles sold at the prices given in Example 5.2a were as follows:

Year	1980	1983	1986	1989	1992
Quantity	200	220	180	300	420

Construct index numbers based on 1980 and interpret.

● **SOLUTION**

The quantity in the chosen base period 1980, $Q_0 = 200$.

a) For 1983, $Q_n = 220$.

$$Q = \frac{Q_n}{Q_0}(100) = \frac{220}{200}(100) = 1.10(100) = 110$$

The quantity in 1983 was 10% higher than in 1980.

b) For 1986, $Q_n = 180$.

$$Q = \frac{Q_n}{Q_0}(100) = \frac{180}{200}(100) = 0.90(100) = 90$$

The quantity in 1986 was 10% lower than in 1980.

c) For 1989, $Q_n = 300$.

$$Q = \frac{Q_n}{Q_0}(100) = \frac{300}{200}(150) = 1.50(100) = 150$$

The quantity in 1989 was 50% higher than in 1980.

d) For 1992, $Q_n = 420$.

$$Q = \frac{Q_n}{Q_0}(100) = \frac{420}{200}(100) = 2.10(100) = 210$$

The quantity in 1992 was 110% higher than in 1980.

C. Simple Value Indexes

The formula for constructing a **simple value index** is

$$V = \frac{V_n}{V_0}(100)$$

where $V_0 = P_0 Q_0$ is the value in the chosen base period;
$V_n = P_n Q_n$ is the value in any other given period.

○ **EXAMPLE 5.2c**

Construct value indexes for the prices and quantities given in examples 5.2a and 5.2b respectively and interpret.

● **SOLUTION**

The value in the chosen base period 1980,

$$V_0 = (P_0)(Q_0) = (20)(200) = 4000$$

a) For 1983, $V_n = (P_n)(Q_n) = (22)(220) = 4840$.

$$V = \frac{V_n}{V_0}(100) = \frac{4840}{4000}(100) = 1.21(100) = 121$$

The value in 1983 was 21% higher than in 1980.

b) For 1986, $V_n = (P_n)(Q_n) = (25)(180) = 4500$.

$$V = \frac{V_n}{V_0}(100) = \frac{4500}{4000}(100) = 1.125(100) = 112.5$$

The value in 1986 was 12.5% higher than in 1980.

c) For 1989, $V_n = (P_n)(Q_n) = (20)(300) = 6000$.

$$V = \frac{V_n}{V_0}(100) = \frac{6000}{4000}(100) = 1.50(100) = 150$$

The value in 1989 was 50% higher than in 1980.

d) For 1992, $V_n = (P_n)(Q_n) = (18)(420) = 7560$.

$$V = \frac{V_n}{V_0}(100) = \frac{7560}{4000}(100) = 1.89(100) = 189$$

The value in 1992 was 89% higher than in 1980.

SECTION 5.3 **Aggregate Indexes**

A. Unweighted Aggregate Indexes

An *unweighted aggregate price index* is obtained by summing the prices of a number of commodities and comparing the sums over time.

UNWEIGHTED AGGREGATE PRICE INDEX, $P = \dfrac{\sum P_n}{\sum P_0}(100)$

○ **EXAMPLE 5.3a**

Use the following information to compute the aggregate commodity price index for 1992 with base period 1981.

Commodity	Price 1981	Price 1992
Loaf of bread	$0.80	$1.60
Kg of beef	$10.00	$30.00
Car	$8000.00	$10 000.00

● **SOLUTION**

$$\Sigma P_0 = 0.80 + 10.00 + 8000.00 = 8010.80$$

$$\Sigma P_n = 1.60 + 30.00 + 10\ 000.00 = 10\ 031.60$$

$$P = \frac{\Sigma P_n}{\Sigma P_0}(100) = \frac{10\ 031.60}{8010.80}(100) = 1.25226(100) = 125.2$$

Note   Unweighted aggregate price indexes tend to be misleading, since all commodities in the group are given the same weight. A more appropriate approach to the construction of aggregate price indexes takes quantities into account.

B. Weighted Aggregate Indexes (Laspeyres Method)

Weighted aggregate indexes are obtained by allowing for quantities as well as prices. While there are several methods of constructing weighted aggregate indexes, the most frequently used method is referred to as the *Laspeyres method*.

In this method, an aggregate price index is obtained by using the base-year quantities to measure the relative change in prices.

$$P = \frac{\Sigma P_n Q_0}{\Sigma P_0 Q_0}(100)$$

A weighted aggregate quantity index is similarly obtained by using base-year prices to measure relative change in quantities.

$$Q = \frac{\Sigma P_0 Q_n}{\Sigma P_0 Q_0}(100)$$

A weighted aggregate value index to measure relative change in value can be constructed by using the formula

$$V = \frac{\Sigma P_n Q_n}{\Sigma P_0 Q_0}(100)$$

Note   In all three formulas the *denominator* is $\Sigma P_0 Q_0$.

○ EXAMPLE 5.3b

Use the following information to compute weighted aggregate price, quantity and value indexes for 1992 using 1981 as base year.

	1981		1992	
	Price	Quantity	Price	Quantity
Commodity	P_0	Q_0	P_n	Q_n
Bread (loaves)	$0.80	400	$1.60	600
Beef (kg)	$1.00	100	$30.00	80
Car	$8000.00	1	$10 000.00	1

● SOLUTION

The formulas require the calculation of the products

$$P_0 Q_0, \quad P_n Q_0, \quad P_0 Q_n, \quad P_n Q_n$$

as shown below.

$P_0 Q_0$	$P_n Q_0$	$P_0 Q_n$	$P_n Q_n$
0.80(400) = 320	1.60(400) = 640	0.80(600) = 480	1.60(600) = 960
10.00(100) = 1 000	30.00(100) = 3 000	10.00(80) = 800	30.00(80) = 2 400
8 000.00(1) = 8 000	10 000.00(1) = 10 000	8 000.00(1) = 8 000	10 000.00(1) = 10 000
$\sum P_0 Q_0$ = 9 320	$\sum P_n Q_0$ = 13 640	$\sum P_0 Q_n$ = 9 280	$\sum P_n Q_n$ = 13 360

Now compute the index numbers by substituting the sum of the products determined above in the appropriate formula.

$$P = \frac{\sum P_n Q_0}{\sum P_0 Q_0}(100) = \frac{13\ 640}{9\ 320}(100) = 146.4$$

Prices in 1992 were 46.4% higher than in 1981.

$$Q = \frac{\sum P_0 Q_n}{\sum P_0 Q_0}(100) = \frac{9280}{9320}(100) = 99.6$$

Quantities in 1992 were 0.4% lower than in 1981.

$$V = \frac{\sum P_n Q_n}{\sum P_0 Q_0}(100) = \frac{13\ 360}{9\ 320}(100) = 143.3$$

The value in 1992 was 43.3% more than in 1981.

EXERCISE 5.3

Note   Unless otherwise specified, the base period is the chronologically earliest time period.

1. The following information has been compiled about a group of commodities:

Commodity	1985		1991	
	Price	Quantity	Price	Quantity
Bread (loaf)	$0.80	2000	$1.40	3600
Milk (litre)	$0.90	600	$1.20	800
Butter (kg)	$3.60	400	$4.80	300

a) Compute simple price, quantity and value indexes for each of the commodities listed.

b) Compute an unweighted aggregate price index for the group of commodities.

c) Compute weighted aggregate price, quantity and value indexes for the group of commodities.

2. From the following data, compute
   a) simple price, quantity and value indexes for each item;
   b) an unweighted aggregate price index for the items;
   c) weighted aggregate price, quantity and value indexes.

Commodity	1981		1991	
	Price	Quantity	Price	Quantity
Automobile	$12 000	20	$20 000	25
Oil (barrel)	$15	1 200	$20	2 400
Transformers	$750	60	$600	80
Tires	$80	300	$200	200

SECTION 5.4 # Special-Purpose (Composite) Indexes

Special-purpose (composite) indexes are created by a combination of business and/or economic indicators selected to measure trends in specific areas of concern.

○ EXAMPLE 5.4a

An index of general business activity is to be constructed from the following data (base 1987).

Indicator	1987	1992	Weight
Sales ($ millions)	60	126	40%
Index of employment	120	150	25%
Units of output (000)	100	120	15%
Unemployment (000)	60	45	20%

● **SOLUTION**

First compute the simple index for each indicator; then weigh the resulting index by the percent weight and add the weighted values.

Indicator	Simple index	Weighted value
Sales	$I = \dfrac{126}{60}(100) = 210$	$210(0.40) = 84.00$
Employment	$I = \dfrac{150}{120}(100) = 125$	$125(0.25) = 31.25$
Output	$I = \dfrac{120}{100}(100) = 120$	$120(0.15) = 18.00$
Unemployment	$I = \dfrac{45}{60}(100) = 75$	$75(0.20) = \underline{15.00}$
		General business index = 148.25

General business activity in 1992 was 48.3% higher than in 1987.

EXERCISE 5.4

1. Construct a special cost-of-living index from the following data, using 1986 as base.

	Index		
Item	1986	1992	Weight
Food	120	180	35%
Rent	110	135	10%
Clothing	112	95	17%
Utilities	130	170	8%
Miscellaneous	104	120	30%

2. Construct an index of business progress from the following information (base 1980).

Indicator	1980	1990	Weight
Assets ($000)	$3 400	$20 400	40%
Number of members	1 774	5 140	25%
Capital ($000)	$9	$238	15%
Reserves ($000)	$12	$737	20%

SECTION 5.5 The Consumer Price Index and Its Uses

A. The Consumer Price Index (CPI)

The **consumer price index** (CPI) is the most widely accepted indicator of changes in the overall price level of goods and services. The Canadian consumer price index is currently based on 1986 price levels and is published monthly by Statistics Canada.

For example, a consumer price index of 125.5 in April 1991 indicates that price level has increased 25.5% relative to the base year (1986) price level.

The main uses of the CPI include the determination of the *purchasing power* of the Canadian dollar and the computation of *real income.*

B. Purchasing Power of the Dollar

The **purchasing power of the dollar** is the *reciprocal* of the consumer price index; that is,

$$\text{PURCHASING POWER OF DOLLAR} = \frac{\$1}{\text{CPI}}(100)$$

○ **EXAMPLE 5.5a**
Determine the purchasing power of the dollar for a CPI of 200, and interpret.

● **SOLUTION**

$$\text{PURCHASING POWER OF DOLLAR} = \frac{\$1}{\text{CPI}}(100) = \frac{\$1}{200}(100) = \$0.50$$

This means that the current dollar is worth half of the base-year (1986) dollar. Assuming, for example, that the change in the price level of bread corresponds to the change in the CPI and that $10.00 would buy 10 loaves in 1986, you would be able to buy only 5 loaves when the CPI is 200.

○ **EXAMPLE 5.5b**
Compute the purchasing power of the Canadian dollar from 1986 to 1991.

● **SOLUTION**
The Canadian consumer price indexes for the selected years according to Statistics Canada sources were as follows:

Year	1986	1987	1988	1989	1990	1991 (April)
CPI	100.0	104.8	108.6	114.0	119.5	125.5

The calculations by means of the formula are shown below.

Year	CPI	Computation $\dfrac{\$1}{CPI}(100)$	Purchasing power of dollar
1986	100.0	$\dfrac{\$1}{100.0}(100)=$	$1.00
1987	104.8	$\dfrac{\$1}{104.8}(100)=$	$0.95
1988	108.6	$\dfrac{\$1}{108.6}(100)=$	$0.92
1989	114.0	$\dfrac{\$1}{114.0}(100)=$	$0.88
1990	119.5	$\dfrac{\$1}{119.5}(100)=$	$0.84
1991(Apr)	125.5	$\dfrac{\$1}{125.5}(100)=$	$0.80

C. Computing Real Income

The CPI can be used to correct the effect of inflation on income by adjusting nominal income (income stated in current dollars) to **real income** (income stated in base-period dollars).

$$\text{REAL INCOME} = \frac{\text{INCOME IN CURRENT DOLLARS}}{\text{CPI}}(100)$$

○ **EXAMPLE 5.5c**

Anne's salary increased from $28 600 in 1986 to $33 200 in 1988 and $36 400 in 1990. Given 1986 as the base year for the Canadian CPI and an index of 108.6 in 1988 and 119.5 in 1990,
a) determine Anne's real income in 1988 and 1990;
b) comment on the changes in stated income compared with the changes in real income.

● **SOLUTION**

a) Real income in 1988 $= \dfrac{33\ 200}{108.6}(100) = \$30\ 571$

   Real income in 1990 $= \dfrac{36\ 400}{119.5}(100) = \$30\ 460$

b) To compare nominal income with real income it is useful to determine income changes in absolute and relative terms, as shown below.

Year	1986	1988	1990
Nominal income	$28 600	$33 200	$36 400
Percent	100.0%	116.1%	127.3%
$ increase		$4 600	$7 800
Percent increase		16.1%	27.3%
Real income	$28 600	$30 571	$30 460
Percent	100.0%	106.9%	106.5%
$ increase		$1 971	$1 860
Percent increase		6.9%	6.5%

While Anne's income in 1988 has increased 16.1% over her 1986 income, her purchasing power, reflected by her real 1988 income, has increased less than 7% over the four-year period. Over the next two years her nominal income has increased 27.3% over her 1986 income. Her real income, however, has increased 6.5%, indicating that her real income actually declined from 1988 to 1990.

EXERCISE 5.5

1. Given the information below,
   a) determine the purchasing power of the dollar for 1989;
   b) determine Jack's real income for 1991;
   c) comment on Jack's purchasing power in 1991 relative to 1989.

Year	CPI	Jack's annual income
1986	100.0	$40 000
1989	114.0	$44 000
1991	125.5	$52 000

2. Jane's monthly income was $1200 in 1987 and $2200 in 1991. Given that the CPI for 1987 was 104.8 and 125.5 for 1991,
   a) determine the purchasing power of the dollar for 1991;
   b) determine Jane's real income for 1987;
   c) comment on Jane's purchasing power in 1991 relative to 1987.

SECTION 5.6 Shifting the Base and Splicing

A. Shifting the Base

The base of an existing index number series can be shifted to a more current year to bring the series up-to-date or to make it comparable with other series of index numbers.

The process of **shifting the base** is accomplished by dividing each index number in the old series by the index of the newly designated base year starting with the new base year.

This process was used by Statistics Canada to update the Canadian CPI for the years 1981 to 1985, with 1981 as the new base (see Example 5.6a below).

○ **EXAMPLE 5.6a**

Given the Canadian consumer price indexes for 1981 to 1985 with base year 1971 (see column 2 below), obtain the updated CPI numbers by shifting the base to 1981.

● **SOLUTION**

Year	Old CPI with base year 1971	Computation to shift base	New CPI with base year 1981
1981	236.9	$\frac{236.9}{236.9}(100) =$	100.0
1982	262.5	$\frac{262.5}{236.9}(100) =$	110.8
1983	277.6	$\frac{277.6}{236.9}(100) =$	117.2
1984	289.7	$\frac{289.7}{236.9}(100) =$	122.3
1985	301.3	$\frac{301.3}{236.9}(100) =$	127.2

B. Splicing

Splicing is a procedure used to create a continuous series of index numbers from two separate series. For this, the two series of numbers must have an overlap for one year. Normally, the year of overlap is the base year for the spliced series.

○ **EXAMPLE 5.6b**

The table below contains two aggregate index number series for a group of commodities. The first series covers the time period 1985 to 1990 with base year 1985; the second series is a revised series for the time period 1990 to 1994 with base year 1990. Splice the two series with 1990 as base year.

Year	1985	1986	1987	1988	1989	1990	1991	1992	1993	1994
Old series	100.0	106.4	115.9	121.4	126.3	137.8				
Revised series						100.0	105.3	112.8	119.3	130.0

● **SOLUTION**

To splice the two series compute the so-called index quotient for the overlap year.

$$\text{INDEX QUOTIENT} = \frac{\text{INDEX NUMBER OF REVISED SERIES FOR NEW BASE YEAR}}{\text{INDEX NUMBER OF OLD SERIES FOR NEW BASE YEAR}}$$

$$= \frac{100.0}{137.8} = 0.72569$$

Now multiply each index number in the old series by the index quotient to obtain the spliced series index numbers for the years 1985 to 1989:

Year	Computation	Spliced values
1985	100.0(0.72569) =	72.6
1986	106.4(0.72569) =	77.2
1987	115.9(0.72569) =	84.1
1988	121.4(0.72569) =	88.1
1989	126.3(0.72569) =	91.7

Spliced series:

Year	1985	1986	1987	1988	1989	1990	1991	1992	1993	1994
Index numbers	72.6	77.2	84.1	88.1	91.7	100.0	105.3	112.8	119.3	130.0

EXERCISE 5.6

1. The following data represent index series of the average prices of common shares of two companies for the time period 1986 to 1991.

Year	1986	1987	1988	1989	1990	1991
Company A	154.3	167.9	184.7	200.9	204.3	195.3
Company B	129.3	134.6	144.8	167.8	179.6	184.3

Shift the base of both series to 1986 two make the two series comparable.

2. The assets (in millions of dollars) of Peel Credit Union for the 10-year period 1981 to 1990 were as follows:

1981	1982	1983	1984	1985	1986	1987	1988	1989	1990
4.4	5.2	6.8	7.9	9.5	11.1	13.0	15.0	18.1	20.4

a) Show the growth in assets in relative terms by creating an index number series with base year 1981.

b) Create a new series for the period 1986 to 1990 by shifting the base to 1986.

3. The following series is an index of reserves for the period 1982 to 1986 for Peel Credit Union:

1982	1983	1984	1985	1986
115.0	357.8	1865.6	3603.4	4983.3

Splice the above series with the corresponding series for the period 1986 to 1990:

1986	1987	1988	1989	1990
100.0	121.0	164.1	249.3	372.3

4. The Canadian consumer price indexes for the 10-year period 1982 to 1991 with base years 1981 and 1986 as applicable are listed below. Splice the two series using 1986 as base year.

Year	1982	1983	1984	1985	1986	1987	1988	1989	1990	1991
Old	110.8	117.2	122.3	127.2	134.0					
New					100.0	104.8	108.6	114.0	119.5	125.5

REVIEW EXERCISE

1. Grand Motors, a Mazda dealership, has reported sales statistics for the Mazda Miata as follows:

June 1, 1989–May 31, 1990		June 1, 1990–May 31, 1991	
Base price	Quantity	Base price	Quantity
$14 500	20	$17 500	30

a) Determine the simple price index, quantity index and value index for the Miata sales.
b) Interpret the meaning of the indexes in part **a)**.

2. The following data show crude oil production in Alberta for the years 1984 and 1987.

	1984		1987	
Commodity	Price	Quantity (millions)	Price	Quantity (millions)
Crude oil production (m³)	$209.96	60.020	$142.01	55.257

a) Compute the simple price, quantity and value indexes for crude oil production for 1987 based on 1984.
b) Interpret the indexes computed in part **(a)**.

3. For the data below, compute the following for 1992 based on 1990:
a) simple price, quantity, and value indexes for each item;
b) an unweighted aggregate price index for the group of items;
c) weighted aggregate price, quantity and value indexes.

Item	1990 Price	1990 Quantity	1992 Price	1992 Quantity
Movie pass	$7.50	50	$9.00	42
Large popcorn	$1.75	70	$2.50	60
Bus ticket	$1.05	40	$1.25	30

4. For the data below, determine
   a) simple price, quantity and value indexes for each commodity;
   b) unweighted aggregate price indexes for the group of commodities;
   c) price, quantity and value indexes by the Laspeyres method.

Month	Commodity A Price	Commodity A Quantity	Commodity B Price	Commodity B Quantity	Commodity C Price	Commodity C Quantity
April	$0.54	240	$3.99	4	$2.30	50
May	$0.49	260	$3.49	7	$3.10	38

5. Using 1989 as base for the data below, compute
   a) the weighted aggregate price index for 1990;
   b) the weighted aggregate quantity index for 1991;
   c) value indexes for 1990 and 1991 by the Laspeyres method.

			Imports into Canada (First Quarter)			
	1991		1990		1989	
Commodity	Price	Quantity (000)	Price	Quantity (000)	Price	Quantity (000)
Diamonds (carats)	$632.00	32.1	$713.55	39.6	$573.40	22.2
Gold (grams)	$14.20	16 450.8	$13.72	10 055.1	$12.77	8 486.8
Silver (grams)	$0.13	32 993.2	$0.20	54 948.0	$0.18	10 646.6

6. For the data below, determine
   a) the Laspeyres price index for 1989 (base year 1987);
   b) the 1989 quantity index by the Laspeyres method (base year 1988);
   c) the weighted aggregate value index for 1989 (base year 1987).

			Canadian Exports (in petajoules)			
	1987		1988		1989	
Commodity	Price (one)	Quantity (billions)	Price (one)	Quantity (billions)	Price (one)	Quantity (billions)
Petroleum	$3 729	7.0	$2 821	6.5	$3 242	6.9
Natural gas	2 460	2.6	2 214	3.0	4 763	3.0
Electricty	7 361	1.2	8 411	0.9	10 606	0.7

7. For the information below,
   a) use 1989 as the base period to compute the weighted aggregate price,

quantity and value indexes for 1990;

b) interpret the indexes computed in **(a)**.

Residential Mortgage Loans Made by Canadian Financial Institutions				
	Average amount		Quantity	
Type	1989	1990	1989	1990
New construction	$91 419	$93 694	113 445	114 949
Existing property	$72 537	$70 764	681 772	619 144

8. Use the data below to

a) compute weighted aggregate price, quantity and volume indexes for April 15, 1992 (April 15, 1991 = 100);

b) interpret the price index computed in **a)**.

	Price of stock		Volume traded	
Company	91−04−15	92−04−15	91−04−15	92−04−15
A	$34.50	$39.00	10 000	15 000
B	17.00	11.25	40 000	30 000
C	7.00	7.25	20 000	10 000

9. Construct an index series (1984 = 100) for the following data:

Year	1984	1985	1986	1987	1988	1989	1990
Number of passengers (000)	3859	3968	3988	4126	4550	4642	4766

10. Construct an index series (1986 = 100) for the following data:

Canadian Gross Domestic Product at Current Prices						
Year	1984	1985	1986	1987	1988	1989	1990
GDP ($ billions)	445.5	479.3	506.5	551.3	603.3	651.6	677.9

11. A shoe store compiled the following information about its sales for 1991 and 1992.

	Average price (per pair)		Total sales	
Shoe group	1990	1992	1990	1992
Women's	$130	$144	$162 240	$182 592
Men's	$96	$110	$89 856	$99 660
Children's	$48	$56	$29 952	$37 184

a) Determine the number of pairs of shoes sold for each year by shoe group.

b) Calculate the price, quantity and value indexes for 1992 by the Laspeyres method (1990 = 100).

12. The purchasing records of Northern Grey Building Supplies for 1989 and 1992 showed the following:

Item	Number of units purchased		Total cost	
	1989	1992	1989	1992
Hammers	210	140	$1 785	$1 113
Screwdrivers	1 200	860	$2 700	$1 677
Wrenches	3 800	4 020	$17 670	$18 291

a) Determine the average cost per item for each year.
b) Compute the weighted aggregate price, quantity and value indexes for 1992 based on 1989.
c) Interpret the results in (b).

13. Construct a special-purpose index for April 1991 based on April 1990 to measure interest rate movements in the Canadian economy from the following data. Interpret your results.

	Treasury bill rate	Long-term government bond rate	Corporate bond rate	Residential mortgage rate
April 1990	13.55%	11.54%	12.56%	13.67%
April 1991	9.08%	9.85%	10.96%	11.26%
Weight	25%	10%	30%	35%

14. Compute an index of savings for one of the Maritime provinces for 1991, based on 1986, from the following data:

	Savings Rate as a Percentage of Gross National Product		
	Household	Corporate	Government
1986	9.5%	1.6%	1.5%
1991	10.9%	1.4%	0.9%
Weight	30%	20%	50%

15. For the data below construct a special-purpose index for 1991 (1987 = 100) and interpret the meaning of the index.

	Time lost in work stoppages (000s of person-days)	Interest on public debt ($ millions)	Federal government expenditures ($ millions)
1987	3 984	45 965	235 669
1990	5 154	63 371	295 345
Weight	30%	60%	10%

16. For the following data construct a leading indicator (special-purpose index) of economic activity for 1990, based on 1989, and interpret your results.

	Department store sales ($ millions)	Unemployment rate	Chartered bank loans ($ millions)
1989	13 914	7.5%	112 920
1990	14 184	8.1%	129 527
Weight	45%	20%	35%

17. Calculate the consumer price index for 1989 given the following sub-group indexes (1986 = 100).

	Durable goods	Semi-durable goods	Non-durable goods	Services
1989 index	111.8	114.1	112.8	115.9
Weight	15.97%	9.98%	29.22%	44.83%

18. Calculate the consumer price index for 1990, based on 1986, given the indexes for the following sub-groups:

	Durable goods	Semi-durable goods	Non-durable goods	Services
1986	100.0	100.0	100.0	100.0
1990	112.5	117.3	119.8	122.4
Weight	15.97%	9.98%	29.22%	44.83%

19. The Canadian consumer price indexes for 1987 and 1989 are 104.8 and 114.0 respectively, based on 1986. Calculate the purchasing power of the dollar for the two years relative to 1986.

20. The CPIs for 1983 and 1985 were 117.2 and 127.2 respectively, based on 1981. Compute the purchasing power of the dollar for the two years relative to 1981.

21. Purchases of life insurance in Canada during 1988 and 1989 were as follows:

	Purchases ($000)	CPI (1986 = 100)
1988	148 090	108.6
1989	163 559	114.0

Restate the purchases of life insurance in real terms using 1986 as the base period.

22. During salary negotiations for 1991 for the faculties of the Ontario Colleges of Applied Arts and Technology, the following background information was available to the negotiating teams.

Year	Maximum salary	CPI (1986 = 100)
1988	$50 767	108.6
1989	$53 318	114.0
1990	$56 517	119.5

a) Express the salaries for the three years in real 1986 terms.
b) Comment on the results.

23. Statistics for Bell Canada are shown below.

Year	1986	1987	1988	1989	1990
Revenue ($ millions)	6103.2	6372.1	6623.6	7272.9	7654.7
CPI	100.0	104.8	108.6	114.0	119.5

a) Determine Bell Canada's revenue, in real terms, from 1986 to 1990, using 1986 as base year.

b) Compute the percent change in real revenue from year to year.

c) Compute the percent change in actual revenue from year to year.

d) Compare the percent changes determined in (b) and (c) and comment on the changes.

24. The average selling prices for townhouses and two-storey houses in Burlington for April 1991 were

Townhouses	$140 000
Two-storey houses	$182 000

Using the consumer price index of 125.5 for April 1991, adjust the selling prices to 1986 dollars.

25. For the following three items, shift the base to 1989.

Raw Material Price Indexes (1986 = 100)			
Year	Animal products	Mineral fuels	Non-ferrous metals
1989	101.1	98.2	127.6
1990	105.7	117.6	114.7

26. For the following data, shift the base to 1989.

Construction Price Indexes for March (1986 = 100)			
Year	New housing	Wages	Land prices
1989	140.7	112.9	157.5
1990	146.7	114.0	169.6
1991	130.3	119.8	157.3

27. Canadian meat-processing shipments for the period 1980 to 1990 (even years) were as follows:

Year	1980	1982	1984	1986	1988	1990
Value ($ millions)	6944	7927	8277	8531	8730	9129

a) Show the industry growth by creating an index series, using 1980 as the base year.

b) Create a new series for the period 1980 to 1990 by shifting the base to 1986.

28. Revenue from long-distance calls originating in Canada for the period 1986 to 1990 was as follows:

Year	1986	1987	1988	1989	1990
Revenue ($ millions)	5817	6055	6312	6791	7143

a) Show the growth in revenue in relative terms by creating an index number series with base year 1986.

b) Create a new series beginning with 1988 as base.

29. For the following data, show each country's capacity in relative terms by indexing both series to 1989.

Uncoated Mechanical Printing Paper Capacity by Country (as of June 1990, forecasted to 1993, in 000s tonnes)					
Year	1989	1990	1991	1992	1993
Canada	1810	2030	2270	2320	2390
U.S.	1675	1640	1705	1730	1730

30. The following data are index series of revenue for Canada's two largest steel companies, Stelco and Dofasco, for the period 1985 to 1990.

Year	1985	1986	1987	1988	1989	1990
Stelco	128.1	126.3	134.0	142.7	144.7	110.6
Dofasco	104.9	101.8	113.8	125.4*	131.8*	121.3*

*(excluding Algoma Steel results)

Shift the base for both companies to 1985 to compare their current growth.

31. Splice the following two consumer price index series using 1986 as base year.

1981	1982	1983	1984	1985	1986	1987	1988	1989	1990
72.3	80.2	84.8	88.5	92.0	95.8				
					104.2	108.8	113.1	118.8	124.5

32. The following is an index series of wage levels for a wholesaler for the period 1984 to 1988:

1984	1985	1986	1987	1988
103.1	109.9	116.2	124.5	133.3

Splice the foregoing series with the corresponding series for the period 1988 to 1992:

1988	1989	1990	1991	1992
116.9	122.0	127.7	134.0	138.5

SELF-TEST

1. The following data represent retail prices of 2-L ice cream containers for the period 1987 to 1992:

Year	1987	1988	1989	1990	1991	1992
Price	$4.75	$4.99	$5.25	$5.50	$5.99	$6.25

a) Construct a simple price index using 1988 as the base year.
b) Compute the percent change in price from 1988 to 1992.
c) Compute the percent change in price from 1990 to 1992.
d) Explain why indexes can make analysis easier and faster.

Questions **2** through **7** are based on the following data:

Year	Price of beer (case of 24)	Consumer price index	Annual wages of Mr. Key
1986	$17.50	100.0	$32 000
1988	20.25	108.6	34 200
1989	22.25	114.0	36 400
1990	23.75	119.5	38 000

2. Calculate the price index of beer for 1990 based on 1986.

3. Determine Mr. Key's real income for 1989 based on 1986.

4. Can Mr. Key purchase more or fewer goods and services in 1989 compared with 1986? Explain.

5. Compute the purchasing power of the consumer dollar for 1990 compared with 1986.

6. Compare the price index of beer for 1990 with the consumer price index for the same year and comment on the relative changes.

7. Price indexes for Country A for the years 1986, 1988, 1989 and 1990 are 131.0, 150.2, 163.4 and 169.6 respectively. Is the rate of inflation higher in Canada or in country A?

8. a) Construct weighted aggregate price, quantity, and value indexes by the Laspeyres method for 1992 based on 1990.
   b) Briefly interpret the three indexes computed in (a).

	1990		1992	
Item	Price per unit	Number of units	Price per unit	Number of units
W	$3000	20	$4000	25
X	6	1200	8	800
Y	150	60	200	80
Z	3	1500	5	2000

9. A special-purpose index is to be constructed for the financial sector of the Canadian economy. Four key series have been selected, and the base is to be 1990.

Item	Weight	1990	1992
TSE 300 Composite Index	20%	2380.8	2295.1
Money supply ($ billions)	10%	228	204
Consumer credit ($ billions)	30%	110	98
Money turnover index	40%	98.0	90.8

a) Compute the special-purpose index for 1992.
b) Interpret the index and comment on how it might be used.

10. The following data represent index series of net income of two companies for the time period 1987 to 1992.

Year	1987	1988	1989	1990	1991	1992
Company A – old	166.9	183.7	199.1			
– new			100.0	108.8	110.6	105.7
Company B	118.0	123.2	129.9	137.4	147.8	149.6

a) Splice the two series for Company A (1989 = 100).
b) Shift the base to 1987 to make the two series comparable.
c) Comment on the net income performance of Company A relative to Company B.

Key Terms

Summary of Formulas

1. Simple Indexes:

a) Price:
$$P = \frac{P_n}{P_0}(100)$$

b) Quantity:
$$Q = \frac{Q_n}{Q_0}(100)$$

c) Value:
$$V = \frac{P_n Q_n}{P_0 Q_0}$$

2. UNWEIGHTED AGGREGATE PRICE INDEX, $P = \dfrac{\sum P_n}{\sum P_0}(100)$

3. Weighted Aggregate Indexes (Laspeyres Method):

a) Price:
$$P = \frac{\sum P_n Q_0}{\sum P_0 Q_0}(100)$$

b) Quantity:
$$Q = \frac{\sum P_0 Q_n}{\sum P_0 Q_0}(100)$$

c) Value:
$$V = \frac{\sum P_n Q_n}{\sum P_0 Q_0}(100)$$

4. PURCHASING POWER OF A DOLLAR $= \dfrac{\$1}{\text{CPI}}(100)$

5. REAL INCOME $= \dfrac{\text{INCOME IN CURRENT DOLLARS}}{\text{CPI}}(100)$

CHAPTER 6

Time-Series Analysis

Introduction

Forecasting or making predictions about the future is an integral part of the business planning process. A first step to looking at the future is to analyze the past. Business data collected over a number of time periods (days, months, quarters, years) provide the basis for such analysis. This so-called time-series analysis provides the basic methods for detecting trends and patterns in the data and may lead to a better understanding of events.

Objectives

Upon completion of this chapter you will be able to do the following:
Given an appropriate time series,
1. determine the linear trend equation using the least squares method;
2. construct the least squares trend line (line of best fit);
3. predict the value associated with a particular future time period;
4. determine and graph the percent of trend (cyclical relative);
5. compute and graph moving averages;
6. compute, graph and interpret seasonal indexes;
7. deseasonalize data.

SECTION 6.1 ## Components of a Time Series

A time series is a set of values observed over a period of time such as the following sales data compiled for a number of years:

Year	1985	1986	1987	1988	1989	1990	1991
Sales ($ millions)	8	10	11	14	18	20	19

The analysis of such historical data is useful in identifying past and current trends and as an aid in forecasting.

The basic time-series model identifies four time-related components:

1. **Secular trend** — the long-term tendency of economic and business data to increase or decrease.

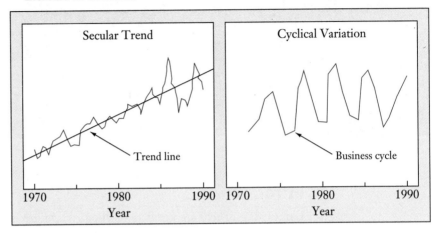

2. **Cyclical variation** — the up-and-down fluctuations associated with the business cycle.
3. **Seasonal variation** — the annually recurring seasonal patterns present in monthly or quarterly data.

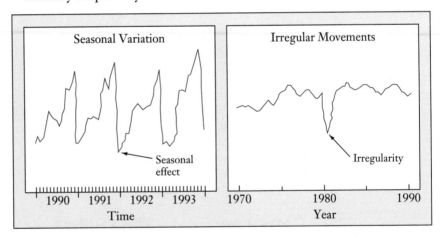

4. **Irregular movements** — the unpredictable changes resulting from un-expected events such as political upheaval, labour problems, catastrophic events.

Secular Trend Analysis

A. Introduction

Secular trend is the most often analyzed component of a time series. Secular trend analysis is concerned with describing historical patterns and is useful in projecting persistent trends. Depending on the data, straight-line or curvilinear models are available to perform the analysis. This chapter will examine only the linear model known as the least squares method of fitting a straight line.

B. The Basic Least Squares Method

The long-term trend of many business data approximates a straight line. The **least squares method** is used to determine the equation of the straight line that best fits such data.

The equation of that straight line is of the form

$$y_p = a + bx$$

where y_p is the *projected* (computed) value of the y variable for a selected time period x;

a is the value of y_p when $x = 0$;

b is the *slope* of the straight line and represents the average increase (when $b > 0$) or decrease (when $b < 0$) in y_p for one time unit;

x is the value assigned to a selected time period.

For a given time series, the specific straight-line equation can be determined by computing the values of a and b from the following:

$$b = \frac{n(\sum xy) - (\sum x)(\sum y)}{n(\sum x^2) - (\sum x)^2}$$

$$a = \frac{\sum y}{n} - b\left(\frac{\sum x}{n}\right)$$

○ **EXAMPLE 6.2a**

Use the least squares method to fit the trend line for the following time series:

Year	1985	1986	1987	1988	1989	1990	1991
Sales ($ millions)	8	10	11	14	18	20	19

● **SOLUTION**

Step 1   Draw a scatter diagram.

A *scatter diagram* (see Figure 6.1) is a visual representation of the time series and indicates whether the use of the straight-line method is appropriate for the given time series. Furthermore, the diagram is useful for graphically fitting the straight line to the data.

To draw the diagram, use the *horizontal* axis (x axis) for the *time* periods and the *vertical* axis (y axis) for the *quantitative* values associated with the time periods.

FIGURE 6.1   Scatter Diagram of Output Sales (1985–1991)

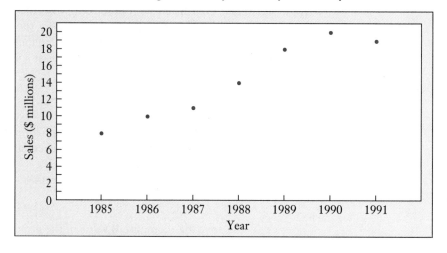

Step 2   Determine the values $\sum x, \sum y, \sum xy, \sum x^2$.

To compute the numerical values of a and b, we must first determine the values $\sum x, \sum y, \sum xy$ and $\sum x^2$.

In a time series the x values are assigned to the time periods, while the y values are the quantitative values associated with the time periods. However, rather than use the numerals identifying the years (1985, 1986, 1987, etc), assign the first year (1985) the x value 0; 1986 becomes 1, 1987 becomes 2 and so on, as shown in Table 6.1. The values in the y column are the sales figures associated with each of the years. The values in the xy column are obtained by multiplication and the values in the x^2 column are found by squaring the respective x values.

TABLE 6.1   Determination of $\sum x, \sum y, \sum xy, \sum x^2$

Year	x	y	xy	x^2
1985	0	8	(0)(8) = 0	(0)(0) = 0
1986	1	10	(1)(10) = 10	(1)(1) = 1
1987	2	11	(2)(11) = 22	(2)(2) = 4
1988	3	14	(3)(14) = 42	(3)(3) = 9
1989	4	18	(4)(18) = 72	(4)(4) = 16
1990	5	20	(5)(20) = 100	(5)(5) = 25
1991	6	19	(6)(19) = 114	(6)(6) = 36
$n = 7$	$\sum x = 21$	$\sum y = 100$	$\sum xy = 360$	$\sum x^2 = 91$

$\sum x, \sum y, \sum xy, \sum x^2$ are obtained by adding the values in the respective columns; n represents the number of observed values.

Step 3   Substitute the values obtained in Table 6.1 in the formulas for a and b:

$$b = \frac{7(360) - (21)(100)}{7(91) - 21^2}$$

$$= \frac{2520 - 2100}{637 - 441}$$

$$= \frac{420}{196}$$

$$= 2.1428571$$

$$a = \frac{100}{7} - (2.1428571)\left(\frac{21}{7}\right)$$

$$= 14.285714 - 6.4285713$$

$$= 7.8571427$$

Step 4   Substitute the computed numerical values of a and b in the general equation $y_p = a + bx$ to obtain the trend line equation

$$y_p = 7.857 + 2.143\,x$$

Step 5   Plot the trend line equation on the scatter diagram drawn in Step 1. To plot the line, substitute values for x in the trend line equation $y_p = 7.857 + 2.143\,x$ and compute the corresponding values of y_p; for example:

for $x = 0$, $y_p = 7.857 + 2.143(0) = 7.857$
for $x = 6$, $y_p = 7.857 + 2.143(6) = 7.857 + 12.858 = 20.715$

Now plot the two points (0,7.86) and (6,20.72) and join the two points to draw the graph of the trend line equation, as shown in Figure 6.2.

FIGURE 6.2 Scatter Diagram with Trend Line and Trend Line Projection — Basic Method

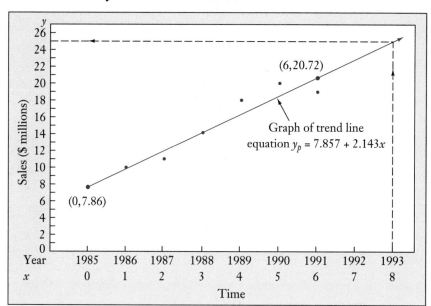

Step 6 Forecast future values.

The graph of the trend line equation or the equation itself can be used to project sales into the future.

For 1993 the projected sales are obtained graphically by drawing a vertical line from the x axis to the graph of the trend line. From there draw a horizontal line to intersect the y axis at approximately 25.

The exact value can be calculated by substituting $x = 8$ in the trend line equation:

$$y_p = 7.857 + 2.143(8) = 7.857 + 17.144 = 25.001$$

EXERCISE 6.2

1. For the following time series,
   a) draw a scatter diagram;
   b) use the least squares method to obtain the trend line equation and plot the graph of equation;
   c) predict sales for (i) 1991 and (ii) 1993.

Year	1984	1985	1986	1987	1988	1989	1990
Sales	10	11	12	14	12	13	15

2. For the following time series,
   a) draw a scatter diagram;

b) obtain the trend line equation by the least squares method;
c) plot the graph of the trend equation;
d) predict production for **(i)** 1993 and **(ii)** 1995.

Year	1986	1987	1988	1989	1990	1991
Production	86	80	72	60	54	48

The Coded Least Squares Method

A. *Formulas for Computing* a *and* b

When using the least squares method, simplification is possible by coding the time periods in such a way that $\sum x = 0$.
When $\sum x = 0$,

the equation $a = \dfrac{\sum y}{n} - b\dfrac{\sum x}{n}$ simplifies to

$$a = \frac{\sum y}{n};$$

the equation $b = \dfrac{n(\sum xy) - (\sum x)(\sum y)}{n(\sum x^2) - (\sum x)^2}$ simplifies to

$$b = \frac{\sum xy}{\sum x^2}$$

The use of the coded method centres around the condition that $\sum x = 0$. In creating this condition, two cases must be considered:
1. when the number of observations n is an odd number;
2. when the number of observations n is an even number.

B. *Using the Coded Method When n Is an Odd Number*

To make $\sum x = 0$ when n is an odd number, assign to the *middle* year the value $x = 0$. The years after the middle years are numbered $x = 1, 2, 3$, etc., while the years before the middle year are numbered $x = -1, -2, -3$, etc., as shown in Example 6.3a below.

○ **EXAMPLE 6.3a**
For the time series used in Example 6.2a, determine the trend line equation by the coded method.

● **SOLUTION**

Step 1 Construct the scatter diagram (as done for Example 6.2a).

Step 2   Assign the value $x = 0$ to the middle year 1988. 1989 becomes $x = 1$, 1990 becomes $x = 2$, 1991 becomes $x = 3$, 1987 becomes $x = -1$, 1986 becomes $x = -2$ and 1985 becomes $x = -3$.

Step 3   Determine the values $\sum x, \sum y, \sum xy, \sum x^2$, as shown in Table 6.2 below.

TABLE 6.2   **Computation of the Values $\sum x, \sum y, \sum xy, \sum x^2$ (When n Is an Odd Number)**

Year	x	y	xy	x^2
1985	-3	8	$(-3)(8) = -24$	$(-3)(-3) = 9$
1986	-2	10	$(-2)(10) = -20$	$(-2)(-2) = 4$
1987	-1	11	$(-1)(11) = -11$	$(-1)(-1) = 1$
1988	0	14	$(0)(14) = 0$	$(0)(0) = 0$
1989	1	18	$(1)(18) = 18$	$(1)(1) = 1$
1990	2	20	$(2)(20) = 40$	$(2)(2) = 4$
1991	3	19	$(3)(19) = 57$	$(3)(3) = 9$
$n = 7$	$\sum x = 0$	$\sum y = 100$	$\sum xy = 60$	$\sum x^2 = 28$

Step 4   Use the formulas to compute a and b.

$$a = \frac{\sum y}{n} = \frac{100}{7} = 14.28571$$

$$b = \frac{\sum xy}{\sum x^2} = \frac{60}{28} = 2.142857$$

Step 5   The trend line equation by the coded method is

$$y_p = 14.286 + 2.143\, x$$

FIGURE 6.3   **Scatter Diagram with Trend Line and Trend Line
Projection — Coded Method**

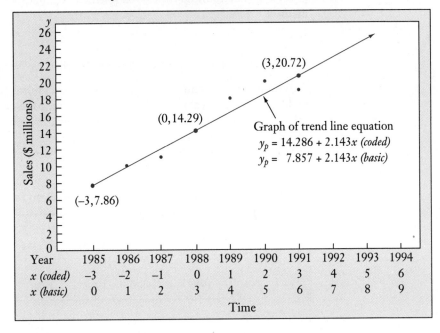

Step 6

For $x = 0$, $y_p = 14.286 + 2.143(0)$
              $= 14.286$

For $x = -3$, $y_p = 14.286 + 2.143(-3)$
              $= 14.286 - 6.429$
              $= 7.857$

For $x = 3$, $y_p = 14.286 + 2.143(3)$
              $= 14.286 + 6.429$
              $= 20.715$

Step 7   Sales projection for 1993.
The x value for 1993 by the coded method is $x = 5$.

$$y_p = 14.286 + 2.143(5) = 14.286 + 10.715 = 25.001$$

C. Comparison of the Two Methods When n Is an Odd Number

Plotting the points obtained by substituting in the coded trend line equation in the scatter diagram (see Figure 6.3) and joining the points results in the same graph as shown in Figure 6.2 (basic method). While the value of $b = 2.142857$

is the same in both equations, the value of a is different because the origin $(x = 0)$ has been shifted from 1985 for the basic method to 1988 for the coded method.

D. Using the Coded Method When n Is an Even Number

A complication in coding arises when n is an even number, since there are two middle years. The value $x = 0$ is located *between* the two middle years. The value $x = -1$ is assigned to the chronologically earlier of the two middle years and the value $x = 1$ is assigned to the following year. This means that the difference between the two middle years is two time units (see Figure 6.4).

FIGURE 6.4 Coding the Years When *n* Is an Even Number

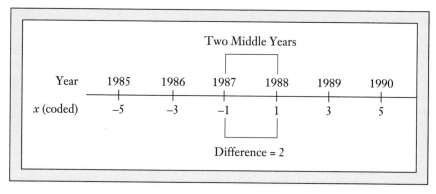

For consistency, the other years must be numbered $-3, -5, -7$, etc., and $3, 5, 7$, etc.

○ **EXAMPLE 6.3b**
For the following time series,
a) determine the trend line equation by (i) the basic method and (ii) the coded method;
b) draw a scatter diagram and plot the trend line equation.

Year	1982	1983	1984	1985	1986	1987	1988	1989	1990	1991
Units	7	8	6	5	6	4	5	3	2	3

● **SOLUTION**
a) i) Determining the trend line equation by the basic method:

Year	x	y	xy	x^2
1982	0	7	0	0
1983	1	8	8	1
1984	2	6	12	4
1985	3	5	15	9
1986	4	6	24	16
1987	5	4	20	25
1988	6	5	30	36
1989	7	3	21	49
1990	8	2	16	64
1991	9	3	27	81
$n = 10$	$\sum x = 45$	$\sum y = 49$	$\sum xy = 173$	$\sum x^2 = 285$

$$b = \frac{n(\sum xy) - (\sum x)(\sum y)}{n(\sum x^2) - (\sum x)^2}$$

$$= \frac{10(173) - (45)(49)}{10(285) - 45^2}$$

$$= \frac{1730 - 2205}{2850 - 2025}$$

$$= \frac{-475}{825}$$

$$= -0.5757576$$

$$a = \frac{\sum y}{n} - b\left(\frac{\sum x}{n}\right)$$

$$= \frac{49}{10} - (-0.5757576)\left(\frac{45}{10}\right)$$

$$= 4.9 + 2.5909092$$

$$= 7.4909092$$

The trend equation $y_p = 7.4910 - 0.5758\,x$.

ii) Determining the trend line equation by the coded method:

Year	x	y	xy	x^2
1982	-9	7	-63	81
1983	-7	8	-56	49
1984	-5	6	-30	25
1985	-3	5	-15	9
1986	-1	6	-6	1
1987	1	4	4	1
1988	3	5	15	9
1989	5	3	15	25
1990	7	2	14	49
1991	9	3	27	81
$n = 10$	$\sum x = 0$	$\sum y = 49$	$\sum xy = -95$	$\sum x^2 = 330$

$$a = \frac{\sum y}{n} = \frac{49}{10} = 4.9000$$

$$b = \frac{\sum xy}{\sum x^2} = \frac{-95}{330} = 0.2879$$

The trend equation $y_p = 4.9000 - 0.2879\,x$

b) Scatter diagram and plotting the trend line:

Basic Method

$y_p = 7.4910 - 0.5758\,x$

for 1982 when $x = 0$,

$$y_p = 7.4910$$

for 1990 when $x = 8$,

$$y_p = 7.4910 - 0.5758(8)$$
$$= 7.4910 - 4.6064$$
$$= 2.8846$$

Coded Method

$y_p = 4.9000 - 0.2879\,x$

for 1982 when $x = -9$,

$$y_p = 4.9000 - 0.2879(-9)$$
$$= 4.9000 + 2.5911$$
$$= 7.4911$$

for 1990 when $x = 7$,

$$y_p = 4.9000 - 0.2879(7)$$
$$= 4.9000 - 2.0153$$
$$= 2.8847$$

FIGURE 6.5 Scatter Diagram with Trend Line

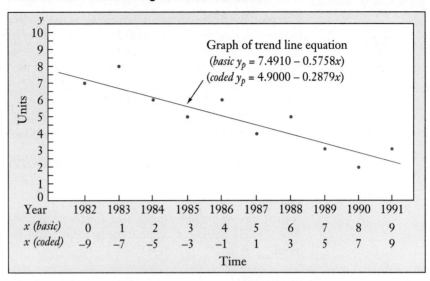

Graph of trend line equation
($basic\ y_p = 7.4910 - 0.5758x$)
($coded\ y_p = 4.9000 - 0.2879x$)

Year	1982	1983	1984	1985	1986	1987	1988	1989	1990	1991
x (basic)	0	1	2	3	4	5	6	7	8	9
x (coded)	–9	–7	–5	–3	–1	1	3	5	7	9

Time

EXERCISE 6.3

1. Use the coded method to obtain the trend line equation for Question **1** in Exercise 6.2.

2. Use the coded method to obtain the trend line equation for Question **2** in Exercise 6.2.

3. For the following time series,

Year	1983	1984	1985	1986	1987	1988	1989	1990
Debt	21	19	16	17	15	11	12	9

   a) draw a scatter diagram;
   b) use the coded method to obtain the trend line equation;
   c) plot the trend line;
   d) predict the amount of debt for **(i)** 1992 and **(ii)** 1994.

4. For the following time series,
   a) draw a scatter diagram;
   b) use the coded method to obtain the trend line equation;
   c) plot the trend line;
   d) predict the volume for **(i)** 1993 and **(ii)** 1996.

Year	1985	1986	1987	1988	1989	1990	1991
Volume	60	85	113	135	152	180	205

5. ***Data disk exercise*** Using the enclosed data disk and file name TIME, determine the sales trend line (variable name SALES).

SECTION 6.4 Percent of Trend

The **percent of trend**, also referrred to as the **cyclical relative**, is a concept used to describe the deviation of the observed values from the computed trend line. It has two important uses in business applications:

1. It permits us to see the size of the deviations of the individual values in the time series.
2. It can be used to determine the length of the business cycle that is used in computing moving averages (see Section 6.5).

The percent of trend (cyclical relative) is computed by means of the formula

$$\text{PERCENT TREND (or CYCLICAL RELATIVE)} = \frac{y}{y_p}(100)$$

○ **EXAMPLE 6.4a**

For the time series in Example 6.2a, the trend equation was determined to be $y_p = 7.857 + 2.143\,x$. Compute the percent trend and graph the percent of trend.

● **SOLUTION**

First compute y_p for each value of x by substituting in the trend equation as shown in Table 6.3. Then divide each observed value y by the corresponding value y_p and multiply by 100 to obtain the percent of trend values.

Note The following observations can be made about the percent of trend:
1. When $y > y_p$, the percent of trend > 100.
2. When $y = y_p$, the percent of trend $= 100$.
3. When $y < y_p$, the percent of trend < 100.

TABLE 6.3   Computation of Percent Trend

Year	x	Observed values y	Computed trend values $y_p = 7.857 + 2.143\,x$	Computation of the percent of trend $\frac{y}{y_p}(100)$
1985	0	8	$7.857 + 2.143(0) = 7.857$	$\frac{8}{7.857}(100) = 101.8$
1986	1	10	$7.857 + 2.143(1) = 10.000$	$\frac{10}{10.000}(100) = 100.0$
1987	2	11	$7.857 + 2.143(2) = 12.243$	$\frac{11}{12.243}(100) = 90.6$
1988	3	14	$7.857 + 2.143(3) = 14.286$	$\frac{14}{14.286}(100) = 98.0$
1989	4	18	$7.857 + 2.143(4) = 16.249$	$\frac{18}{16.249}(100) = 109.6$
1990	5	20	$7.857 + 2.143(5) = 18.572$	$\frac{20}{18.572}(100) = 107.7$
1991	6	19	$7.857 + 2.143(6) = 20.715$	$\frac{19}{20.715}(100) = 91.7$

To represent the observations over time in terms of their variation from the trend, a graph can be constructed by plotting the percent of trend against the time periods with the trend line set at 100% (see Figure 6.6).

FIGURE 6.6   Graph of Percent of Trend

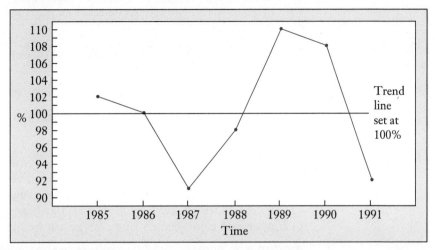

The graph points out the ups and downs associated with the cyclical component of the time series much better than Figure 6.2, which shows the actual time series and the trend line. The *relative* variation over time can be observed by noting the extent to which the plotted line deviates from the trend line set at 100%.

EXERCISE 6.4

1. Compute and graph the percent of trend for Question **3** in Exercise 6.3.
2. Compute and graph the cyclical relative for Question **4** in Exercise 6.3.

SECTION 6.5 Cyclical Analysis — The Moving Average Method

The **moving average method** can be used to smooth cyclical fluctuations in order to provide a better picture of the overall long-term trend in a time series. The results of the smoothing procedure are affected by the time period selected for computing the averages.

While the selection of the time period is somewhat arbitrary, the period chosen should be a whole number that approximates the average length of a cycle in the time series or is a multiple of the length of the cycle. A rough indicator of the length of a cycle can be obtained from the data by counting the number of time periods from one peak to another peak.

○ **EXAMPLE 6.5a**

The annual sales (in $ millions) of a product line are given below for the time period 1970 to 1991.

1970	1971	1972	1973	1974	1975	1976	1977	1978	1979	1980
6	5	6	7	10	9	8	12	13	11	14

1981	1982	1983	1984	1985	1986	1987	1988	1989	1990	1991
12	12	13	16	17	15	18	19	18	17	20

a) Plot the data.
b) Fit a seven-year moving average to the data and plot the results on the graph produced in part (a).

● **SOLUTION**

a) The data can be plotted as shown in Figure 6.7. The graph shows that there are peaks every three to four years, indicating a business cycle of three to four years in length. For a multiple cycle (*two-year* cycle) a six-year, seven-year, or even eight-year moving average may be appropriate.

FIGURE 6.7   Annual Sales Chart with Seven-Year Moving Average

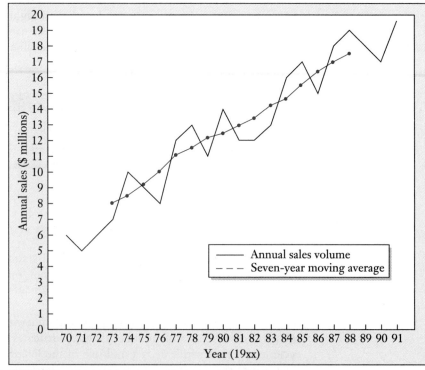

SYGRAPH®

b) Computation of a seven-year moving average.

Details of the computations are shown in Table 6.4. A step-by-step explanation of the computation follows.

Step 1   Determine the seven-year moving totals.

The first such total is obtained by adding the annual sales figures for the first seven years, 1970 to 1976 inclusive ($6 + 5 + 6 + 7 + 10 + 9 + 8 = 51$). The total (51) is then written opposite the *middle* year 1973, as shown in Table 6.4.

The next total can be obtained by adding the next group of seven years, 1971 to 1977 inclusive ($5 + 6 + 7 + 10 + 9 + 8 + 12 = 57$). This total is located against the new middle year 1974.

Successive totals can be obtained in a similar manner, e.g.,

1972 to 1978, $6 + 7 + 10 + 9 + 8 + 12 + 13 = 65$;
1973 to 1979, $7 + 10 + 9 + 8 + 12 + 13 + 11 = 70$;
1974 to 1980, $10 + 9 + 8 + 12 + 13 + 11 + 14 = 77$;

and so on, until the last possible seven-year total for the group of seven years from 1985 to 1991 ($17 + 15 + 18 + 19 + 18 + 17 + 20 = 124$) is obtained and located against the middle year 1988.

TABLE 6.4   **Computation of the Seven-Year Moving Average**

Year	Annual sales	Seven-Year moving totals		Seven-year moving averages	
1970	6				
1971	5				
1972	6				
1973	7	$6 + 5 + 6 + 7 + 10 + 9 + 8 =$	51	$51 \div 7 =$	7.29
1974	10	$51 - 6 + 12 =$	57	$57 \div 7 =$	8.14
1975	9	$57 - 5 + 13 =$	65	$65 \div 7 =$	9.29
1976	8	$65 - 6 + 11 =$	70	$70 \div 7 =$	10.00
1977	12	$70 - 7 + 14 =$	77	$77 \div 7 =$	11.00
1978	13	etc.	79	etc.	11.29
1979	11		82		11.71
1980	14		87		12.43
1981	12		91		13.00
1982	12		95		13.57
1983	13		99		14.14
1984	16		103		14.71
1985	17		110		15.71
1986	15		116		16.57
1987	18		120		17.14
1988	19		124		17.71
1989	18				
1990	17				
1991	20				

The computation of the successive seven-year totals after the first such total has been obtained can be simplified by deducting the first sales figure in the previous group of seven from that total and adding the sales figure for the year following the last year in that group.

For example, to obtain the 1971 to 1977 total, deduct the sales figure for the dropped year 1970 (6) from the total (51) and add the sales figure for the added year 1977 (12); that is, $51 - 6 + 12 = 57$.

For 1972 to 1978, $57 - 5 + 13 = 65$;
for 1973 to 1979, $65 - 6 + 11 = 70$;
for 1974 to 1980, $70 - 7 + 14 = 77$;

and so on, until the last possible grouping 1985 to 1991, $120 - 16 + 20 = 124$, is reached.

Step 2   Compute the seven-year moving average.
The seven-year moving averages are determined by dividing the successive seven-year moving totals by the number of sales figures in each total (that is, divide by 7) as shown in Table 6.4.

Step 3   Plot the seven-year moving averages.
Each average can now be plotted against the middle year. The average 7.29 is plotted against 1973; 8.15 against 1974; and so on, until the last average 17.71 is plotted against 1988.

When the points are joined, a graphical picture of the smoothing effect of this method is obtained (see Figure 6.7).

○ **EXAMPLE 6.5b**
For the data used in Example 6.5a, fit a six-year moving average and plot the results.

● **SOLUTION**
The computation of six-year moving averages is similar to the computation of seven-year averages except that six sales figures are included in each total. The average values are then obtained by dividing by 6.

Care must be taken in locating the totals and the averages. Since there is *no single* middle year for an even number of years, the totals and averages are located halfway between the two middle years in each group of six, as shown in Table 6.5.

The same approach is taken when plotting the averages in the sales chart. Each average is plotted as a point located between the respective middle years, as shown in Figure 6.8.

TABLE 6.5 Computation of the Six-Year Moving Average

Year	Annual sales ($ millions)	Six-year moving totals	Six-year moving averages
1970	6		
1971	5		
1972	6		
		← 6 + 5 + 6 + 7 + 10 + 9 = 43	43 ÷ 6 = 7.17
1973	7		
		← 43 − 6 + 8 = 45	45 ÷ 6 = 7.50
1974	10		
		← 45 − 5 + 12 = 52	52 ÷ 6 = 8.67
1975	9		
		59	9.83
1976	8		
		63	10.50
1977	12		
		67	11.17
1978	13		
		70	11.67
1979	11		
		74	12.33
1980	14		
		75	12.50
1981	12		
		78	13.00
1982	12		
		84	14.00
1983	13		
		85	14.17
1984	16		
		91	15.17
1985	17		
		98	16.33
1986	15		
		103	17.17
1987	18		
		104	17.33
1988	19		
		107	17.83
1989	18		
1990	17		
1991	20		

FIGURE 6.8 Annual Sales Chart with Six-Year Moving Average

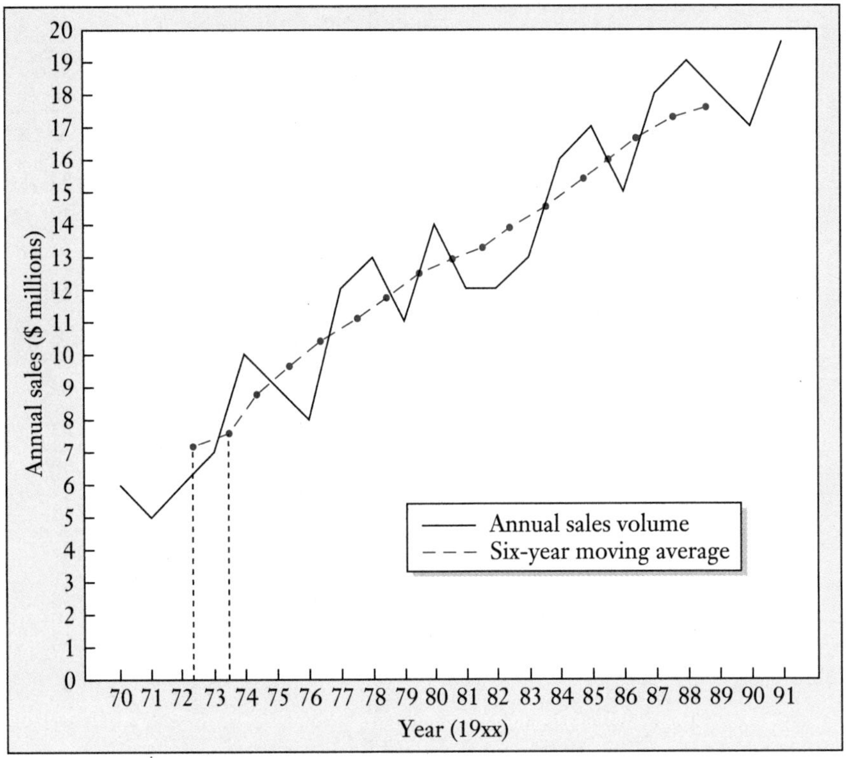

EXERCISE 6.5

1. Plot the following data, compute a five-year moving average and plot the results.

1980	1981	1982	1983	1984	1985	1986	1987	1988	1989	1990	1991
10	8	9	10	14	12	12	13	16	19	9	9

2. The following tabulation summarizes the yearly average cost per unit to produce a steel frame. Graph the data, compute a seven-year moving average and graph the results.

Year	1973	1974	1975	1976	1977	1978	1979	1980	1981	1982
$ Unit cost	281	280	337	406	402	452	432	582	598	618

Year	1983	1984	1985	1986	1987	1988	1989	1990	1991	1992
$ Unit cost	577	602	629	602	657	775	878	893	905	898

3. Plot the following data, compute a four-year moving average and plot the results.

1979	1980	1981	1982	1983	1984	1985
7	10	8	9	10	14	12

1986	1987	1988	1989	1990	1991
12	14	16	17	14	11

4. Graph the following data, compute a six-year moving average and graph the results.

1978	1979	1980	1981	1982	1983	1984	1985
97	109	76	84	102	113	76	83

1986	1987	1988	1989	1990	1991	1992
105	117	75	86	102	106	77

5. *Data disk exercise* Using the enclosed data disk and file name TIME, plot the five-year moving sales average (variable name SALES).

SECTION 6.6 Seasonal Analysis

A. Introduction

Business activity in many industries is subject to identifiable seasonal variation during the annual business cycle due to supply and demand factors. These recurring seasonal patterns are best described by a set of seasonal indexes. While different industries have different recurring seasonal patterns, most of these variations can be accommodated by either monthly or quarterly indexes.

The most common approach to obtaining such sets of seasonal indexes for a time series is the use of a moving-average method. Of the various methods available, the most commonly used is the **ratio-to-moving-average method**. For quarterly data, a four-quarter moving average is employed, while a 12-month moving average is appropriate for monthly data.

The aim of the method is to produce a seasonal index that can be used to measure the seasonal variation in the data. As for any index, the base is 100. The difference between 100 and the computed index for a particular month or quarter indicates the seasonal effect on the data.

A sales index of 150 for a particular month indicates that sales for that month were 50 percent higher than would be expected without any seasonal effect. Similarly, a sales index of 80 for a particular quarter indicates that sales for that quarter were 20 percent lower than could be expected without any seasonal effect.

The method requires much computational work performed in a six-step procedure that is designed to eliminate the long-term trend, the cyclical variation and irregular movements. The method is illustrated in the following examples.

B. The Ratio-to-Moving-Average Method — Monthly Data

○ **EXAMPLE 6.6a**
The monthly sales (in $ millions) of Custom Golf Inc. for the time period 1989 to 1991 are listed below.

Year	Jan	Feb	Mar	Apr	May	June	July	Aug	Sept	Oct	Nov	Dec
1989	1	1	2	4	6	7	6	5	7	3	1	3
1990	2	3	3	5	7	9	8	7	9	4	2	4
1991	3	2	4	6	9	11	9	8	10	5	3	5
1992	4	3	5	8	11	14	12	10	13	7	4	4

a) Plot the monthly data for each year in the form of a multiple line graph.
b) Obtain a set of monthly seasonal indexes using the ratio-to-moving-average method.

● **SOLUTION**
a) A graph of the data is useful as a visual aid in identifying the existing seasonal pattern, as illustrated in Figure 6.9.

FIGURE 6.9 Monthly Sales of Custom Golf Inc., 1989–1992

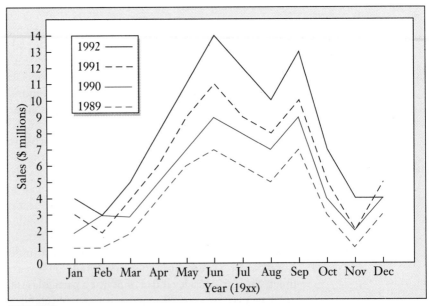

SYGRAPH®

TABLE 6.6   Computation of Centred 12-Month Moving Average and Specific Seasonal Index

Year	Month	Sales ($ millions)	12-month moving totals	12-month moving averages (÷12)	Centred 12-month moving averages	Specific seasonal indexes
1989	January	1				
	February	1				
	March	2				
	April	4				
	May	6				
	June	7				
			46	3.833		
	July	6			3.875	154.8
			47	3.917		
	August	5			4.000	125.0
			49	4.083		
	September	7			4.125	169.7
			50	4.167		
	October	3			4.208	71.3
			51	4.250		
	November	1			4.292	23.3
			52	4.333		
	December	3			4.417	67.9
			54	4.500		
1990	January	2			4.583	43.6
			56	4.667		
	February	3			4.750	63.2
			58	4.833		
	March	3			4.917	61.0
			60	5.000		
	April	5			5.042	99.2
			61	5.083		
	May	7			5.125	136.6
			62	5.167		
	June	9			5.208	172.8
			63	5.250		
	July	8			5.292	151.2
			64	5.333		
	August	7			5.292	132.3
			63	5.250		
	September	9			5.292	170.1
			64	5.333		
	October	4			5.375	74.4
			65	5.417		
	November	2			5.500	36.4
			67	5.583		
	December	4			5.667	70.6
			69	5.750		

b) ***Step 1***   Compute the 12-month moving *totals* as explained in Section 6.5. First total the sales for the 12 months in 1989 (1 + 1 + 2 + 4 + 6 + 7 + 6 + 5 + 7 + 3 + 1 + 3 = 46). Since the number of months is an even number, locate the total (46) between June and July, as shown in Table 6.6.

TABLE 6.6 Continued

Year	Month	Sales ($ millions)	12-month moving totals	12-month moving averages (÷12)	Centred 12-month moving averages	Specific seasonal indexes
1991	January	3			5.792	51.8
			70	5.833		
	February	2			5.875	34.0
			71	5.917		
	March	4			5.958	67.1
			72	6.000		
	April	6			6.042	99.3
			73	6.083		
	May	9			6.125	146.9
			74	6.167		
	June	11			6.209	177.2
			75	6.250		
	July	9			6.292	143.0
			76	6.333		
	August	8			6.375	149.4
			77	6.417		
	September	10			6.458	154.8
			78	6.500		
	October	5			6.583	76.0
			80	6.667		
	November	3			6.750	44.4
			82	6.833		
	December	5			6.958	71.9
			85	7.083		
1992	January	4			7.250	55.2
			89	7.417		
	February	3			7.500	40.0
			91	7.583		
	March	5			7.708	64.9
			94	7.833		
	April	8			7.917	101.0
			96	8.000		
	May	11			8.042	136.8
			97	8.083		
	June	14			8.042	174.1
			96	8.000		
	July	13				
	August	10				
	September	13				
	October	7				
	November	4				
	December	6				

The total for the next group of 12 months can be obtained by deducting the sales figure for January 1989 from the previous total of 46 and adding the sales figure for January 1990; that is, $46 - 1 + 2 = 47$. This total is located between July and August.

The next total is $47 - 1 + 3 = 49$, located between August and September, followed by $49 - 2 + 3 = 50$, located between September and October, and so on, until the total for the last possible group of 12 months is found as $97 - 5 + 4 = 96$ and located between June 1992 and July 1992.

Step 2   Compute the 12-month moving *averages* by dividing each of the 12-month moving totals by 12, and list the averages next to the 12-month moving totals, as shown in Table 6.6.

Step 3   Obtain the *centred* 12-month averages. This can be done by averaging two successive averages obtained in Step 2. To illustrate the procedure, an excerpt from Table 6.6 is reproduced in the following diagram, which shows the calculation of the first two centred averages.

Year	Month	Sales ($ millions)	12-month moving totals	12-month moving averages ($\div 12$)	Centred 12-month moving averages
1989					
	June	7			
			46	3.833	$\dfrac{3.833 + 3.917}{2} = 3.875$
	July	6			
			47	3.917	$\dfrac{3.917 + 4.083}{2} = 4.000$
	August	5			
			49	4.083	
	September	7			

The procedure converts the 12-month moving averages to centred 12-month moving averages aligned with the appropriate month. The first centred average of 3.875 is associated and aligned with July 1989. The next centred average of 4.000 is associated and aligned with August 1989.

The remaining centred averages are determined in a similar manner. The last possible centred average, associated and aligned with June 1992

$$= \frac{8.083 + 8.000}{2} = 8.042.$$

Note   For a 12-month moving average, centred averages are not available for the first or last six months of the time series.

Step 4   Convert the *original* data associated with each centred average to indexes. This is done by dividing the original data by the corresponding centred average and multiplying by 100.

The first of these so-called *specific seasonal indexes* is obtained by dividing the sales figure for July 1989 by the corresponding centred average; that is, $\dfrac{6}{3.87}(100) = 154.8$. The next such index is $\dfrac{5}{4.000}(100) = 125.0$ for August 1989. The remaining specific seasonal indexes are calculated in a similar manner until the last such index possible is determined for June 1992 as $\dfrac{14}{8.042}(100) = 174.1$.

Step 5   Determine the *modified mean* for each month. This is accomplished by first preparing a table of the specific seasonal indexes by month, as shown in Table 6.7.

Not surprisingly, there are various methods of computing modified means. In this text the modified means are obtained from the table by eliminating the *lowest* and the *highest* seasonal index for each month and computing the arithmetic mean of the remaining indexes for each month.

TABLE 6.7   **Determination of Modified Means and Seasonal Indexes**

Month	Specific seasonal indexes				Modified means	Seasonal indexes
	1989	1990	1991	1992		
January	–	43.6	51.8	55.2	51.8	51.7
February	–	63.2	34.0	40.0	40.0	40.0
March	–	61.0	67.1	64.9	64.9	64.8
April	–	99.2	99.3	101.0	99.3	99.2
May	–	136.6	146.9	136.8	136.8	136.6
June	–	172.8	177.2	174.1	174.1	173.9
July	154.8	151.2	143.0	–	151.2	151.0
August	125.0	132.3	149.4	–	132.3	132.1
September	169.7	170.1	154.8	–	169.7	169.5
October	71.3	74.4	76.0	–	74.4	74.3
November	23.3	36.4	44.4	–	36.4	36.4
December	67.9	70.6	71.9	–	70.6	70.5
				Totals	1201.5	1200.0

In our case only three seasonal indexes are available for each month. Since two of the indexes are dropped, the averaging calculation is simplified to listing the remaining third value as the modified mean.

For the month of January, dropping the low of 43.6 and the high of 55.2, the modified mean is 51.8. For the month of February, dropping 34.0 and 63.2, the modified mean is 40.0, and so on, as listed in Table 6.7.

Step 6   Compute the *seasonal indexes*.

Since there are 12 months and the average for the 12 months is 100, the sum of the 12 modified means should be 1200.0. However, this is not likely to be the case for any time series because of rounding and the omission of the highest and the lowest specific indexes in Step 5.

To eliminate the discrepancy between the actual sum of the modified means and 1200.0, a *correction factor* is used to adjust the modified means to obtain the seasonal indexes for the 12 months.

$$\text{Correction factor} = \frac{1200.0}{\text{Sum of the modified means}}$$
$$= \frac{1200.0}{1201.5} = 0.99875$$

The seasonal indexes for the 12 months are now computed by multiplying each modified mean by the correction factor.

For the month of January, the index = 0.99875(51.8) = 51.7;
for the month of February, the index = 0.99875(40.0) = 40.0;

and so on, as listed in the last column in Table 6.7.

C. The Ratio-to-Moving-Average Method — Quarterly Data

○ **EXAMPLE 6.6b**

For the following series of quarterly sales,
a) plot the quarterly data in the form of a line graph;
b) obtain a set of seasonal indexes.

Year	Quarter I	Quarter II	Quarter III	Quarter IV
1988	4	17	18	7
1989	8	21	24	10
1990	9	26	27	13
1991	12	33	36	15
1992	11	32	34	14

● **SOLUTION**

a) Plot the quarterly data.

FIGURE 6.10   Quarterly Sales 1988–1992

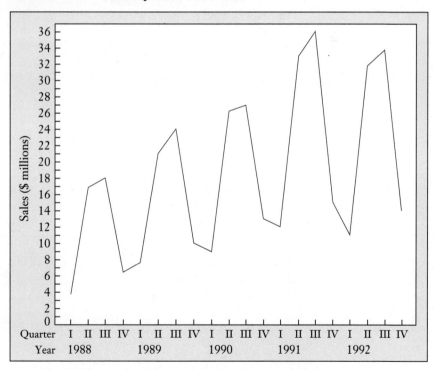

b)

   Step 1   Compute the four-quarter moving totals.

The first such total = $4 + 17 + 18 + 7 = 46$, located between the second and third quarters of 1988; followed by $46 - 4 + 8 = 50$, located between the third and fourth quarter of 1988; and so on until the last possible moving total is determined as $92 - 15 + 14 = 91$, located between the second and third quarters of 1992, as shown in Table 6.8.

Step 2   Compute the four-quarter moving averages by dividing each of the four-quarter moving totals by four, as shown in Table 6.8.

Step 3   Determine the centred four-quarter moving averages by averaging successive pairs of quarterly moving averages. The centred quarterly averages are now aligned with the appropriate quarter.

Step 4   Calculate the specific seasonal indexes for each quarter by dividing the original quarterly sales data by the corresponding centred moving averages. The first such index is obtained as $\dfrac{18}{12}(100) = 150.0$ and is associated with the third quarter of 1988.

The next index $= \dfrac{7}{13}(100) = 53.8$ and is associated with the fourth quarter of 1988. Compute all other indexes in a similar manner until the last such index is found as $\dfrac{32}{22.875}(100) = 139.9$, associated with the second quarter of 1992.

TABLE 6.8 Centred Four-Quarter Moving Averages and Specific Seasonal Indexes

Year	Quarter	Sales ($ millions)	4-quarter moving totals	4-quarter moving averages (÷4)	Centred 4-quarter moving averages	Specific seasonal indexes
1988	I	4				
	II	17				
			46	11.50		
	III	18			12.000	150.0
			50	12.50		
	IV	7			13.000	53.8
			54	13.50		
1989	I	8			14.250	56.1
			60	15.00		
	II	21			15.375	136.6
			63	15.75		
	III	24			15.875	151.2
			64	16.00		
	IV	10			16.625	60.2
			69	17.25		
1990	I	9			17.625	51.1
			72	18.00		
	II	26			18.375	141.5
			75	18.75		
	III	27			19.125	141.2
			78	19.50		
	IV	13			20.375	63.8
			85	21.25		
1991	I	12			22.375	53.6
			94	23.50		
	II	33			23.750	138.9
			96	24.00		
	III	36			23.875	150.8
			95	23.75		
	IV	15			23.625	63.5
			94	23.50		
1992	I	11			23.250	47.3
			92	23.00		
	II	32			22.875	139.9
			91	22.75		
	III	34				
	IV	14				

Step 5 Determine the modified means by first listing the specific quarterly seasonal indexes, as shown in Table 6.9.

TABLE 6.9   **Determination of the Modified Quarterly Means and the Seasonal Index for Each Quarter**

Quarter	Specific seasonal indexes					Modified means	Seasonal indexes
	1988	1989	1990	1991	1992		
I	–	56.1	51.1	53.6	47.3	52.35	51.9
II	–	136.6	141.5	138.9	139.9	139.40	138.0
III	150.0	151.2	141.2	150.8	–	150.40	148.9
IV	53.8	60.2	63.8	63.5	–	61.85	61.2
					Totals	404.00	400.0

Drop the lowest and the highest index for each quarter, and average the remaining indexes for each quarter. Since we have four specific seasonal indexes for each quarter, the averaging procedure is performed by adding the two remaining indexes for each quarter and dividing by two.

$$\text{The modified mean for the first quarter} = \frac{51.1 + 53.6}{2}(100) = 52.35;$$

$$\text{for the second quarter the modified mean} = \frac{138.9 + 139.9}{2}(100) = 139.40;$$

$$\text{for the third quarter} = \frac{150.0 + 150.8}{2}(100) = 150.40;$$

$$\text{and for the fourth quarter} = \frac{60.2 + 63.5}{2}(100) = 61.85.$$

Step 6   Compute the seasonal indexes.

Since there are four quarters, the sum of the modified means should be 400.0. However, the four modified means listed add up to 404.0. The appropriate correction factor = $400.0 \div 404.0 = 0.99010$. The seasonal index for the first quarter = $52.35(0.99010) = 51.9$. In a similar manner, the remaining modified means, when adjusted by the correction factor, become the seasonal indexes as shown in Table 6.9.

D.   *Seasonally Adjusted (Deseasonalized) Time Series*

A series of seasonal indexes shows the extent of the seasonal variation in a time series. These seasonal fluctuations are related to regular annual events such as climate changes (the four seasons), vacation periods and legal holidays, and to regular production and retail sales associated with them. These fluctuations can be eliminated by computing a series of seasonally adjusted values as follows:

$$\text{SEASONALLY ADJUSTED VALUE} = \frac{\text{ACTUAL VALUE}}{\text{SEASONAL INDEX}}(100)$$

○ **EXAMPLE 6.6c**

Adjust the quarterly data used in Example 6.6b for seasonal variation.

● **SOLUTION**

The seasonally adjusted values are calculated as shown in Table 6.10 using the data given for Example 6.6b and the seasonal indexes listed in Table 6.9.

TABLE 6.10 Seasonally Adjusted Quarterly Sales for 1988–1992

Data for Example 6.6b			Seasonal indexes from Table 6.9	Seasonally adjusted quarterly sales ($ millions)
Year	Quarter	Sales ($ millions)		
1988	I	4	51.9	$(4 \div 51.9)(100) =$ 7.71
	II	17	138.0	$(17 \div 138.0)(100) = 12.32$
	III	18	148.9	$(18 \div 148.9)(100) = 12.09$
	IV	7	61.2	$(7 \div 0.612) = 11.44$
1989	I	8	51.9	$(4 \div 0.519) = 15.41$
	II	21	138.0	15.22
	III	24	148.9	16.11
	IV	10	61.2	16.34
1990	I	9	51.9	$(9 \div 0.519) = 17.34$
	II	26	138.0	18.84
	III	27	148.9	18.12
	IV	13	61.2	21.24
1991	I	12	51.9	23.12
	II	33	138.0	23.91
	III	36	148.9	24.16
	IV	15	61.2	24.51
1992	I	11	51.9	21.19
	II	32	138.0	23.19
	III	34	148.9	22.83
	IV	14	61.2	22.88

Note The purpose of adjusting for seasonal variation is to compare values of a time series at different points in time. Seasonally adjusted values permit us to determine whether differences in the original data can be explained by the regular seasonal fluctuations.

To illustrate, for 1992 the drop in third-quarter sales of $34 million to $14 million in the fourth quarter looks quite significant. However the corresponding seasonally adjusted values (what sales would have been without seasonal effects) of $22.83 million for the third quarter and $22.88 million for the fourth quarter indicate that there was very little change.

EXERCISE 6.6

1. Sales for the Outlet Retail Store for the period July 1, 1987, to June 30, 1992, were as follows:

	Quarter			
Year	I	II	III	IV
1987	–	–	4	8
1988	3	6	5	10
1989	4	8	6	13
1990	6	10	7	16
1991	5	9	4	12
1992	4	9	–	–

a) Use the ratio-to-moving-average method to compute the seasonal indexes.

b) Compute the seasonally adjusted values.

2. Quarterly shoe sales (in $ millions) in Canada for the period October 1, 1986, to September 30, 1991, were as follows:

Year	I	II	III	IV
1986				360
1987	269	380	370	471
1988	287	405	396	507
1989	299	439	423	520
1990	386	510	452	556
1991	281	423	388	

a) Use the ratio-to-moving-average method to determine the seasonal index series.

b) Compute the seasonally adjusted quarterly shoe sales.

3. The following sales figures (in $ millions) have been compiled for a chain of restaurants for the period 1988 to 1992:

Year	Jan	Feb	Mar	Apr	May	Jun
1988	12	10	12	14	13	13
1989	12	11	15	13	14	14
1990	14	12	14	15	15	15
1991	15	13	15	16	17	18
1992	16	14	16	18	18	19

Year	Jul	Aug	Sep	Oct	Nov	Dec
1988	11	12	12	15	17	27
1989	12	11	11	16	20	32
1990	13	15	14	16	19	31
1991	16	19	17	20	23	36
1992	17	21	18	21	26	40

a) Use the ratio-to-moving-average method to compute the seasonal indexes.

b) Compute the seasonally adjusted sales figures.

4. The following table shows the value of building permits in index form (1986 = 100) issued in Canada by month for the period 1986 to 1991:

Year	Jan	Feb	Mar	Apr	May	Jun
1988	123.2	130.8	123.1	124.9	147.6	135.3
1989	158.5	160.0	154.6	144.2	137.7	136.9
1990	153.2	141.7	142.7	123.0	118.5	126.3
1991	86.0	84.8	83.2	97.2	104.5	108.6

Year	Jul	Aug	Sep	Oct	Nov	Dec
1988	137.7	125.3	133.4	124.5	130.4	143.2
1989	146.0	154.8	142.3	136.1	137.8	143.8
1990	107.7	105.3	100.4	105.2	91.2	78.2
1991	114.4	108.1	122.0	111.7	115.6	106.5

a) Determine the seasonal index series using the ratio-to-moving-average method.

b) Compute the seasonally adjusted monthly indexes.

SECTION 6.7 **Irregular Movements**

In business the practice of forecasting or predicting is based on past events and anticipated future events. As with any prediction, the best-laid plans do not always work out due to unanticipated events.

Unpredictable changes can result from events such as

1. political events (dissolution of the USSR, reunification of East and West Germany);
2. labour problems (Canadian postal strikes);
3. catastrophic events (tornado in Edmonton, *Exxon Valdez* oil spill).

In spite of the difficulties inherent in the forecasting process, business cannot afford *not* to devote time, energy and resources to predicting the future.

REVIEW EXERCISE

1. Calgary International Airport moves millions of passengers yearly, as shown below:

Year	1983	1984	1985	1986	1987	1988	1989	1990
Number of passengers (millions)	3784	3859	3968	3988	4126	4550	4642	4719

a) Plot the data in a scatter diagram.

b) Compute the trend line equation using the basic least squares method, and plot the line.

c) Determine the average annual increase in passenger traffic at Calgary International Airport since 1983.

    d) Estimate the number of passengers for **(i)** 1992 and **(ii)** 1994.

2. Increased global competition and weak market prices were a factor in reducing exports of Canadian beef during the period 1985 to 1989, as shown below:

Year	1985	1986	1987	1988	1989
Exports (millions kg)	114	100	87	81	93

    a) Draw a scatter diagram.
    b) Obtain the least squares equation by the basic method and plot the trend line.
    c) Estimate Canadian beef exports for **(i)** 1991 and **(ii)** 1993.
    d) What was the average annual change in beef exports?

3. Share prices for a company listed on the Vancouver Stock Exchange have eroded, as summarized below:

Year	1986	1987	1988	1989	1990	1991	1992
Share price	$9.75	$7.00	$5.50	$4.56	$4.00	$3.60	$3.32

    a) Draw a scatter diagram.
    b) Calculate the least squares equation using the coded method and plot the trend line.
    c) Predict the share price for 1993.

4. Canadian energy consumption, measured in petajoules, has increased significantly over the period shown below:

Year	1985	1986	1987	1988	1989	1990
Energy consumption (petajoules)	8717	8802	9049	9543	9729	9998

    a) Construct a scatter diagram.
    b) Determine the trend line equation, using the coded method, and plot the trend line.
    c) Predict the level of energy consumption for **(i)** 1991 and **(ii)** 1993.

5. Warranty repairs for J.R. Electronics Canada were as follows:

1985	1986	1987	1988	1989	1990	1991	1992
2929	2633	2418	2163	1916	1711	1549	1434

    a) Construct a scatter diagram.
    b) Determine the trend line equation and plot the line.
    c) Forecast warranty repairs for **(i)** 1994 and **(ii)** 1996.

6. Flook Transport's records indicate that damage to items shipped has increased over the period shown below:

1984	1985	1986	1987	1988	1989	1990	1991
62	69	79	90	102	116	136	153

a) Represent the data on a scatter diagram.
b) Obtain the least squares equation and plot the trend line.
c) Predict the number of items damaged for (i) 1993 and (ii) 1995.

7. From the trend line equation $y_p = 550 + 30x$, where y represents annual unit sales and 1986 is the year of origin, determine
   a) the expected sales for 1986;
   b) the average annual change in unit sales;
   c) the estimated sales for 1997.

8. Using the least squares equation $y_p = 60 - 3x$, with 1990 as the year of origin, determine
   a) the trend value for 1984;
   b) the average annual change;
   c) the predicted trend value for 1994.

9. Government spending for the years 1988 to 1992, in billions of dollars, was 66, 73, 78, 87 and 99 respectively. The least squares equation best describing the data is $y_p = 64.6 + 8x$, with 1988 as the year of origin.
   a) Plot the observed values on a scatter diagram and join the points plotted by straight line segments.
   b) Plot the trend line on the scatter diagram.
   c) Compute the trend values for the years 1988 to 1992.
   d) Compute the percent of trend for the time period.
   e) Construct a graph depicting the percent of trend.

10. The equation $y_p = 5.6 - 1.4x$, origin 1989, is the least squares equation for the following time series:

Year	1987	1988	1989	1990	1991
y	9	7	4	5	3

   a) Plot the observed values y and join the points.
   b) Plot the trend line.
   c) Compute the trend value for each year.
   d) Calculate the cyclical relatives.
   e) Construct a graph of the cyclical relatives.

11. Total assets of the Canadian Broadcasting Corporation (CBC) for the period 1984 to 1990 are shown below:

Year	1984	1985	1986	1987	1988	1989	1990
Total assets ($ millions)	639	691	731	755	799	869	941

   a) Determine the trend line equation.
   b) Construct a scatter diagram and plot the trend line.
   c) Estimate the corporation's total assets for 1993.
   d) Compute the percent of trend and draw a graph showing the percent of trend.

12. Members of the House of Commons vote themselves annual salary and benefit increases. These changes are reflected in the annual payroll shown below for the period 1985 to 1990.

Year	1985	1986	1987	1988	1989	1990
Payroll ($ millions)	47 145	48 650	53 539	55 949	60 701	64 248

a) Construct a scatter diagram.
b) Determine the trend line equation and plot the line.
c) Predict the payroll for 1995.
d) Calculate and graph the cyclical relatives.

13. The number of corporate tax returns processed by a tax accountant for the period 1976 to 1991 are listed below:

1976	1977	1978	1979	1980	1981	1982	1983
65	51	65	70	60	97	69	50

1984	1985	1986	1987	1988	1989	1990	1991
42	46	41	37	24	22	36	40

a) Compute the five-year moving averages.
b) Plot the time series and the moving averages.

14. For the following data,
a) calculate six-year moving averages;
b) plot the time series and the moving averages.

1972	1973	1974	1975	1976	1977	1978	1979	1980	1981
170	180	205	205	180	195	210	230	245	215

15. Over the years, sales of a special pump used in hospital operating rooms have increased, as shown in the data below:

1970	1971	1972	1973	1974	1975	1976	1977
123	143	153	157	152	187	196	199
1978	1979	1980	1981	1982	1983	1984	1985
190	226	244	248	222	257	262	274
1986	1987	1988	1989	1990	1991	1992	
280	288	324	309	305	327	362	

a) Determine the computed trend and the percent of trend.
b) Construct a graph of the percent of trend and estimate from it the length of the business cycle for the pump.
c) Fit a moving average to the data based on your results in (b) .
d) Plot the sales data and the moving averages on a separate graph.

16. Annual sales ($000) for Connex Variety Stores are given below:

1973	1974	1975	1976	1977	1978	1979	1980	1981	1982
52.8	88.6	78.7	72.1	88.0	96.9	85.1	87.8	108.1	130.8

1983	1984	1985	1986	1987	1988	1989	1990	1991
79.4	61.5	70.3	71.4	50.8	48.2	54.7	62.3	48.3

a) Compute a four-year moving average.

b) Plot the sales data and the four-year moving averages.

17. Monthly restaurant sales in the clubhouse of a local vacation resort offering golfing and skiing were as follows:

Month	1989	1990	1991	1992
January	31 800	32 100	34 500	37 800
February	40 200	40 500	42 900	47 400
March	29 100	28 800	26 700	27 900
April	19 500	16 500	19 500	22 800
May	16 200	19 200	21 900	24 300
June	21 900	22 200	20 700	24 300
July	27 900	30 300	33 900	23 700
August	31 500	31 800	34 200	31 600
September	25 500	22 500	25 500	
October	15 300	14 400	19 500	
November	8 100	8 700	10 500	
December	10 200	8 400	11 700	

a) Determine the twelve-month moving average.

b) Use the ratio-to-moving-average method to compute the monthly seasonal indexes.

c) Adjust the original data for seasonal variation.

18. Sales (in $000) of the Play-With-Me Toy Store for the three-year period 1990 to 1992 are listed below:

Month	1990	1991	1992
January	500	400	600
February	600	700	500
March	700	800	700
April	1000	1100	1000
May	900	900	900
June	600	800	700
July	700	800	900
August	500	400	400
September	800	700	900
October	1000	1000	1100
November	1200	1300	1300
December	1500	1700	1600

a) Compute the monthly seasonal indexes using the ratio-to-moving-average method.

b) Determine the seasonally adjusted values.

19. The following data summarize the number of business failures in Canada over a four-year period:

Year	Quarter			
	I	II	III	IV
1	717	693	693	543
2	651	502	495	565
3	616	497	411	546
4	554	477	532	636

a) Compute the quarterly moving average.
b) Construct a graph of the time series and plot the moving averages on it.
c) Use the ratio-to-moving average method to determine the quarterly seasonal indexes.
d) Compute the seasonally adjusted values.

20. Unemployment rates (in percents) in Canada for a recent four-year period are shown below:

Year	Quarter			
	I	II	III	IV
1	5.3	3.2	1.7	2.4
2	5.6	3.8	2.9	5.0
3	9.4	6.6	4.5	6.0
4	9.3	5.7	3.7	5.5

a) Calculate the quarterly seasonal indexes using the ratio-to-moving average method.
b) Construct a graph showing the original data and the moving averages.
c) Adjust the data for seasonal fluctuation.

SELF-TEST

1. A historical study of the hourly earnings at Clarkson Manufacturing Company yielded the following data:

1984	1985	1986	1987	1988	1989	1990	1991	1992
11.36	12.34	13.40	14.54	15.98	16.98	17.66	19.46	19.38

   a) Determine the trend line equation.
   b) Construct a scatter diagram and plot the trend line.
   c) Predict the hourly earnings for **(i)** 1994 and **(ii)** 1996.

2. Sales of records for the B & B Audio Entertainment Centre have declined over a period of years due to introduction of new products

Year	1987	1988	1989	1990	1991	1992
Sales ($000)	560	552	538	522	517	497

   a) Compute the equation of the least squares trend line.
   b) Determine the average change in sales of records over the time period 1987 to 1992.
   c) Forecast sales for 1995.
   d) Compute the cyclical relatives.
   e) Construct a graph of the percent of trend.

3. The growth of consumer credit in Canada can be described by the following trend model:

$$y_p = 1\ 600\ 000 + 110\ 000x$$

   where y_p is the annual dollar value of consumer credit;

   x represents year under consideration (origin is 1990).
   a) What is the average expected annual change in consumer credit?
   b) What was the trend value for 1988?
   c) Predict consumer credit for 1995.

4. The following number of building starts were recorded over the last 12-year period:

Year	1	2	3	4	5	6
Starts	32 064	14 473	45 727	43 237	23 875	8 410
Year	7	8	9	10	11	12
Starts	40 437	39 678	33 815	17 788	55 413	48 856

   a) Compute the four-year moving average.
   b) Construct a graph showing the original data and the moving average.

5. The Bank of Canada reported spending by Canadians travelling abroad (in $ millions) as follows:

Year	Quarter I	Quarter II	Quarter III	Quarter IV
1987	8 672	8 628	8 792	9 220
1988	9 152	9 504	9 724	10 148
1989	10 072	10 632	10 948	11 212
1990	11 480	11 792	12 216	12 356

a) Use the ratio-to-moving-average method to compute the quarterly seasonal indexes.

b) Compute the seasonally adjusted values.

Key Terms

Cyclical relative   140
Cyclical variation   128, 142
Irregular movement   128, 160
Least squares method   129
Moving average method   142
Percent of trend   140
Ratio-to-moving-average method   148
Seasonal variation   128, 147
Secular trend   128, 129

Summary of Formulas

1. General Trend Line Equation:

$$y_p = a + bx$$

2. Formulas for Computing a and b:

a) Basic method:

$$a = \frac{\sum y}{n} - b\left(\frac{\sum x}{n}\right)$$

$$b = \frac{n(\sum xy) - (\sum x)(\sum y)}{n(\sum x^2) - (\sum x)^2}$$

b) Coded method (when time periods coded so that $\sum x = 0$):

$$a = \frac{\sum y}{n}$$

$$b = \frac{\sum xy}{\sum x^2}$$

3. PERCENT OF TREND (CYCLICAL RELATIVE) = $\dfrac{y}{y_p}(100)$

4. SEASONALLY ADJUSTED VALUE = $\dfrac{\text{ACTUAL VALUE}}{\text{SEASONAL INDEX}}(100)$

CHAPTER 7

An Introduction to Probability

Introduction

In the first six chapters of this text we have dealt with the basic ideas of descriptive statistics. We have discussed the various ways of presenting data in tables, charts and graphs, and of reducing large bodies of data to a few summarizing statistics such as the mean and the standard deviation.

In chapters 8 to 13 we will be concerned with inferential statistics. This is the area of statistics that concerns itself with drawing conclusions about a population by analyzing samples drawn from that population. To deal with inferential statistics we must first become familiar with the basic concepts of probability.

Objectives

Upon completion of this chapter you will be able to

1. differentiate between the classical, relative frequency and subjective approaches to probability theory;
2. explain the terms random, experiment, outcome and event, and define the probability of an event;
3. construct and use Venn diagrams to illustrate and define the concepts of complementary, mutually exclusive and joint events, and to deal with the union and intersection of events;
4. use Venn diagrams to solve simple problems;
5. explain and use the basic rules of probability;
6. use tree diagrams to solve problems involving conditional and joint probabilities;
7. use counting rules to solve more complex problems.

SECTION 7.1 Approaches to Probability

A. The Classic View of Probability

Probability theory started with the study of games of chance such as playing cards or throwing dice. The phrase "games of chance" implies that the result of a particular roll of a pair of dice or the drawing of a particular card from an unmarked deck of cards is uncertain.

In the context of probability, the rolling of a die or the drawing of a card from a well-shuffled deck of cards, or any other activity for which the result depends on chance alone, is called a **random experiment**.

Consider the experiment of tossing a "fair" coin. The result of the experiment can go two ways — "head" or "tail," each of which is equally likely to happen. If the "head" shows, it is called the **Event** (Head); if the "tail" shows, it is called the Event (Tail). The probability of the Event (Head) = 0.50, as is that of the Event (Tail).

Similarly, the experiment of rolling a die has six possible events, one of which must occur, as diagrammed in Figure 7.1.

FIGURE 7.1 **The Six Events of Rolling One Die**

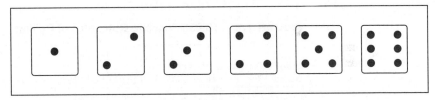

Provided the die is "fair," each of the six events is equally likely to occur. The probability of a particular event occurring = $\frac{1}{6}$.

A more complex situation arises when we use a pair of dice. In the experiment of rolling a pair of dice the *event of interest* is the total number of spots showing. There are 11 possible events — 2 spots, 3 spots, 4 spots and so on to 12 spots.

To illustrate the experiment let us use a pair of coloured dice — one red die and one black die. Now consider the Event (Four Spots). As indicated in Figure 7.2, this event is made up of the three possible arrangements (**outcomes**) of the two dice.

FIGURE 7.2 The Three Possible Outcomes of Rolling Two Dice for the Event (Four Spots)

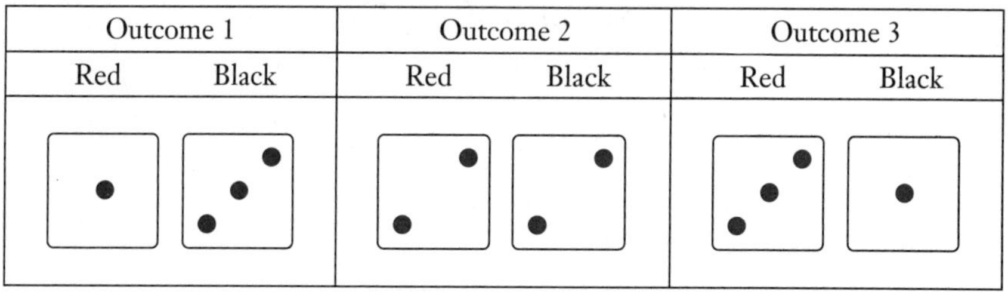
Outcome 1
Red Black
Outcome 2
Red Black
Outcome 3
Red Black

The 11 possible events (2 spots, 3 spots, 4 spots and so on to 12 spots) are illustrated in Figure 7.3. As shown, there are 36 possible arrangements of the two dice; that is, there are 36 possible outcomes to the experiment of rolling two dice.

The number of outcomes that make up a particular event are referred to as *favourable* outcomes. For example, the three outcomes that result in the Event (Four Spots) are referred to as the favourable outcomes for this event. Similarly, there are four favourable outcomes for the Event (Nine Spots).

FIGURE 7.3 The 36 Possible Outcomes of Rolling Two Dice

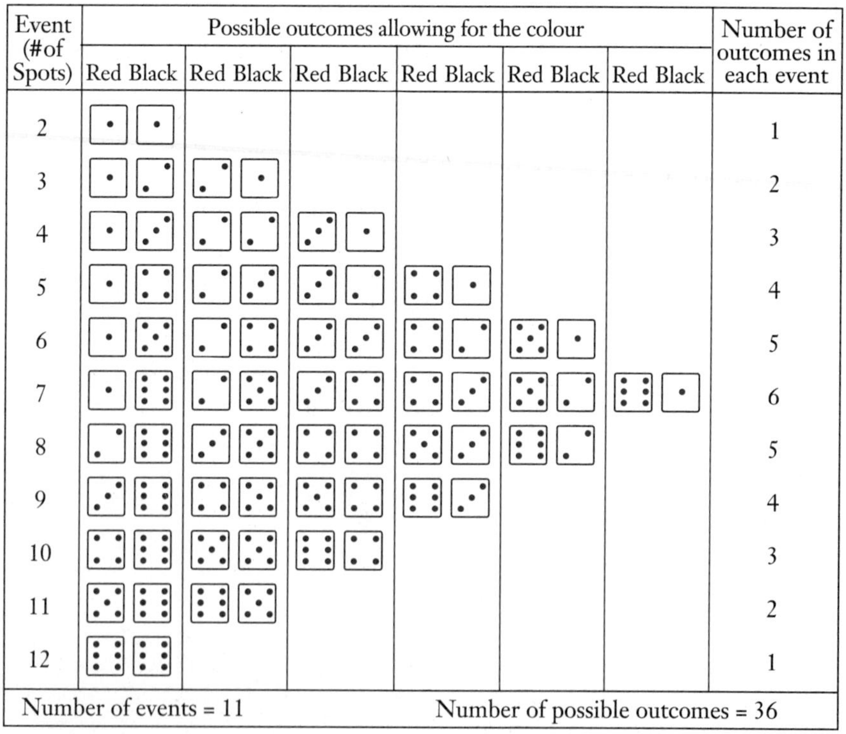
Event (#of Spots)
Possible outcomes allowing for the colour
Red Black Red Black Red Black Red Black Red Black Red Black
Number of outcomes in each event
2 1
3 2
4 3
5 4
6 5
7 6
8 5
9 4
10 3
11 2
12 1
Number of events = 11
Number of possible outcomes = 36

Since there is a total of 36 different outcomes, the probability of the Event (Four Spots) = $\frac{3}{36} = \frac{1}{12} = 0.0833$, while the probability of the Event (Nine Spots) = $\frac{4}{36} = 0.1111$.

In general,

$$\text{PROBABILITY OF AN EVENT} = \frac{\text{NUMBER OF FAVOURABLE OUTCOMES}}{\text{TOTAL NUMBER OF POSSIBLE OUTCOMES}}$$

For the 11 events in Figure 7.3, the probabilities are

$$\frac{1}{36}, \frac{2}{36}, \frac{3}{36}, \frac{4}{36}, \frac{5}{36}, \frac{6}{36}, \frac{5}{36}, \frac{4}{36}, \frac{3}{36}, \frac{2}{36}, \frac{1}{36}.$$

○ **EXAMPLE 7.1a**

What is the probability of drawing an ace from a well-shuffled, unmarked deck of 52 cards?

● **SOLUTION**

Since the total number of cards is 52, there are 52 *equally likely outcomes* on a draw of a single card.

There are four aces in the deck. The Event (Drawing an Ace) consists of the four possible equally likely outcomes of drawing the ace of clubs, the ace of diamonds, the ace of hearts or the ace of spades.

$$P(\text{EVENT = DRAWING AN ACE})$$
$$= \frac{\text{NUMBER OF FAVOURABLE OUTCOMES}}{\text{TOTAL NUMBER OF POSSIBLE OUTCOMES}}$$
$$= \frac{4}{52} = 0.0769$$

B. *The Relative Frequency Concept of Probability*

There are many situations in which the outcomes of an experiment are not equally likely to occur. Whether your car will start or not on any given morning are not equally likely events. Similarly, rain, snow or sunny skies are not equally likely events on any particular day in any place in Canada. In such situations the *relative frequency* of the occurrence of the event in the past is taken as its probability of occurrence.

The relative frequency interpretation of probability is based on historical data and can be defined as the *proportion of the time that the event will occur over the long run*.

For example, if you use public transit to get to work each day, you will eventually get a "feel" for its reliability in terms of whether the system gets you to work on time "every time," "nearly every time" or "about half the time" and so on.

If you were to count the total number of trips to work over a long period of time and the number of times you were late because of delays in the transportation system, you could put a numerical value to your "feeling" about the reliability of the system. Suppose, for example, that out of 240 working days you were late 12 times due to delays in the system. Your long-run probability of being late would be

$$P(\text{EVENT} = \text{LATE TO WORK}) = \frac{\text{NUMBER OF TIMES LATE}}{\text{NUMBER OF WORKING DAYS}}$$
$$= \frac{12}{240} = 0.05 = 5\%$$

C. The Concept of Subjective Probability

Both the classical concept of probability and the relative frequency approach to probability are objective in that they rely on counting the frequency of occurrence of outcomes. In the classical case this can be done theoretically; in the relative frequency approach it is done by an actual count of what has happened.

In the subjective approach to probability the individual assigns his or her own *personal estimate* of the chance that an event may happen. This estimate will be based on the person's previous experience with similar situations. This subjective approach is useful in situations where neither theoretical data nor actual data are available.

EXERCISE 7.1

1. How many outcomes are possible for the following experiments:
   a) rolling a single die once?
   b) rolling a single die twice?

2. How many outcomes are possible for
   a) tossing a single coin once?
   b) tossing a single coin twice?
   c) tossing a single coin three times?

3. State which of the following selection methods are random:
   a) drawing a number out of a hat;
   b) choosing the first name in the phone book under each letter of the alphabet;
   c) spinning an equally divided wheel;
   d) drawing the winning entries for a contest from a barrel.

4. Is asking a person to pick a number between 1 and 10 a random selection method? Explain.

5. How many events are possible for
   a) a single coin tossed once?
   b) a single coin tossed twice?
   c) a single coin tossed three times?

6. How many events are possible for the experiment of rolling three dice?

7. What is the probability of drawing a face card (ace, king, queen, jack) from a deck of 52 cards?

8. Determine the probability of drawing from a deck of 52 cards
   a) a red card;
   b) a club;
   c) the queen of spades.

9. During the past semester Jenna Rak was late for classes 4 times out of 80 scheduled classes. What is the probability that Jenna will arrive for classes on time?

10. As part of the basketball team tryouts, each player was asked to get as many baskets possible in one minute. Mike has so far scored on 18 attempts and missed 7. What is the probability that Mike will miss the next attempt?

11. A computer-controlled lathe requires periodic maintenance or it will break down. Maintenance records of similar models disclosed the following information:

Production time (days)	0 to 14	15 to 29	30 to 44	45 to 59
Probability of breakdown	2%	5%	30%	95%

   a) What is the probability of the lathe working at least 25 days without breakdown?
   b) After how many days of operation should periodic maintenance be performed if we are willing to accept a 5% risk of a breakdown?

12. The following data represent regional mortgage lending by Canada's chartered banks:

Region	Mortgage loans ($ millions)	
	Residential	Non-residential
Eastern Canada	71 878	5 224
Western Canada	28 024	2 378

   a) What is the probability that the next dollar granted by a chartered bank for a mortgage will be in western Canada?
   b) What is the probability that the next mortgage will be a non-residential mortgage in eastern Canada?

SECTION 7.2 Venn Diagrams

A. Basic Concepts

The basic components of probability are the outcomes associated with some process such as a game of chance, a weather forecast, the price movements of a particular stock or the number of rejects in a manufacturing process. The possible outcomes of a process can be graphically represented by diagrams named after their developer, John Venn.

A **Venn diagram** starts with an *enclosed* area that contains all possible outcomes of the process. This area is called the **sample space**. In the case of tossing a coin, the sample space consists of two outcomes: "Head" or "Tail." When one die is rolled, the sample space consists of the six possible outcomes: "One Spot," "Two Spots" and so on to "Six Spots." The enclosed area can then be subdivided into parts representing the event or events under consideration.

B. Representative Diagrams

The following set of diagrams are Venn diagrams illustrating basic concepts of probability.

Diagram 1 Complementary Events

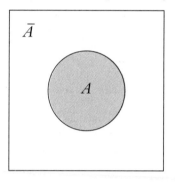

The area within the rectangle represents the sample space. The circular area within the sample space represent Event A. The remainder of the rectangular area contains all other events. These events are referred to as the **complement** of A, represented by the symbol $\overline{A}$ (read as "Not A").

For example, when a card is drawn from a deck of 52, if we define Event A as "Draw an Ace," the complement of A is $\overline{A}$ = "All Other Cards."

Diagram 2 Mutually Exclusive Events

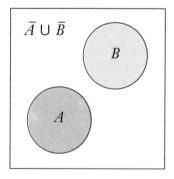

The areas within the two circles marked A and B represent two events that cannot occur at the same time — they are **mutually exclusive** events. The remaining events in the sample space are represented by the symbol $(\overline{A} \cup \overline{B})$ meaning "Not A and Not B."

For example, when one card is drawn, the events "Draw an Ace" or "Draw a King" cannot happen at the same time — they are *mutually exclusive* events.

Diagram 3 Joint Events

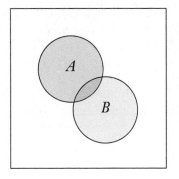

Joint events are events that involve two or more attributes that occur at the same time. Joint events are represented by overlapping areas. For example, the events "Draw an Ace" and "Draw a Diamond" are joint events and overlap because one of the four aces is the ace of diamonds.

The **union** of two events A and B consists of all the outcomes included in either Event A or Event B or in both. It is indicated by the shading of the areas A and B and is denoted by the symbol $(A \cup B)$.

Diagram 4A Union of Mutually Exclusive Events

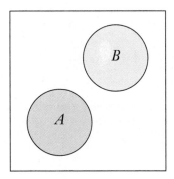

If Event A = Draw an Ace and Event B = Draw a King, the union of the two events represents the idea of drawing either an ace or a king. Because the two events are mutually exclusive, the areas A and B do not overlap, as shown in Diagram 4A.

Diagram 4B Union of Joint Events

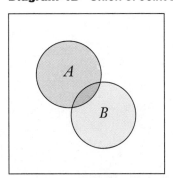

For Event A = Draw an Ace and Event B = Draw a Diamond, the union of the two events $(A \cup B)$ represents the idea of drawing either an ace or a diamond. Since both events may happen when the ace of diamonds is drawn, the areas overlap, as shown in Diagram 4B.

Diagram 5   Intersection of Events

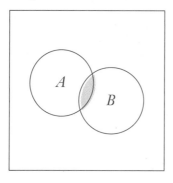

The **intersection** of two events A and B, denoted by the symbol $(A \cap B)$, is indicated by the shading in the area of overlap. It consists of all outcomes that are included in both A and B. Intersection is not possible for mutually exclusive events.

For example, the intersection of the events A (Draw an Ace) and B (Draw a Diamond) is the Event (Draw the Ace of Diamonds), since this is the only outcome that fits the two required attributes.

C. *Examples Using Venn Diagrams*

○ **EXAMPLE 7.2a**

Consider the experiment of drawing a single card at random from a deck of 52 cards. The outcome is favourable if the draw meets the following three attributes:

a) The card drawn must be red.
b) It must be a diamond.
c) It must be an ace.

Represent the event of drawing a card in the form of a Venn diagram, and assign the number of possible outcomes for each area.

● **SOLUTION**

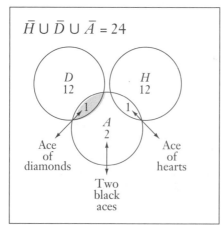

First draw a rectangle representing the sample space of the 52 possible outcomes of drawing a single card.

Partition the sample space according to the three attributes required for a favourable outcome. Draw two non-intersecting circles representing the events D (Diamond) and H (Heart) respectively. The union of the two circles $(D \cup H)$ represents all 26 red cards. The remaining sample space represents the 26 black cards.

Now draw a third circle to represent the Event A (Ace). Since there are two red aces, the ace of diamonds and the ace of hearts, this circle must overlap the two other circles, as shown in the diagram. The area now outside the three circles represents the remaining black cards. The event determined by the three attributes is the Event (Draw the Ace of Diamonds).

The number of possible outcomes associated with each area are as follows:

1 (ace of diamonds) for the *intersection* of the circles A and D; $(A \cap D) = 1$.
1 (ace of hearts) for the *intersection* of the circles A and H; $(A \cap H) = 1$.
2 (the two black aces) for the remaining area of circle A.
12 (the remaining diamonds) for the remaining area of circle D.
12 (the remaining hearts) for the remaining area of circle H.
24 (the remaining black cards) for the area of the sample space outside the three circles. Not H not D and not A, denoted by $(\overline{H} \cup \overline{D} \cup \overline{A}) = 24$.

The total $1 + 1 + 2 + 12 + 12 + 24 = 52$ represents the sum of the possible outcomes of the experiment of drawing one card from a deck of 52.

○ **EXAMPLE 7.2b**

Of the 100 students enrolled in a business program, 52 take statistics. There are 60 female students; 45 of them take accounting, 30 take statistics, and 20 take both courses. 32 male students take accounting and 18 take both accounting and statistics.

Draw a Venn diagram to represent the information and determine
a) the number of female students taking neither accounting nor statistics;
b) the number of male students taking statistics only;
c) the number of students taking accounting only;
d) the number of students taking neither accounting nor statistics.

● **SOLUTION**

First draw a rectangle to represent the sample space for the 100 students. To partition the area inside the rectangle, draw three overlapping circles to represent three attributes in the problem — female, accounting, statistics. The remaining attribute, "male," is represented by the remaining area in the rectangle surrounding the three circles.

To assign numbers to the various areas, start with the area where the largest number of attributes overlap. In this case it is the number of female students (20) taking both accounting and statistics.

a) Determine the number of female students taking neither accounting nor statistics.

Step 1   Locate the area of intersection of the three circles and insert the number 20.

Step 2   Since 45 of the female students take accounting and 20 of them also take statistics, the number of female students taking accounting only is $45 - 20 = 25$. Insert the number 25 in the remaining part of the intersection of circles F and A.

Step 3   Since 30 of the female students take statistics and 20 also take accounting, the number of female students taking statistics only is $30 - 20 = 10$. Insert the number 10 in the remaining part of the intersection of circles F and S.

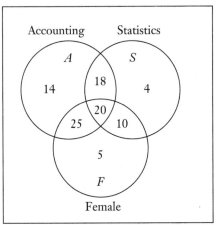

Step 4 Since there are 60 female students and 20 + 25 + 10 = 55 are shown in the areas of intersection with the circles representing accounting and statistics, insert the number 5 in the remaining part of the circle *F*. This means the number of female students taking neither accounting nor statistics is 5.

b) Determine the number of male students taking statistics only.

Step 5 Insert 18 (the number of male students taking both accounting and statistics) into the remaining area of intersection of circles *A* and *S*.

Step 6 Since 52 students take statistics, 20 + 10 = 30 of whom are female, and 18 are male who also take accounting, insert 4 into the remaining area of circle *S*. The number of male students taking statistics only is 4.

c) Determine the number of students taking accounting only.

Step 7 Insert the number 14 into the remaining area of circle *A* since, out of 32 male students taking accounting, 18 also take statistics. Since there are also 25 female students taking accounting only, the total number of students taking accounting only is 18 + 25 = 39.

d) Determine the number of students taking neither accounting nor statistics.

Step 8 We know from the Venn diagram that 14 + 18 + 4 = 36 male students take either accounting or statistics or both courses. Since there are 40 male students enrolled, 40 − 36 = 4 of them take neither accounting nor statistics. From the diagram we also know that 5 female students take neither course. The total number of students taking neither accounting nor statistics is 9.

EXERCISE 7.2

1. What is the sample space for a single coin tossed
   a) once?
   b) twice?
   c) three times?
2. What is the sample space for a game of chess?
3. Which of the following pairs of events are mutually exclusive?
   a) A salesperson takes an order for widgets as the company's widget production line breaks down.
   b) The prime lending rate increases and the value of the Canadian dollar increases.

c) Russell and Kerry are both hired for the same job.

d) On two rolls of a die a four occurs and the sum is four.

4. Which of the following events are not mutually exclusive?

a) The price of a company's stock rises while the company's net income decreases.

b) A posted voters list has William Thomas listed as a male and a female.

c) A retired person collects Canada Pension while paying into the Canada Pension Plan from part-time employment.

d) Cutnife Equipment sales decrease while selling expenses increase.

5. A local basketball team has 15 players, 12 of whom are right-handed. There are 5 players over six feet tall, one of whom is left-handed. Draw a Venn diagram and determine

a) the number of right-handed players under six feet;

b) the number of left-handed players under six feet.

6. The Kay Bank has 12 000 accounts, of which 7000 are chequing accounts, 6000 are savings accounts, and 4000 are both. Draw a Venn diagram and determine

a) the probability that a randomly selected account will be a savings or chequing account or both;

b) the number of accounts that are neither savings nor chequing accounts.

SECTION 7.3 The Basic Rules of Probability

A. *Sum of the Probability of an Event and Its Complement*

The probability of an event is usually denoted by the symbol P(Event). In general, the probability of an Event (A) is given by

$$P(A) = \frac{\text{NUMBER OF FAVOURABLE OUTCOMES}}{\text{TOTAL NUMBER OF POSSIBLE OUTCOMES}}$$

Since the number of favourable outcomes cannot be smaller than zero or greater than the total number of possible outcomes, the value of the fraction

$$\frac{\text{NUMBER OF FAVOURABLE OUTCOMES}}{\text{TOTAL NUMBER OF POSSIBLE OUTCOMES}}$$

must lie between 0 and 1; that is, $0 \leq P(A) \leq 1$. Similarly, for the complement of Event (A), denoted by the symbol $P(\overline{A})$, $0 \leq P(\overline{A}) \leq 1$.

In any experiment the outcome of an event must be either favourable or its complement,

$$P(A) + P(\overline{A}) = 1$$

That is, the SUM OF THE PROBABILITIES IN ANY SAMPLE SPACE = 1.

$P(A) = 1$ means that Event (A) is certain to happen and implies that $P(\overline{A}) = 0$.

For example, in the toss of a two-headed coin the Event (A = Head) is certain to happen; that is, $P(A) = 1$. The complement $\overline{A}$ = Tail cannot happen; that is, $P(\overline{A}) = 0$.

In the roll of one die, which has six equally possible outcomes, the Event (A = Six Spots) has the probability $P(A) = \dfrac{1}{6}$. The complementary Event ($\overline{A}$ = Not a Six Spot) has the probability $P(\overline{A}) = \dfrac{5}{6}$. Again, $P(A) + P(\overline{A}) = \dfrac{1}{6} + \dfrac{5}{6} = 1$.

B. The Special Rule of Addition for Mutually Exclusive Events

If two or more events are mutually exclusive, the probability that one or any of the others will occur is the sum of their individual probabilities of occurrence. For two mutually exclusive events A and B, the addition is denoted by

$$P(A \text{ or } B) = P(A \cup B) = P(A) + P(B)$$

This special rule can be expanded for any number of mutually exclusive events. In the case of three such events, A, B and C, the addition is denoted by

$$P(A \text{ or } B \text{ or } C) = P(A \cup B \cup C) = P(A) + P(B) + P(C)$$

○ **EXAMPLE 7.3a**
Determine the probability of drawing either a black ace or any king from a deck of 52 cards on a single draw.

● **SOLUTION**

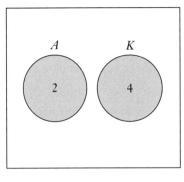

Let the Event (A) = Black Ace and the Event (K) = King. As represented in the Venn diagram, the two events are mutually exclusive.

The probability of drawing a black ace, $P(A) = \dfrac{2}{52}$.

The probability of drawing a king, $P(K) = \dfrac{4}{52}$.

The probability of drawing either a black ace or any king is

$$P(A \text{ or } K) = P(A \cup K) = P(A) + P(K)$$

$$= \frac{2}{52} + \frac{4}{52} = \frac{6}{52} = 0.1154$$

○ **EXAMPLE 7.3b**

Determine the probability of drawing either the ace of clubs, a red king, or any queen from a deck of 52 cards.

● **SOLUTION**

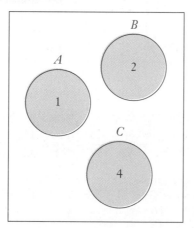

Let Event (A) = Ace of Clubs, Event (B) = Red King and Event (C) = Queen.

The probability of drawing the ace of clubs, $P(A) = \frac{1}{52}$.

The probability of drawing a red king, $P(B) = \frac{2}{52}$.

The probability of drawing a queen, $P(C) = \frac{4}{52}$.

Since the three events are mutually exclusive,

$$P(A \text{ or } B \text{ or } C) = P(A \cup B \cup C)$$

$$= P(A) + P(B) + P(C)$$

$$= \frac{1}{52} + \frac{2}{52} + \frac{4}{52}$$

$$= \frac{7}{52} = 0.1346$$

C. The General Rule of Addition

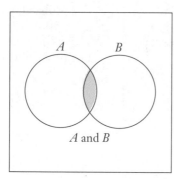

A and B

For two events that are *not* mutually exclusive, the probability that one or the other event will occur is the sum of their individual probabilities of occurrence diminished by the probability of both events occurring at the same time.

$$P(A \text{ or } B) = P(A \cup B)$$

$$= P(A) + P(B) - P(A \text{ and } B)$$

$$= P(A) + P(B) - P(A \cap B)$$

○ **EXAMPLE 7.3c**
Determine the probability of drawing an ace or a heart from a deck of 52.

● **SOLUTION**
Let Event (A) = Ace and Event (B) = Heart.
   The two events are not mutually exclusive since both attributes are met if the ace of hearts is drawn.

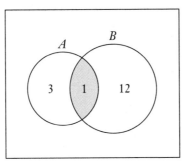

The probability of drawing an ace,
$P(A) = \dfrac{4}{52}$.
   The probability of drawing a heart,
$P(B) = \dfrac{13}{52}$.
   The probability of drawing the ace of hearts, $P(A \text{ and } B) = \dfrac{1}{52}$.

$$P(A \text{ or } B) = P(A) + P(B) - P(A \text{ and } B)$$
$$= \frac{4}{52} + \frac{13}{52} - \frac{1}{52} = \frac{16}{52} = 0.3077$$

Note   The same result can be obtained from the Venn diagram by adding the numbers assigned to the areas and dividing by the total number of possible outcomes, $\dfrac{3 + 1 + 12}{52} = \dfrac{16}{52}$.

○ **EXAMPLE 7.3d**
The Deerfield Golf Club has 250 playing members. Fifty of them also play tennis and 75 play squash. Of the 75 squash players, 20 also play tennis. Compute the probability that a club member plays either squash or tennis.

● **SOLUTION**

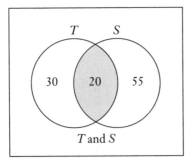

Let Event (T) = Tennis and Event (S) = Squash.
   The probability of playing tennis,
$P(T) = \dfrac{50}{250}$.
   The probability of playing squash,
$P(S) = \dfrac{75}{250}$.

The probability of playing tennis and squash, $P(T \text{ and } S) = \dfrac{20}{250}$.

$$P(T \text{ or } S) = P(T) + P(S) - P(T \text{ and } S)$$

$$= \frac{50}{250} + \frac{75}{250} - \frac{20}{250} = \frac{105}{250} = 0.42$$

Note The general rule for addition $P(A) + P(B) - P(A \text{ and } B)$ reduces to the special rule of addition for mutually exclusive events $P(A) + P(B)$, since in this case $P(A \text{ and } B) = 0$.

D.   *The Special Rule of Multiplication for Mutually Independent Events*

Two events are mutually independent if the probability of the occurrence of one event is not affected by the occurrence of the other event. For two independent events A and B, the probability that both events will occur is found by multiplying their probabilities:

$$P(A \text{ and } B) = P(A \cap B) = P(A) \cdot P(B)$$

Note The symbol $\cdot$ is used to indicate multiplication.
This rule applies to any number of independent events.

$$P(A \text{ and } B \text{ and } C \text{ and } \ldots) = P(A \cap B \cap C \cap \ldots)$$
$$= P(A) \cdot P(B) \cdot P(C) \cdot \ldots$$

○ **EXAMPLE 7.3e**
What is the probability of getting two heads in consecutive flips of a balanced coin?

● **SOLUTION**
Let Event (A) = First Flip and Event (B) = Second Flip.

The probability of getting a head on the first flip, $P(A) = \dfrac{1}{2}$.

The probability of getting a head on the second flip, $P(B) = \dfrac{1}{2}$.

Since the two events are mutually independent, the probability of getting two heads on consecutive flips

$$P(A \text{ and } B) = P(A \cap B) = P(A) \cdot P(B)$$

$$= \left(\frac{1}{2}\right)\left(\frac{1}{2}\right) = \frac{1}{4} = 0.25$$

This result can be verified by listing all possible outcomes of consecutive tosses of one coin as shown below:

Toss	Possible outcomes			
First	Head	Head	Tail	Tail
Second	Head	Tail	Tail	Head

The outcome "head,head" is one of four possible outcomes. It has one chance in four to occur. Its probability of occurrence is 0.25.

○ **EXAMPLE 7.3f**

What is the probability of getting two aces in consecutive draws of one card from a deck of 52 if the first card is replaced and the deck is reshuffled before the second draw is made? (This process is referred to as sampling with replacement.)

● **SOLUTION**

Let Event (A) = First Draw and Event (B) = Second Draw.

The probability of the first card being an ace, $P(A) = \dfrac{4}{52}$.

The probability of the second card being an ace, $P(B) = \dfrac{4}{52}$.

The probability of drawing two aces in consecutive draws with replacement,

$$P(A \text{ and } B) = P(A \cap B) = P(A) \cdot P(B)$$

$$= \left(\frac{4}{52}\right)\left(\frac{4}{52}\right) = 0.0059172;$$

that is, approximately 6 chances in 1000, or once in 167 draws.

E. Conditional Probability

So far we have assumed that any item chosen was drawn from the total population. Sometimes, however, we already have some information about the event under study. In such cases the size of the population to be considered is reduced.

For example, suppose we want to determine the probability that the top card in a deck of 52 cards is either the king of hearts or the king of diamonds. If we also know that the top card in the deck is a red card, then we need no longer consider the total population of cards in the deck but can confine our attention to that part of the deck that consists of the 26 red cards.

In problems such as this we are specifically interested in the probability of the occurrence of one event given that some other event has already occurred or will occur. This is referred to as **conditional probability** and is denoted by $P(A \mid B)$. The symbol $P(A \mid B)$ is read as "the probability of A given B" and means the probability that Event A will occur given that Event B has occurred.

In general,

$$P(A \mid B) = \frac{P(A \text{ and } B)}{P(B)} = \frac{P(A \cap B)}{P(B)}$$

$$= \frac{\text{Probability of Events } A \text{ and } B \text{ Occurring Together}}{\text{Probability of Event } B}$$

Similarly, the symbol $P(B \mid A)$ is read as "the probability of B given A" and means the probability that Event B will occur given that Event A has occurred.

$$P(B \mid A) = \frac{P(A \text{ and } B)}{P(A)}$$

$$= \frac{\text{Probability of Events } A \text{ and } B \text{ Occurring Together}}{\text{Probability of Event } A}$$

In our example, the probability that the top card of the deck is a red king, given that the top card is known to be red, would be written as

$$P(\text{Red King} \mid \text{Red}) = \frac{P(\text{Red and King})}{P(\text{Red})} = \frac{\dfrac{2}{52}}{\dfrac{26}{52}} = \frac{2}{26}$$

○ **EXAMPLE 7.3g**

The manager of Shirts Unlimited has determined that out of 100 buying customers 52 will buy a shirt, 44 will buy a tie, and 30 will buy both. Determine the probability that
a) a customer who has bought a shirt will also buy a tie;
b) a customer who has bought a tie will also buy a shirt.

● **SOLUTION**

Let $P(S)$ be the probability that a customer buys a shirt.

$$P(S) = \frac{52}{100} = 0.52$$

Let $P(T)$ be the probability that a customer buys a tie.

$$P(T) = \frac{44}{100} = 0.44$$

The probability that a customer buys a shirt and a tie is

$$P(S \text{ and } T) = P(S \cap T) = \frac{30}{100} = 0.30$$

a) For those customers who buy a shirt, the probability that they will also buy a tie is

$$P(T \mid S) = \frac{P(T \text{ and } S)}{P(S)} = \frac{0.30}{0.52} = 0.5769$$

This means that approximately 58% of the customers who buy a shirt will also buy a tie.

b) For those customers who buy a tie the probability that they also buy a shirt is

$$P(S \mid T) = \frac{P(T \text{ and } S)}{P(T)} = \frac{0.30}{0.44} = 0.6818$$

Approximately 68% of the customers who buy a tie will also buy a shirt.

F. *The General Rule of Multiplication*

The general rule of multiplication applies to events that are not mutually independent and is derived from the formulas for conditional probability.

The two formulas

$$P(A \mid B) = \frac{P(A \text{ and } B)}{P(B)} \quad \text{and} \quad P(B \mid A) = \frac{P(A \text{ and } B)}{P(A)}$$

can be rearranged to

$$P(A \text{ and } B) = P(B) \cdot P(A \mid B) \text{ and } P(A \text{ and } B) = P(A) \cdot P(B \mid A)$$

In general, $P(A \text{ and } B) = P(B) \cdot P(A \mid B) = P(A) \cdot P(B \mid A)$. This means that for two events that are not mutually independent the probability that both will occur is given by

$$\begin{pmatrix} \text{Probability} \\ \text{that one event} \\ \text{will occur} \end{pmatrix} \quad \text{times} \quad \begin{pmatrix} \text{The conditional probability that the} \\ \text{other event will occur given that the} \\ \text{first event has occurred or will occur} \end{pmatrix}$$

Note   It does not matter which event is called the first event or the second event, or which event is labelled Event A or Event B.

○ EXAMPLE 7.3h

What is the probability of getting two aces in consecutive draws of one card from a deck of 52 if the first card drawn is not replaced before the second draw is made? (This is known as sampling without replacement.)

● SOLUTION

Let Event (A) = First Draw and Event (B) = Second Draw.

The probability of the first card being an ace, $P(A) = \dfrac{4}{52}$.

Since the first card drawn is not replaced, 51 cards are left in the deck. Also, if the first card drawn was an ace, only three aces remain. Events A and B are no longer independent.

The probability that the second drawn will be an ace given that the first card was an ace becomes the conditional probability $P(B \mid A) = \dfrac{3}{51}$.

The probability that both cards drawn will be aces

$$P(A \text{ and } B) = P(A) \cdot P(B \mid A)$$

$$= \left(\frac{4}{52}\right)\left(\frac{3}{51}\right) = 0.0045249$$

This means there are about 45 chances in 10 000, or one in 222, of drawing two aces without replacement.

○ **EXAMPLE 7.3i**

An order of 25 transistor radios delivered to a store contains 5 defective radios. If the radios are selected at random, what is the probability that, of the first two selected,
a) both will be defective?
b) none will be defective?
c) one will be defective?

● **SOLUTION**

a) The probability that the first radio selected is defective,

$$P(A) = \frac{5}{25} = \frac{1}{5}.$$

The selection of the second radio is from 24 and if the first radio was defective, 4 defective radios are left. The probability of the second radio also being defective,

$$P(B) = \frac{4}{24} = \frac{1}{6}.$$

Since the two events are mutually independent, the probability that both radios are defective,

$$P(A \text{ and } B) = P(A) \cdot P(B)$$

$$= \left(\frac{1}{5}\right)\left(\frac{1}{6}\right) = \frac{1}{30} = 0.0333.$$

b) The number of radios that are not defective is 20. The probability that the first radio selected is not defective, $P(A) = \dfrac{20}{25} = \dfrac{4}{5}$.

The selection of the second radio is from 24 and if the first radio was not defective, 19 non-defective radios are left: $P(B) = \dfrac{19}{24}$.

Since the two events are mutually independent, the probability that both radios are non-defective,

$$P(A \text{ and } B) = P(A) \cdot P(B)$$

$$= \left(\frac{4}{5}\right)\left(\frac{19}{24}\right) = \frac{76}{120} = 0.6333.$$

c) The situation in which only one of the first two selected radios is defective can occur in two ways:
   i) The first radio selected is defective. In this case, the second radio must be non-defective.
   ii) The first radio selected is non-defective. In this case, the second radio must be defective.
   This means the second event depends on the outcome of the first event. The events are not independent.

For (i), the probability that the first radio is defective, $P(A) = \frac{5}{25} = \frac{1}{5}$.

The probability that the second radio is non-defective, given that the first radio is defective,

$$P(B \mid A) = \frac{20}{24} = \frac{5}{6}.$$

The joint probability that the first radio selected is defective while the second radio is non-defective,

$$P(A \text{ and } B) = P(A) \cdot P(B \mid A)$$

$$= \left(\frac{1}{5}\right)\left(\frac{5}{6}\right) = \frac{5}{30} = \frac{1}{6} = 0.1667.$$

Similarly, for ii), the probability that the first radio selected is non-defective, $P(B) = \frac{20}{25} = \frac{4}{5}$.

The probability that the second radio is defective, given that the first radio is non-defective,

$$P(A \mid B) = \frac{5}{24}.$$

The joint probability that the first radio selected is non-defective while the second radio is defective,

$$P(A \text{ and } B) = P(B) \cdot P(A \mid B)$$

$$= \left(\frac{4}{5}\right)\left(\frac{5}{24}\right) = \frac{1}{6} = 0.1667.$$

The combined probability of the two possible ways of selecting one defective radio and one non-defective radio = $0.1667 + 0.1667 = 0.3334$.

Note   The three conditions stated in (a), (b) and (c) cover all possible outcomes of selecting the first two radios. The sum of the probabilities for the three outcomes should be one. This is the case since $0.0333 + 0.6333 + 0.3334 = 1.0000$.

EXERCISE 7.3

1. If a student was late for 4 of the last 20 classes, what is the probability that the student will be on time for the next class?

2. A deck of cards was rescued from an angry dog who had torn 13 cards. What are the chances of not being dealt a torn card?

3. What is the probability of drawing the following cards from a well-shuffled deck of 52 cards?
   a) a red ace or a black queen;
   b) a black card or a red card;
   c) a black card or a red jack;
   d) a five or a face card (ace, king, queen or jack).

4. A finance department has been equipped with 21 computers and printers. Fourteen of the printers are dot matrix and three are laser printers. What is the probability that an employee will be given a computer with
   a) a laser printer?
   b) a dot matrix printer?
   c) a laser printer or a dot matrix printer?

5. Twelve student awards were given by the Faculty of Business. Seven of the recipients were women and 5 of the 12 students started their program as mature students. Four of the mature students were female. What is the probability that an award recipient chosen at random would be a mature student or female?

6. Eighteen sales representatives are being evaluated to determine which one should be promoted to manager. Four of the candidates have worked in the eastern region and 12 have post-secondary education. Two of the representatives with experience in the eastern region have post-secondary education. What is the probability that a randomly promoted individual will have worked in the eastern region or have post-secondary education?

7. A coin is tossed three times. What is the probability that a "tail" will appear all three times?

8. The probability that an office supplies sales representative following up a magazine response card will make a sale is 0.60. Given that a representative has three independent leads, what is the probability that
   a) none will buy?
   b) all three will buy?
   c) one will buy?

9. A small software developer displays and demonstrates dental and optical research software programs at trade shows. From previous shows it is known that out of 20 visitors to the booth, 8 will buy the dental program, 6 will buy the optical program and 2 will buy both. What is the probability that a visitor buying a dental program will also buy an optical program?

10. The office manager for Bete Shoes has found in the past that 40% of internal stationery requests include pens, 74% include binders and 18% include both pens and binders. Determine the probability that an internal order will include pens if binders are known to be on that order.

11. A box contains 15 white and 5 black marbles. What is the probability that, of the first two marbles picked,
    a) both will be black?
    b) both will be white?
    c) the first will be black and the second will be white?
    d) one will be black and one will be white?

12. A shipment of 200 steel bars is known to contain 15 defective bars. If two bars are randomly selected for inspection, what is the probability that one or two bars will be defective?

SECTION 7.4 Tree Diagrams

A **tree diagram** is a convenient method of graphically showing all possible outcomes of the events of more complex problems involving conditional and joint probabilities.

○ **EXAMPLE 7.4a**

Draw a tree diagram to show all possible outcomes of tossing a coin three times and determine the probabilities of the Event(Number of Heads).

● **SOLUTION**

First toss. Since one toss of a coin has two possible outcomes, the tree diagram is started by drawing two branches from a point on the left side of the diagram, as shown in Figure 7.4. The endpoints of the two branches are marked H (for "head") and T (for "tail") to represent the possible outcomes for the first toss of the coin.

Second toss. The second toss has the same two outcomes. To allow for each possible outcome, two branches are drawn from the endpoint of the branch marked H. The endpoints of these two new branches are marked H and T respectively to indicate the possible outcomes of the second toss, given that the outcome of the first toss was H (head). Similarly, two branches are drawn from the endpoint marked T. The endpoints of these two new branches are marked H and T respectively to indicate the possible outcomes of the second toss, given that the outcome of the first toss was T (tail).

Third toss. Two branches are drawn from each of the four endpoints resulting from the second toss to indicate the possible outcomes of the third toss. The eight endpoints are marked H, T, H, T, H, T, H, T, as shown in Figure 7.4.

The eight possible outcomes of tossing a coin three times are now listed under the heading of possible outcomes (HHH, HHT, HTH, etc.). The number of heads in each outcome in each of the eight outcomes are shown in the last column of Figure 7.4.

FIGURE 7.4 **Tree Diagram for Three Tosses of a Coin**

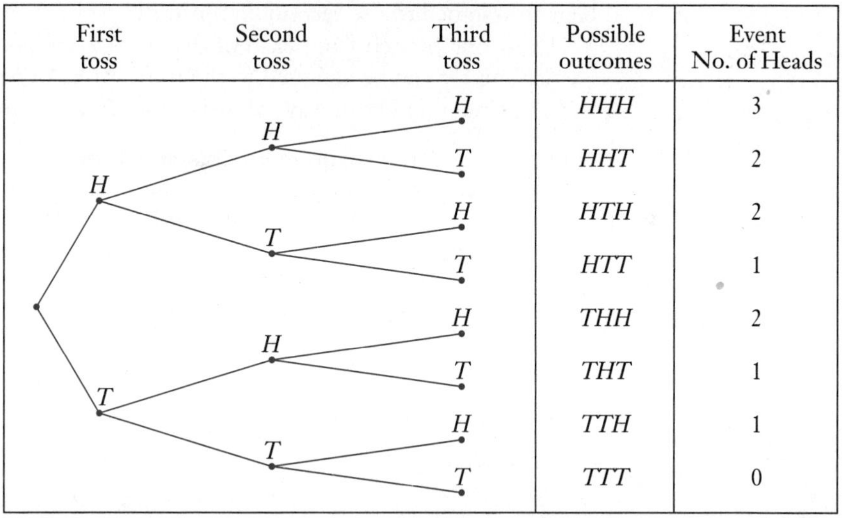

First toss	Second toss	Third toss	Possible outcomes	Event No. of Heads
		H	HHH	3
	H	T	HHT	2
H		H	HTH	2
	T	T	HTT	1
		H	THH	2
	H	T	THT	1
T		H	TTH	1
	T	T	TTT	0

The resulting events (Number of Heads) and their frequency and probability of occurrence can now be determined, as shown in Table 7.1 below.

TABLE 7.1 **Probabilities of Tossing a Coin Three Times**

Event No. of Heads	Frequency of occurrence	Probability of occurrence (relative frequency)
Three heads	1	$P(3 \text{ heads}) = \dfrac{1}{8} = 0.125$
Two heads	3	$P(2 \text{ heads}) = \dfrac{3}{8} = 0.375$
One head	3	$P(1 \text{ head}) = \dfrac{3}{8} = 0.375$
Zero heads	1	$P(0 \text{ heads}) = \dfrac{1}{8} = 0.125$
Total	8	1.000

○ **EXAMPLE 7.4b**

The 24-Hour Taxi Service employs three drivers and uses three cars. Determine the number of possible ways of assigning the three drivers to the three cars.

● **SOLUTION**

Let the drivers be identified as *D1, D2, D3* and the cars as *A, B, C* respectively.

Draw three branches from the starting point to indicate that a car can be assigned to the first driver (whether it be *D1*, *D2* or *D3* does not matter) in three ways.

Draw two branches from each of the three endpoints to indicate that a car can be assigned to the second driver in two ways.

Draw one branch from each of the six new endpoints to indicate that the remaining car can be assigned to the third driver in only one way.

The total number of ways of assigning three cars to three drivers is 6.

FIGURE 7.5   Tree Diagram For Example 7.4b

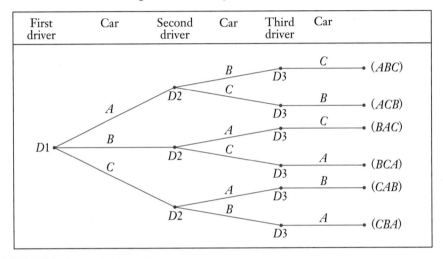

EXERCISE 7.4

1. A panel of consumers was asked to rate two products as "excellent," "average" or "poor." Determine the number of possible outcomes by drawing a tree diagram.

2. A sales representative must call four customers to find out if they will be ordering for the second quarter (yes or no). Construct a tree diagram to determine the number of ways in which the representative can be answered.

3. Third-year accounting students at a college face the task of choosing two courses, ranked by preference, from four courses offered. Use a tree diagram to determine in how many ways a student can make a choice of two courses ranked by order of preference.

4. A club has four candidates to fill the offices of president, treasurer and secretary. Draw a tree diagram to show in how many different ways the offices can be filled.

Counting Rules

For complex problems involving many outcomes, the listing of all possible outcomes may not be feasible. To deal with such complex problems, a variety of counting schemes are available, three of which are considered in this section.

A. Multiplication Schemes

Rule 1 If any one of m mutually exclusive events can occur on each of n trials, the number of possible outcomes is m^n.

○ **EXAMPLE 7.5a**

Determine the number of possible outcomes of tossing a coin three times (see also Example 7.4a).

● **SOLUTION**

The number of possible events, $m = 2$;
the number of trials, $n = 3$;
the number of possible outcomes, $m^n = 2^3 = 8$.

This is the number of different outcomes obtained by drawing the tree diagram in Example 7.4a.

○ **EXAMPLE 7.5b**

Determine the number of possible outcomes of rolling a die five times.

● **SOLUTION**

The number of possible events, $m = 6$;
the number of trials, $n = 5$;
the number of possible outcomes, $m^n = 6^5 = 7776$.

Rule 2 If there are m possible outcomes for one event and n possible outcomes for another event, the total number of outcomes for the two events is $(m)(n)$.

This rule can be extended to three or more events. For three events with m, n and p possible outcomes respectively, the total number of possible outcomes is $(m)(n)(p)$.

○ **EXAMPLE 7.5c**

Determine the number of different ways in which four consumers can choose among three new car models.

● **SOLUTION**

The number of consumers, $m = 4$;
the number of possible ways of selecting a new car model for each consumer, $n = 3$;
the total number of possible ways in which four consumers can choose a car,
$$mn = (4)(3) = 12.$$

○ **EXAMPLE 7.5d**
A task force consisting of 1 director, 1 vice-president, 2 senior administrators and 3 senior technical experts is to be selected from a company's management staff of 7 directors, 3 vice-presidents, 6 senior administrators and 10 senior technical experts. In how many ways can the task force be selected?

● **SOLUTION**
The number of ways of selecting a director, $m = 7$;
the number of ways of selecting a vice-president, $n = 3$;
the number of ways of selecting the first administrator, $p = 6$;
the number of ways of selecting the second administrator, $q = 5$;
the number of ways of selecting the first senior expert, $r = 10$;
the number of ways of selecting the second senior expert, $s = 9$;
the number of ways of selecting the third senior expert, $t = 8$.
The number of ways of selecting the task force is given by
$(m)(n)(p)(q)(r)(s)(t) = (7)(3)(6)(5)(10)(9)(8) = 453\ 600$.

○ **EXAMPLE 7.5e**
How many different car licence plates can be made if a licence plate number consists of three digits followed by three letters?

● **SOLUTION**
There are 10 digits, each of which can occur three times:

$$m = 10, \quad n = 3; \quad m^n = 10^3$$

There are 26 letters, each of which can occur three times:

$$p = 26, \quad q = 3, \quad p^q = 26^3$$

The number of different licence plates is

$$(m^n)(p^q) = (10^3)(26^3) = (1000)(17\ 576) = 17\ 576\ 000.$$

○ **EXAMPLE 7.5f**
In how many ways can a bridge hand be set up to contain four aces and include five spades, four hearts, three diamonds and one club?

● **SOLUTION**
The ace of spades can be selected in one way; the remaining four spades in 12, 11, 10, and 9 ways respectively. The five spades can be selected in $(1)(12)(11)(10)(9) = 11\ 880$ ways.
For each of the ways of selecting spades, the four hearts can be selected in $(1)(12)(11)(10) = 1320$ ways.
For each of the above, the three diamonds can be selected in $(1)(12)(11) = 132$ ways, and the ace of clubs in 1 way.

The total number of ways of assembling the bridge hand

$$= (1)(132)(1320)(11\ 880) = 2\ 069\ 971\ 200.$$

Rule 3   The number of ways in which n objects can be arranged in order is

$$n! = n(n-1)(n-2)(n-3)\ldots(3)(2)(1)$$

where the symbol $n!$, read as "n factorial," denotes the product of the first n whole numbers, and $0!$ is defined to equal 1.

For example,

$$6! = (6)(5)(4)(3)(2)(1) = 720$$
$$15! = (15)(14)(13)(12)\ldots(3)(2)(1) = 130\ 767\ 436\ 800$$

○ **EXAMPLE 7.5g**

A student taking five subjects has bought one textbook for each subject. In how many ways can the books be arranged on a bookshelf?

● **SOLUTION**

The first book can be selected in 5 ways;
after that, the second book can be selected in 4 ways;
after that, the third book can be selected in 3 ways;
after that, the fourth book can be selected in 2 ways;
and the fifth book in 1 way.

The total number of possible arrangements

$$= (5)(4)(3)(2)(1) = 5! = 120.$$

○ **EXAMPLE 7.5h**

How many ordered arrangements can be made from
a) the nine non-zero digits?
b) the 26 letters of the alphabet?
c) the 52 cards in a deck?

● **SOLUTION**

a) The nine non-zero digits can be arranged in $9!$ ways:

$$9! = (9)(8)(7)(6)(5)(4)(3)(2)(1) = 362\ 880$$

b) The 26 letters of the alphabet can be arranged in $26!$ ways:

$$26! = (26)(25)(24)(23)\ldots(3)(2)(1) = 4.0329146(10^{26})$$

c) The deck of 52 cards can be arranged in $52!$ ways:

$$52! = (52)(51)(50)\ldots(3)(2)(1) = 8.0658175(10^{67})$$

B. Permutations

The term **permutation** refers to the number of ways in which a subset of a group of objects can be arranged in order. Each possible arrangement is a permutation.

Examples 7.5g and 7.5h represent the special case of permutation in which the subset consists of all objects in the group. This means the total number of permutations for 5 objects is 5! = 120; for 9 objects it is 9! = 362 880; and for 52 objects it is an astronomical 52! = $8(10^{67})$.

The counting rule for permutations — that is, the number of ways in which x objects selected from an entire group of n objects can be *arranged in order* — is given by

$$_nP_x = \frac{n!}{(n-x)!}$$

where n = the number of objects in the entire group;
$\quad x$ = the number of objects to be selected from the group of n objects;
$\quad P$ = the number of permutations (ordered ways) in which the x selected objects can be arranged.

Note   For the special case in which all objects are included, the permutation formula becomes multiplication Rule 3. In this case, $x = n$ and $_nP_x$ becomes

$$_nP_n = \frac{n!}{(n-n)!} = \frac{n!}{0!} = n! \text{ (since } 0! = 1\text{)}.$$

○ **EXAMPLE 7.5i**

In how many ordered ways can 5 cards be selected from a deck of 52 cards?

● **SOLUTION**

The first card can be selected in 52 ways;
the second card can be selected in 51 ways;
the third card can be selected in 50 ways;
the fourth card can be selected in 49 ways;
the fifth card can be selected in 48 ways.

By using multiplication Rule 2, the total number of ordered ways (permutations) of selecting five cards from a deck of 52 is (52)(51)(50)(49)(48) = 311 875 200.

The permutation formula abbreviates the process to

$$_nP_x = \frac{n!}{(n-x)!}, \text{ where } n = 52 \text{ and } x = 5. \text{ That is,}$$

$$_{52}P_5 = \frac{52!}{(52-5)!}$$

$$= \frac{52!}{47!} = \frac{(52)(51)(50)(49)(48)\,(47)(46)(45)(44)\ldots(3)(2)(1)}{(47)(46)(45)(44)\ldots(3)(2)(1)}$$

$$= \frac{(52)(51)(50)(49)(48)(47!)}{47!} = (52)(51)(50)(49)(48)$$

$$= 311\ 875\ 200$$

○ **EXAMPLE 7.5j**

Determine the number of ordered sets of 13 cards that can be selected from a deck of 52.

● **SOLUTION**

$n = 52; x = 13$

$$_{52}P_{13} = \frac{52!}{(52-13)!}$$

$$= \frac{52!}{39!} = \frac{(52)(51)(50)\ldots(42)(41)(40)(39!)}{39!}$$

$$= (52)(51)(50)(49)\ldots(42)(41)(40) = 4.0531(10^{21})$$

C. Combinations

In determining the number of permutations (ordered arrangements) of n objects taken x at a time, it is important to consider the order in which the objects are selected.

Consider the case of four persons (Brian, Margaret, Pierre and Sheila) running for the positions of president and vice-president of the student council. The person getting the most votes becomes president and the runner-up becomes vice-president.

In this case the order of finish is all important. The number of possible selections are listed below:

Brian-Margaret	Brian-Pierre	Brian-Sheila
Margaret-Brian	Pierre-Brian	Sheila-Brian
Margaret-Pierre	Margaret-Sheila	Pierre-Sheila
Pierre-Margaret	Sheila-Margaret	Sheila-Pierre

The number of possible ordered arrangements (permutations) is 12 and could have been determined using the permutation formula where $n = 4$ and $x = 2$:

$$_nP_x = {}_4P_2 = \frac{4!}{(4-2)!} = \frac{4!}{2!} = \frac{(4)(3)(2!)}{2!} = (4)(3) = 12$$

Now consider the selection of two of the four persons to form a committee of two. In this case, the order of finish (first or second) does not matter. The pairings Brian-Margaret and Margaret-Brian result in the same committee. This means six distinct pairings are possible from the group of four persons.

Arrangements in which the order is of no consequence are called **combinations** of n objects taken x at a time, and are denoted by the symbol $_nC_x$. The number of such combinations is given by the formula

$$_nC_x = \frac{n!}{x!(n-x)!}$$

where n = the number of objects in the entire group;
  x = the number of objects to be selected from the group;
  C = the number of combinations or ways in which x objects can be chosen without regard to order.

In our example, for $n = 4$ and $x = 2$,

$$_nC_x = {_4}C_2 = \frac{4!}{2!(4-2)!} = \frac{4!}{2!(2!)}$$
$$= \frac{(4)(3)(2!)}{(2)(1)(2!)} = \frac{(4)(3)}{(2)(1)} = 6$$

○ **EXAMPLE 7.5k**

Regarding Example 7.5i, how many poker hands consisting of 5 cards can be selected from a deck of 52 cards?

● **SOLUTION**

In a poker hand the order in which the five cards are dealt is of no consequence. We want to determine the number of combinations where $n = 52$ and $x = 5$.

$$_{52}C_5 = \frac{52!}{5!(52-5)!} = \frac{52!}{5!(47!)} = \frac{(52)(51)(50)(49)(48)(47!)}{(5)(4)(3)(2)(1)(47!)}$$
$$= \frac{(52)(51)(50)(49)(48)}{(5)(4)(3)(2)(1)} = \frac{311\ 875\ 200}{120} = 2\ 598\ 960$$

Note The number of combinations = $\dfrac{\text{Number of permutations}}{x!}$; that is, $_nC_x = \dfrac{_nP_x}{x!}$.

○ **EXAMPLE 7.5l**

Regarding Example 7.5j, determine the number of bridge hands (sets of 13 cards) that can be dealt from a deck of 52 cards.

● **SOLUTION**

Since the order in which the cards are dealt does not matter, we want to determine the number of combinations for $x = 52$, $n = 13$.

$$_{52}C_{13} = \frac{52!}{13!(52 - 13)!} = \frac{52!}{13!(39!)} = \frac{(52)(51)(50\ldots(42)(41)(40)}{(13)(12)(11)\ldots(3)(2)(1)}$$

$$= \frac{4.0531(10^{21})}{7.0919(10^9)} = \frac{40.531(10^{20})}{7.0919(10^9)} = 5.7151(10^{11})$$

○ **EXAMPLE 7.5m**

To have the winning number in Lotto 6/49 you must match the six numbers drawn at random. What is the probability of matching the winning set of numbers with one selection of six numbers?

● **SOLUTION**

The order in which the numbers are selected does not matter. $n = 49$; $x = 6$.

$$_{49}C_6 = \frac{49!}{6!(49 - 6)!} = \frac{49!}{6!(43!)} = \frac{(49)(48)(47)(46)(45)(44)}{(6)(5)(4)(3)(2)(1)}$$

$$= \frac{10\ 068\ 347\ 520}{720} = 13\ 983\ 816$$

The probability of having the winning combination with one set of six numbers = $\dfrac{1}{13\ 983\ 816}$ = $7.1511(10^{-8})$.

○ **EXAMPLE 7.5n**

A safety committee of 5 is to be chosen from a group of 10 employees who have indicated their willingness to stand for election to the committee. In how many ways can the committee be made up?

● **SOLUTION**

$n = 10$; $x = 5$. The order in which the candidates are elected does not matter. The number of combinations of 5 persons from a group of 10,

$$_{10}C_5 = \frac{10!}{5!(10 - 5)!} = \frac{10!}{5!(5!)} = \frac{(10)(9)(8)(7)(6)}{(5)(4)(3)(2)(1)} = 252$$

○ **EXAMPLE 7.5o**

Refer to Example 7.3i in which a shipment of 25 radios contained 5 defective ones. In how many ways can 5 non-defective radios and 2 defective radios be selected?

● **SOLUTION**

The order in which the radios are selected from the 20 non-defectives and the 5 defectives does not matter. The number of combinations of selecting 5 non-defective radios from 20 is given by

$$_{20}C_5 = \frac{20!}{5!(20-5)!} = \frac{20!}{5!(15!)} = \frac{(20)(19)(18)(17)(16)}{(5)(4)(3)(2)(1)} = 15\ 504$$

The number of ways of selecting two defective radios from the group of five defective ones is given by

$$_5C_2 = \frac{5!}{2!(5-2)!} = \frac{5!}{2!(3!)} = \frac{(5)(4)}{(2)(1)} = 10$$

The total number of combinations = $(15\ 504)(10) = 155\ 040$.

EXERCISE 7.5

1. A sales representative must call on five customers in the same area. In how many ways can the representative arrange his itinerary?

2. Third-year students at a college must choose three courses, ranked in order of preference, from eight courses offered. In how many ways can a student choose the three courses?

3. Management has offered to let all staff in a department attend courses to become computer literate. Three of the 12 staff will be permitted to attend classes during business hours. In how many ways can the three employees be chosen?

4. In how many ways can 3 marketing representatives be selected from 10 applicants to serve different regions?

5. Six different desks are available for allocation to six distinct office areas. In how many ways can the desks be assigned?

6. A committee of 3 must be formed from 18 managers. In how many ways can the committee be formed?

7. An early-retirement package is to be offered to a group of 20 employees with the stipulation that only 12 of the employees will be able to take advantage of the offer. In how many ways can the offer be accepted by the group of employees?

8. Recent changes in government regulation permit a college's board of governors to have student and faculty representation. In the first election 5 students are running for 2 seats and 15 faculty for 3 seats. In how many ways can the 5 board vacancies be filled?

9. How many four-digit passwords can be made from the numbers 0 through 9?

10. In how many ways can four men and four women make up a special committee looking into safety in the workplace if three persons are to be selected and at least one committee member must be a woman?

REVIEW EXERCISE

1. Given a shopping allowance of $100, how many outcomes are possible to spend the toal amount on the following items?

Shirt	$25	Pants	$50	Shoes	$75
Tie	$25	Sweater	$50		

2. In a recent student election, three candidates ran for two positions. How many outcomes are possible for the election?

3. Consumers were asked to rate products A and B according to the following scale.
   0 – like
   1 – indifferent
   2 – dislike
   How many outcomes can result for rating the two products?

4. A wine tasting required participants to rate three characteristics of the wine as satisfactory or unsatisfactory. How many outcomes are possible for rating the three characteristics?

5. How many events are possible for filling two employment positions according to the sex of the applicants?

6. Three consumers are asked whether or not they like a new cereal. They can respond either yes or no. How many events are possible?

7. If two coins were tossed at the same time, what is the probability that at least one coin would show a head?

8. If a jar contained eight black, seven white, three blue, and two red marbles, what is the probability of
   a) picking a black or white marble?
   b) picking a blue, a red or a white marble?
   c) picking a red marble?

9. A sales representative averages 35 orders for every 100 calls. What is the probability that the representative will fail to get an order on a call?

10. Out of 1000 travellers leaving the arrival area of Airport Y, 185 took taxis. Out of 600 persons leaving the arrival area of Airport Z, 150 took taxis. What is the probability that a person will call for a taxi
    a) at Airport Y?
    b) at Airport Z?
    c) at one of the two airports?

11. The following data were gathered during a study of computer use in Canada:

Age group	Able to use a computer	Not able to use a computer
15–19	164	36
20–24	132	68

a) What is the probability that a Canadian between the ages of 15 and 24 will not be able to use a computer?

$$P(A \cap B') = \frac{132}{400}$$

b) What is the probability that a Canadian between the ages of 20 and 24 will know how to use a computer?

12. The following table presents the major causes of death in Canada:

	Causes of Death			
Sex	Cardiovascular disease	Cancer	Other	Total
Male	395	270	335	1000
Female	434	264	302	1000

a) What is the probability that a Canadian will die from cancer?
b) What is the probability that a male will die from cardiovascular disease?
c) What is the probability that a female will die of cancer or cardiovascular disease?

13. When a statistics instructor was asked to predict class attendance on a sunny day in April if the temperature rose above 21°C, the response was as follows:

Number of students attending	Probability
0–5	5%
5–10	20%
10–15	60%
15–20	10%
Over 20	5%

a) What is the probability that 10 or more students will attend class?
b) What is the probability that fewer than 15 will attend class?

14. An economist for a large Canadian chartered bank has been asked to forecast the exchange rate between the U.S. dollar and the Canadian dollar for the next quarter. Using available econometric models, she feels that 1.19 is twice as likely as 1.21 and that 1.18 is three times as probable as 1.16. She also thinks that the chance of 1.16 occurring is only half of the chance of 1.21 occurring. Based on this information, what is the probability that the exchange rate will be
a) 1.18?
b) 1.19?
c) 1.16?

15. A jar has four red and three blue marbles. What is the sample space for picking
a) one marble?
b) two marbles?

16. Determine the sample space for a student who has to choose two options from three available courses.

17. The manager of a furniture manufacturing company has determined that 20 of the company's products were not chairs, were not assembled and were not on sale. The company makes 8 models of chairs and has 42 products on sale. Five chair models come assembled; two chair models are on sale; one chair model comes assembled and is on sale; 28 products that are not chairs come assembled and are on sale; and two products come assembled, are not on sale and are not chairs. Draw a Venn diagram and determine

a) the number of products carried by the company;

b) the number of models of chairs that come unassembled or are not on sale.

18. In a college parking lot, 450 cars were counted. Registration data showed that 60 cars were made in Canada, 220 cars were red, and 280 cars belonged to students. Thirty of the cars made in Canada and 200 of the red cars belonged to students. Ten red cars were made in Canada and five of the red cars owned by students were made in Canada. Draw a Venn diagram and determine

a) the number of cars in the parking lot that were not red, were not made in Canada or did not belong to students;

b) the number of red cars that did not belong to students;

c) the number of cars belonging to students but not made in Canada.

19. What is the probability of throwing a total of three spots or four spots with a roll of two dice?

20. In American roulette there are 38 slots in which the ball may land. The slots are numbered 00, 0, and 1 through 36. The numbers 00 and 0 are green, the odd numbers are red and the even numbers are black. If a ball rolls randomly into a slot, what is the probability of

a) a five or a six?

b) red or black?

c) green or 27?

d) a number less than 20?

21. A small office is equipped with eight computers, five of which run MYSTAT and three have colour monitors. Of the five computers running MYSTAT, two have colour monitors. What is the probability that a computer chosen at random in this office runs MYSTAT or has a colour monitor?

22. Given that $P(\overline{A}) = 0.40$, $P(B) = 0.30$ and $P(A \cap B) = 0.15$, determine the value of $P(A \cup B)$.

23. Given that $P(A) = 0.60$, $P(B) = 0.30$, $P(\overline{A} \cup \overline{B}) = 0.80$, determine $P(A \cap B)$.

24. Given $P(A) = 0.30$, $P(\overline{B}) = 0.50$ and $P(A \cup B) = 0.10$, determine

a) $P(A \mid B)$;

b) $P(B \mid A)$;

c) $P(A \text{ or } B)$.

25. The following data represent the reaction of men and women over the age of 18 toward a new TV commercial:

	Reaction to TV Commercial		
Gender	Like	Neutral	Dislike
Male	900	200	400
Female	1200	300	1000

a) Determine the probability that a person from this group likes the commercial, given that the person selected is a male.

b) Determine the probability that a person from this group dislikes the commercial, given that the person selected is a female.

c) What are the chances that a person from the group is neutral toward the commercial?

26. A shortlist of applicants for two management trainee positions is made up of four graduates from Kelowna College, three from Assiniboine College, and two from Fredericton College. What is the probability that, of the two successful applicants,
    a) both will be from Kelowna College?
    b) one will be from Assiniboine College and the other from Fredericton College?

27. If two cards are drawn from a well-shuffled deck of 52 cards and the first card drawn is not replaced, what is the probability that
    a) both cards are kings?
    b) the first card is a queen and the second card is a jack?
    c) one card is a queen and one card is a jack?

28. The awarding of a contract for a new government office is to be announced on July 1. The timing of the announcement is conditional on three bids being received. If the submission of the bids are independent events and the probability of the bids being late are 0.10, 0.05 and 0.02 respectively, what are the chances that the announcement will be made as scheduled?

29. Four unrelated stocks on the Alberta Stock Exchange have been judged to increase in price with the following probabilities:

Stock	J	K	L	M
Probability of price increase	0.90	0.95	0.80	0.70

    a) What is the probability that all four stocks will increase in price?
    b) What is the probability that only stocks J and L will increase in price?

30. A jar contains a black, a red and a white marble. Construct a tree diagram to show how two marbles can be chosen.

SELF-TEST

1. The following data represent product acceptance for a group of persons compiled by sex:

	Product Acceptance			
Sex	Excellent	Good	Fair	Poor
Male	75	225	150	25
Female	50	175	225	75

    a) If a person is randomly chosen from this group what is the probability that the person
       i) is male?
       ii) considers the product to be "good"?
       iii) is female and thinks that the product is "poor"?

       iv) is male or thinks the product is "fair"?

  b) Determine the probability of a person from this group rating the product as "excellent," given that the person selected is

     i) female;

     ii) male;

     iii) either male or female.

2. A pyramid-shaped, four-sided die is rolled twice. The four sides have one, two, three or four spots respectively, and the event considered is the sum of the number of spots on the face-down position of the die.

  a) Construct a tree diagram to show all possible outcomes of the experiment.

  b) Determine the possible number of outcomes.

  c) Determine the number of possible events defined above.

  d) Determine the probability of the following events.

     i) three spots;

     ii) four spots or five spots;

     iii) more than six spots.

3. A survey of 700 Canadians showed that 336 were male, 136 had a computer at home and 330 were able to use a computer. Of the 336 males, 200 were able to use a computer, 60 had a computer at home, and 10 had a computer at home and were able to use it. Of the females, 40 were able to use a computer and had one at home.

  a) Construct a Venn diagram to represent the survey data.

  b) How many females were able to use a computer?

  c) How many males had a computer at home but were unable to use it?

4. Two sprinters from a region will be invited to the Canadian track-and-field championships.

  a) If eight sprinters compete in the regional championships and the regional winner runs in the national championships while the runner-up goes as backup, in how many different ways can the two regional runners be selected?

  b) If a rule change allows both the regional winner and the runner-up to compete in the nationals, in how many ways can the two runners be chosen?

5. A local service club is selling 100 tickets at $100 each to raise money for a student exchange program. A draw of three prizes is offered as an incentive to the buyers of the tickets.

  a) What is the probability of winning first prize with the purchase of one ticket if the first-prize winner is the third ticket called?

  b) What is the probability of winning a prize if you purchase one ticket?

  c) What are the chances of winning at least one prize if you purchase three tickets?

Key Terms

Combination   197

Conditional probability   184

Event   169

Summary of Formulas

1.

$$\text{PROBABILITY OF AN EVENT} = \frac{\text{NUMBER OF FAVOURABLE OUTCOMES}}{\text{TOTAL NUMBER OF POSSIBLE OUTCOMES}}$$

2. $0 \leq P(A) \leq 1$

3. $P(A) + P(\overline{A}) = 1$

4. Addition Rules:

a) Special rule of addition (for mutually exclusive events):

$$P(A \text{ or } B) = P(A \cup B) = P(A) + P(B)$$
$$P(A \text{ or } B \text{ or } C \text{ or } \ldots) = P(A \cup B \cup C \cup \ldots) = P(A) + P(B) + P(C) + \ldots$$

b) General rule of addition:

$$P(A \text{ or } B) = P(A) + P(B) - P(A \text{ and } B)$$
$$= P(A) + P(B) - P(A \cap B)$$

5. Multiplication Rules:

a) Special rule of multiplication (for mutually independent events):

$$P(A \text{ and } B) = P(A \cap B) = P(A) \cdot P(B)$$
$$P(A \text{ and } B \text{ and } C \text{ and } \ldots) = P(A \cap B \cap C \cap \ldots) = P(A) \cdot P(B) \cdot P(C) \cdot \ldots$$

b) General rule of multiplication:

$$P(A \text{ and } B) = P(A \cap B) = P(A) \cdot P(B \mid A) = P(B) \cdot P(A \mid B)$$

6. Conditional Probability:

$$P(A \mid B) = \frac{P(A \text{ and } B)}{P(B)} = \frac{P(A \cap B)}{P(B)}$$

$$P(B \mid A) = \frac{P(A \cap B)}{P(A)}$$

7. Counting Rules:

a) Rule 1: If any one of m mutually exclusive events can occur on each of n trials, the number of possible outcomes is m^n.

b) Rule 2: If there are m possible outcomes for one event and n possible outcomes for another event, the total number of outcomes for the two events is $(m)(n)$.

c) Rule 3: The number of ways in which n objects can be arranged in order is

$$n! = n(n-1)(n-2)(n-3)\ldots(3)(2)(1)$$

8. Permutations:

$$_nP_x = \frac{n!}{(n-x)!}$$

9. Combinations:

$$_nC_x = \frac{n!}{(n-x)!\,x!}$$

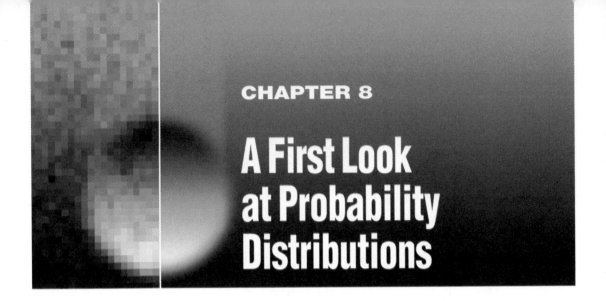

CHAPTER 8

A First Look at Probability Distributions

Introduction

A **probability distribution** is a list of all the outcomes of an experiment together with the probability associated with each outcome. Many business or economic problems can be described by means of probability distributions.

Objectives

Upon completion of this chapter you will be able to
1. understand and define the term probability distribution;
2. distinguish between discrete and continuous probability distributions;
3. compute the expected value (mean), variance and standard deviation of a discrete probability distribution;
4. define the characteristics of the binomial probability distribution;
5. construct and use a binomial distribution for given values of n and p.

SECTION 8.1 Continuous versus Discrete Probability Distributions

In Example 7.4a we created a tree diagram to show all possible outcomes of tossing a coin three times. The frequencies and probabilities (relative frequencies) obtained were shown in Table 7.1 and are reproduced in Table 8.1.

Such a tabulation of the events, together with their associated probabilities, constitutes a probability distribution. The distribution can be represented graphically as shown in Figure 8.1.

208

TABLE 8.1 **Probability Distribution for Example 7.4a**
(Tossing a Coin Three Times)

Event (No. of "Heads")	Frequency of occurrence	Probability of occurrence (relative frequency)
Three heads	1	0.125
Two heads	3	0.375
One head	3	0.375
Zero heads	1	0.125
	Total	1.000

FIGURE 8.1 **Graphical Representation of the Probability**
Distribution for Example 7.4a

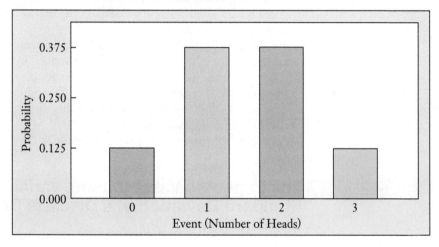

In this context the Event (Number of Heads) is referred to as a **discrete random variable**. The resulting distribution shown in Table 8.1 and Figure 8.1 is called a **discrete probability distribution**. The term *random variable* is used for quantities that result from an experiment and that take on different values by chance. In our example, the random variable "Number of Heads" took the values 0, 1, 2 and 3.

A *discrete* random variable is called discrete because it can only take on clearly defined, *separated* values that are obtained by counting. The distribution is referred to as a *discrete probability distribution* because it consists of a listing of the discrete values that the random variable can assume as well as their associated probabilities.

Other examples of discrete probability distributions are the number of newspapers sold per day from a corner box, the number of vehicles driving into each of the service stations located within a defined geographical area on a particular day, or the distribution of marks on a test for a group of students.

The term **continuous random variable** is usually associated with a measurement that can be made to a desired degree of precision, such as length, height, weight or speed. For example, in downhill skiing the time it takes racers to finish the course is measured to one one-hundredth of a second. Similarly, in the long jump, distance is measured to the nearest centimetre. A tabulation of this type of variable is known as a *continuous frequency distribution* and the tabulation of the associated relative frequencies is referred to as a **continuous probability distribution**.

The following is a summary of important concepts concerning probability distributions in general:

1. A random variable is a variable whose value in a particular experiment is determined purely by chance.
2. A continuous random variable is one that can assume an infinite number of values within a specified interval.
3. A discrete random variable is one that can assume only distinct values within a specified interval.
4. Probability distributions can be either continuous or discrete, depending on the nature of the random variable described by the distribution.
5. The probability distribution of a random variable defines the probability of occurrence of every possible value that the random variable can assume.
6. The probabilities making up the probability distribution are values between 0 and 1, and their sum must equal 1.
7. A probability distribution can be presented in the form of a table or graph.

SECTION 8.2 The Expected Value (Mean), Variance and Standard Deviation of a Discrete Probability Distribution

A. Expected Value (Mean)

Denoted by the symbol $E(x)$, the **expected value** of a random variable x is the *long-run value of* μ. This long-run value is the *mean value* that the variable is expected to attain if the experiment considered is repeated an infinitely large number of times.

To compute the expected value (mean) of a random variable, use the following procedure:

Step 1   Multiply each individual value of the random variable by its corresponding probability of occurrence.

Step 2   Add the products obtained in Step 1. This procedure can be summarized in symbolic terms as

$$E(x) = \mu = \sum x \cdot P(x)$$

where x denotes the individual values that the random variable
can assume;

$P(x)$ is the probability of occurrence of the individual values x;

$E(x)$ is the expected value (mean) of the random variable x.

○ EXAMPLE 8.2a

You are invited to join a game by paying a certain amount for each roll of a die. In return you will be paid an amount equal to the number of spots showing on the upper face of the die when it stops rolling. For example, if you pay $2 to join the game and the six-spot turns up, you get back $6; that is, you win $4. Determine the expected value of the game.

● SOLUTION

The random variable x (Number of Spots) can take the values $x = 1, 2, 3, 4, 5, 6$. Since each outcome for the random variable is equally likely, the probability of each outcome, $P(x) = \dfrac{1}{6}$.

The long-run average value $E(x)$ can now be computed as shown in Table 8.2.

TABLE 8.2   Computation of $E(x)$

Individual value x	Probability $P(x)$	Product $x \cdot P(x)$
1	$\dfrac{1}{6}$	$1\left(\dfrac{1}{6}\right) = \dfrac{1}{6} = 0.1667$
2	$\dfrac{1}{6}$	$2\left(\dfrac{1}{6}\right) = \dfrac{2}{6} = 0.3333$
3	$\dfrac{1}{6}$	$3\left(\dfrac{1}{6}\right) = \dfrac{3}{6} = 0.5000$
4	$\dfrac{1}{6}$	$4\left(\dfrac{1}{6}\right) = \dfrac{4}{6} = 0.6667$
5	$\dfrac{1}{6}$	$5\left(\dfrac{1}{6}\right) = \dfrac{5}{6} = 0.8333$
6	$\dfrac{1}{6}$	$6\left(\dfrac{1}{6}\right) = \dfrac{6}{6} = 1.0000$

$$E(x) = \mu = \sum x \cdot P(x) = \frac{21}{6} = 3.50$$

Note   For the special case in which the probability of occurrence is the same for all values of x, $P(x)$ is a common factor, and the calculation can be simplified to $E(x) = (\sum x)(P(x)) = 21\left(\dfrac{1}{6}\right) = 3.50$.

The result obtained indicates that a person who participates in this game for a long period of time could expect to win on the average $3.50 (even though it is not possible to receive that amount on any single roll of the die). The *expected monetary value* of the game is $3.50, and this is the *maximum* amount anyone should be willing to pay to play the game.

B. The Variance and Standard Deviation of a Discrete Probability Distribution

The concept of expected value is that of a long-run mean over a large number of trials. More often than not, some value other than the computed expected value will occur for the various individual trials. In some cases, such as in Example 8.2a, the expected value ($3.50) *cannot* occur as the outcome of a single trial, since the die can only show the whole numbers 1 to 6.

A measure of the mean difference of the individual values from the expected value is the *variance* of the probability distribution. Denoted by $V(x)$, the variance of a discrete random variable can be computed by multiplying the squared deviations of the individual values from the expected value by the individual probabilities:

$$V(x) = \sigma^2 = \sum (x - \mu)^2 \cdot P(x)$$

○ **EXAMPLE 8.2b**

Determine the variance and the standard deviation for the probability distribution in Example 8.2a.

● **SOLUTION**

The individual values of the random variable are $x = 1, 2, 3, 4, 5, 6$.

The probability of occurrence for all values of x is $P(x) = \dfrac{1}{6}$;

the expected value $E(x) = \mu = 3.50$.

The variance can be computed as shown in Table 8.3.

The variance $V(x) = \sigma^2 = \sum (x - \mu)^2 \cdot P(x) = 2.9166668$;

the standard deviation, $\sigma = \sqrt{2.9166668} = 1.7078252$.

The calculation can be simplified by using the alternate formula

$$\sigma^2 = \sum x^2 \cdot P(x) - \mu^2$$

as shown in Table 8.4.

TABLE 8.3   Computation of Variance

Individual value x	Deviation $(x - \mu)$	Squared deviation $(x - \mu)^2$	$P(x)$	Product $(x - \mu)^2 \cdot P(x)$
1	$1 - 3.50 = -2.50$	6.25	$\dfrac{1}{6}$	$6.25\left(\dfrac{1}{6}\right) = 1.0416667$
2	$2 - 3.50 = -1.50$	2.25	$\dfrac{1}{6}$	$2.25\left(\dfrac{1}{6}\right) = 0.3750000$
3	$3 - 3.50 = -0.50$	0.25	$\dfrac{1}{6}$	$0.25\left(\dfrac{1}{6}\right) = 0.0416667$
4	$4 - 3.50 = 0.50$	0.25	$\dfrac{1}{6}$	$0.25\left(\dfrac{1}{6}\right) = 0.0416667$
5	$5 - 3.50 = 1.50$	2.25	$\dfrac{1}{6}$	$2.25\left(\dfrac{1}{6}\right) = 0.3750000$
6	$6 - 3.50 = 2.50$	6.25	$\dfrac{1}{6}$	$6.25\left(\dfrac{1}{6}\right) = 1.0416667$
				$\sum(x - \mu)^2 \cdot P(x) = 2.9166668$

TABLE 8.4   Simplified Computation of Variance

x	x^2	$P(x)$	$x^2 \cdot P(x)$
1	1	$\dfrac{1}{6}$	$1\left(\dfrac{1}{6}\right) = \dfrac{1}{6}$
2	4	$\dfrac{1}{6}$	$4\left(\dfrac{1}{6}\right) = \dfrac{4}{6}$
3	9	$\dfrac{1}{6}$	$9\left(\dfrac{1}{6}\right) = \dfrac{9}{6}$
4	16	$\dfrac{1}{6}$	$16\left(\dfrac{1}{6}\right) = \dfrac{16}{6}$
5	25	$\dfrac{1}{6}$	$25\left(\dfrac{1}{6}\right) = \dfrac{25}{6}$
6	36	$\dfrac{1}{6}$	$36\left(\dfrac{1}{6}\right) = \dfrac{36}{6}$
	$\sum x^2 = 91$		$\sum x^2 \cdot P(x) = \dfrac{91}{6}$

$$V(x) = \sigma^2 = \sum x^2 \cdot P(x) - \mu^2$$

$$= \frac{91}{6} - (3.50)^2 = 15.1666667 - 12.25$$

$$= 2.9166667$$

For the special case in which $P(x)$ is the same for all values of x, $P(x)$ is a common factor, and the calculation can be simplified to

$$V(x) = \sigma^2 = (\sum x^2) \cdot P(x) - \mu^2$$

$$= 91 \left(\frac{1}{6} \right) - (3.50)^2 = 2.9166667$$

However, in most cases the calculation of the mean and variance of a discrete probability distribution is best done by using a *tabular* format.

○ EXAMPLE 8.2c

In Chapter 7 the experiment of rolling a pair of dice was considered and the possible values of the random variable x (Total Number of Spots), together with the associated probabilities, were listed in Figure 7.1. Using the information from Figure 7.1, compute the mean, the variance and the standard deviation of the probability distribution.

● SOLUTION

The random variable x (Total Number of Spots) is a discrete variable taking the values $x = 2, 3, \ldots, 10, 11, 12$. The corresponding number of outcomes of each event and the probabilities of the events are summarized in Table 8.5 and the calculated values listed.

$$\mu = \sum x \cdot P(x) = 7.00$$

Using the basic method,

$$\sigma^2 = \sum (x - \mu)^2 \cdot P(x) = 5.833333$$

$$\sigma = \sqrt{5.833333} = 2.415229$$

Using the simplified method,

$$\sigma^2 = \sum (x^2 \cdot P(x)) - \mu^2 = 54.833333 - 49.00 = 5.833333$$

$$\sigma = \sqrt{5.833333} = 2.415229$$

TABLE 8.5 **Format for Computing the Mean and Variance of a Discrete Probability Distribution**

x	No. of outcomes	Probability of event $P(x)$	$x \cdot P(x)$	$x^2 \cdot P(x)$	$x - \mu$	$(x - \mu)^2 \cdot P(x)$
2	1	$\frac{1}{36} = 0.027778$	0.055556	0.111111	-5.00	0.694445
3	2	$\frac{2}{36} = 0.055556$	0.166667	0.500000	-4.00	0.888889
4	3	$\frac{3}{36} = 0.083333$	0.333333	1.333333	-3.00	0.750000
5	4	$\frac{4}{36} = 0.111111$	0.555556	2.777778	-2.00	0.444444
6	5	$\frac{5}{36} = 0.138889$	0.833333	5.000000	-1.00	0.138889
7	6	$\frac{6}{36} = 0.166667$	1.666667	8.166667	0.00	0.000000
8	5	$\frac{5}{36} = 0.138889$	1.111111	8.888889	1.00	0.138889
9	4	$\frac{4}{36} = 0.111111$	1.000000	9.000000	2.00	0.444444
10	3	$\frac{3}{36} = 0.083333$	0.833333	8.333333	3.00	0.750000
11	2	$\frac{2}{36} = 0.055556$	0.611111	6.722222	4.00	0.888889
12	1	$\frac{1}{36} = 0.027778$	0.333333	4.000000	5.00	0.694444
Total	36	1.000000	7.000000	54.833333	0.00	5.833333

EXERCISE 8.2

1. The marketing department of B.C. Research has submitted the following information for the purpose of forecasting net income for the next year.

Sales volume	Probability
$1 000 000	0.20
1 200 000	0.30
1 500 000	0.50

Calculate the expected sales volume.

2. The design division of Highrise Construction prepared the following scenarios to estimate the time required for a newly-won contract.

Scenario	Time in days	Probability
Optimistic	100	0.15
Likely	120	0.20
Most likely	150	0.40
Pessimistic	190	0.25

Compute the expected time for the completion of the contract.

3. For the given values of a random variable and the associated probabilities, determine
   a) the mean;
   b) the variance and standard deviation.

x	0	1	2	3	4
$P(x)$	0.10	0.20	0.30	0.20	0.20

4. A commodities trader has assigned the following probabilities for the forward price of a bushel of wheat three months from now.

Price per bushel	Probability
$2.20	0.05
2.30	0.25
2.40	0.35
2.50	0.20
2.60	0.15

a) Compute the expected value for a bushel of wheat three months from now.
b) Compute the standard deviation of the price of wheat.

SECTION 8.3 # The Binomial Distribution

A. Characteristics of the Binomial Distribution

Although a probability distribution can be developed for any random variable, a number of well-known probability distributions are available as models for many business and economic situations. Of these the most widely used discrete probability distribution is the **binomial distribution**.

Many situations have the common characteristic of only two possible outcomes. The toss of a coin has the outcomes "head" or "tail"; the answer to a true or false question is either "correct" or "incorrect"; the decision to participate in a game is either "yes" or "no"; the result of an attempted high jump is either a "success" or a "failure."

The binomial distribution is based on a series of attempts (trials) in which only two outcomes are possible. The following are the essential characteristics of a binomial distribution:

1. The trials are identical.
2. Each trial has only *two* possible outcomes classified as "success" or "failure."
3. The data collected are the result of counting; that is, the distribution is *discrete*.
4. The probability of "success" is the same for each trial. The same is true for the probability of "failure."
5. The trials are independent of each other; that is, the outcome of one trial does not affect the outcome of any other trial.

B. *Formula for Constructing a Binomial Probability Distribution*

For a binomial process the event of interest is "the probability of obtaining exactly x successes in n trials." This probability can be computed by using the formula

$$P(x) = (_nC_x)(p^x)[(1-p)^{n-x}]$$

where n is the number of trials;
$\qquad x$ is the number of successes for the n trials;
$\qquad p$ is the probability of success;
$\quad 1 - p$ is the probability of failure;
$\quad P(x)$ is the probability of x successes in n trials;
$\qquad {}_nC_x$ is the number of combinations.

Since $_nC_x = \dfrac{n!}{x!(n-x)!}$ the formula describing the binomial probability distribution can be written as

$$P(x) = \frac{n!}{x!(n-x)!}(p^x)[(1-p)^{n-x}]$$

C. *Constructing and Using Binomial Distributions*

○ **EXAMPLE 8.3a**
Consider the experiment of eight tosses of a coin for the Event (Number of Heads).
a) Construct the probability distribution.
b) Construct the less-than and more-than cumulative probability distributions.
c) Determine the probabilities of the following outcomes of the eight tosses of the coin:
  i) exactly four heads;
  ii) no more than five heads;
  iii) at least six heads;
  iv) between three and five heads;
  v) fewer than four heads;
  vi) fewer than three heads or more than six heads.

● **SOLUTION**

a) The experiment of tossing a coin eight times meets the requirements of a binomial process:
   1. The number of trials is fixed, $n = 8$.
   2. Each trial has two possible outcomes: "head" or "tail."
   3. The data are discrete (the result of counting).
   4. The probability of success is the same for all trials.
   5. The trials are independent of each other.

   The binomial formula for the probability of success can be used.

   $n = 8$; $p = 0.50$; $(1 - p) = (1 - 0.50) = 0.50$;
   $x = 0, 1, 2, 3, 4, 5, 6, 7$ and 8.
   In this particular case, since $p = (1 - p) = 0.50$,

   $$P(x) = (_nC_x)(p^x)[(1 - p)^{n-x}] = (_8C_x)(0.50^x)[(0.50)^{8-x}]$$

   Since $(0.50^x)(0.50^{8-x}) = 0.50^{x+(8-x)} = 0.50^8 = 0.0039063$,

   $$P(x) = \frac{8!}{x!(8 - x)!}(0.0039063)$$

   The binomial probability distribution for $n = 8$, $p = 0.50$ can now be constructed, as shown in Table 8.6, by evaluating $P(x)$ for the possible values of x.

TABLE 8.6 **Construction of Binomial Distribution**

Column 1	Column 2	Column 3	Column 4
x	Value of $_8C_x = \dfrac{8!}{x!(8-x)!}$	Value of $(p^x)[(1-p)^{n-x}]$ $(0.50^x)(0.50^{8-x})$ $= 0.50^8$	Value of $P(x)$ rounded to four decimals (Column 2 × Column 3)
0	$\dfrac{8!}{(0!)(8!)} = \dfrac{8!}{(1)(8!)} = 1$	0.0039063	$(1)(0.0039063) = 0.0039$
1	$\dfrac{8!}{(1!)(7!)} = \dfrac{8}{1} = 8$	0.0039063	$(8)(0.0039063) = 0.0312$
2	$\dfrac{8!}{(2!)(6!)} = \dfrac{(8)(7)}{(2)(1)} = 28$	0.0039063	$(28)(0.0039063) = 0.1094$
3	$\dfrac{8!}{(3!)(5!)} = \dfrac{(8)(7)(6)}{(3)(2)(1)} = 56$	0.0039063	$(56)(0.0039063) = 0.2188$
4	$\dfrac{8!}{(4!)(4!)} = \dfrac{(8)(7)(6)(5)}{(4)(3)(2)(1)} = 70$	0.0039063	$(70)(0.0039063) = 0.2734$
5	$\dfrac{8!}{(5!)(3!)} = \dfrac{(8)(7)(6)}{(3)(2)(1)} = 56$	0.0039063	$(56)(0.0039063) = 0.2188$
6	$\dfrac{8!}{(6!)(2!)} = \dfrac{(8)(7)}{(2)(1)} = 28$	0.0039063	$(28)(0.0039063) = 0.1094$
7	$\dfrac{8!}{(7!)(1!)} = \dfrac{8}{1} = 8$	0.0039063	$(8)(0.0039063) = 0.0312$
8	$\dfrac{8!}{(8!)(0!)} = 1$	0.0039063	$(1)(0.0039063) = 0.0039$

b) The two *cumulative* probability distributions for $n = 8$, $p = 0.50$, can be determined by computing running totals, as shown in Table 8.7.

TABLE 8.7 **Cumulative Probability Distributions**

x	Probability distribution $P(x)$	Less-than cumulative distribution $P(\leq x)$	More-than cumulative distribution $P(\geq x)$
0	0.0039	0.0039	1.0000
1	0.0312	0.0351	0.9961
2	0.1094	0.1445	0.9649
3	0.2188	0.3633	0.8555
4	0.2734	0.6367	0.6367
5	0.2188	0.8555	0.3663
6	0.1094	0.9649	0.1445
7	0.0312	0.9961	0.0351
8	0.0039	1.0000	0.0039

c) i) From Table 8.6, the probability of the Event (Exactly Four Heads),
$P(4) = 0.2734$.

ii) From Table 8.6, the probability of the Event (No More Than Five Heads),

$$P(\leq 5) = P(0) + P(1) + P(2) + P(3) + P(4) + P(5)$$
$$= 0.0039 + 0.0312 + 0.1094 + 0.2188 + 0.2734 + 0.2188$$
$$= 0.8555$$

or directly from Table 8.7 in the Less-than cumulative probability distribution column, $P(\leq 5) = 0.8555$.

iii) From Table 8.6, the probability of the Event (At Least Six Heads),

$$P(\geq 6) = P(6) + P(7) + P(8)$$
$$= 0.1094 + 0.0312 + 0.0039 = 0.1445$$

or directly from Table 8.7 in the More-than cumulative probability distribution column, $P(\geq 6) = 0.1445$.

iv) The probability of the Event (Between Three and Five Heads),

$$P(3 \text{ or } 4 \text{ or } 5) = P(3) + P(4) + P(5)$$
$$= 0.2188 + 0.2734 + 0.2188 = 0.7110$$

or directly from Table 8.7 in the Less-than cumulative probability distribution column,

$$P(3 \text{ or } 4 \text{ or } 5) = P(\leq 5) - P(\leq 2)$$
$$= 0.8555 - 0.1445 = 0.7110.$$

v) The probability of the Event (Fewer Than Four Heads),

$$P(< 4) = P(\leq 3)$$
$$= P(0) + P(1) + P(2) + P(3)$$
$$= 0.0039 + 0.0312 + 0.1094 + 0.2188 = 0.3633$$

or directly from Table 8.7 in the Less-than cumulative probability column, $P(\leq 3) = 0.3633$.

vi) The probability of the Event (Fewer Than Three Heads or More Than Six Heads),

$$P(< 3 \text{ or } > 6) = P(\leq 2) + P(\geq 7)$$
$$= [P(0) + P(1) + P(2)] + [P(7) + P(8)]$$
$$= [0.0039 + 0.0312 + 0.1094] + [0.0312 + 0.0039]$$
$$= 0.1445 + 0.0351 = 0.1796$$

or directly from Table 8.7, $P(\leq 2) + P(\geq 7) = 0.1445$ (in the Less-than cumulative probability column) + 0.0351 (in the More-than cumulative probability column) = 0.1796.

○ **EXAMPLE 8.3b**

In a complex manufacturing process the probability of producing a satisfactory item is 0.80. If eight items are taken from the production line, determine the probability that this group contains
a) exactly three satisfactory items;
b) at most three satisfactory items;
c) at least three satisfactory items;
d) between four and seven satisfactory items;
e) more than four satisfactory items;
f) fewer than two or more than six satisfactory items.

● **SOLUTION**

The process meets the requirements of a binomial process. In order to answer the questions we need to first construct the probability distribution for

$n = 8; \quad p = 0.80; \quad (1 - p) = (1 - 0.80) = 0.20;$
$x = 0, 1, 2, 3, 4, 5, 6, 7, 8.$

Substituting in the binomial formula $P(x) = ({}_nC_x)(p^x)(1 - p)^{n-x}$, we obtain

$$P(x) = ({}_8C_x)(0.80^x)(0.20^{8-x})$$

$$= \frac{8!}{x!(8-x)!}(0.80^x)(0.20^{8-x})$$

The binomial probability distribution can be constructed as shown in Table 8.8 and the associated cumulative probabilities as shown in Table 8.9.

TABLE 8.8 **Construction of Binomial Distribution, $n = 8$, $p = 0.80$**

Column 1 x	Column 2 Value of ${}_8C_x$	Column 3 Value of $(0.8^x)(0.2^{8-x})$	Column 4 (Col. 2)(Col. 3) to 4 decimals
0	1	$(0.8^0)(0.2^8) = (1)(0.00000256) = 0.00000256$	0.0000
1	8	$(0.8^1)(0.2^7) = (0.8)(0.0000128) = 0.00001024$	0.0001
2	28	$(0.8^2)(0.2^6) = (0.64)(0.000064) = 0.00004096$	0.0011
3	56	$(0.8^3)(0.2^5) = (0.512)(0.00032) = 0.00016384$	0.0092
4	70	$(0.8^4)(0.2^4) = (0.4096)(0.0016) = 0.00065536$	0.0459
5	56	$(0.8^5)(0.2^3) = (0.32768)(0.008) = 0.00262144$	0.1468
6	28	$(0.8^6)(0.2^2) = (0.262144)(0.04) = 0.01048576$	0.2936
7	8	$(0.8^7)(0.2^1) = (0.2097152)(0.2) = 0.04194304$	0.3355
8	1	$(0.8^8)(0.2^0) = (0.16777216)(1) = 0.16777216$	0.1678

TABLE 8.9   Cumulative Probability Distributions, $n = 8$, $p = 0.80$

x	Probability distribution $P(x)$	Less-than cumulative distribution $P(\leq x)$	More-than cumulative distribution $P(\geq x)$
0	0.0000	0.0000	1.0000
1	0.0001	0.0001	1.0000
2	0.0011	0.0012	0.9999
3	0.0092	0.0104	0.9988
4	0.0459	0.0563	0.9896
5	0.1468	0.2031	0.9437
6	0.2936	0.4967	0.7967
7	0.3355	0.8322	0.5033
8	0.1678	1.0000	0.1678

Using the tables,

a) $P(3) = 0.0092$.

b) $P(\leq 3) = P(0) + P(1) + P(2) + P(3)$
   $= 0.0000 + 0.0001 + 0.0011 + 0.0092 = 0.0104$;
   or directly from the Less-than cumulative column $P(\leq 3) = 0.0104$.

c) From the More-than cumulative column, $P(\geq 3) = 0.9988$.

d) $P(4 \text{ or } 5 \text{ or } 6 \text{ or } 7) = P(4) + P(5) + P(6) + P(7)$
   $= 0.0459 + 0.1468 + 0.2936 + 0.3355 = 0.8218$;
   or $P(4 \text{ or } 5 \text{ or } 6 \text{ or } 7) = P(\leq 7) - P(\leq 3)$
   $= 0.8322 - 0.0104 = 0.8218$.

e) $P(> 4) = P(\geq 5) = 0.9437$.

f) $P(< 2 \text{ or } > 6) = P(\leq 1) + P(\geq 7) = 0.0001 + 0.5033 = 0.5034$.

D.   Binomial Tables

In Example 8.3a we constructed a binomial probability distribution for $n = 8$, $p = 0.50$, and in Example 8.3b a similar distribution for $n = 8$, $p = 0.80$. The fact is that a different probability distribution results for every value of n and every value of p. The construction of binomial probability distributions even for small values of n is a tedious arithmetic process.

Because of the usefulness of the binomial distribution, tables have been produced for many values of n and selected values of p.

One such table for $n = 8$ for selected values of p is presented in Table 8.10. The corresponding cumulative probability tables can be constructed without difficulty by listing the running totals for each value of p.

TABLE 8.10 Binomial Probability Table for *n* = 8

p x	0.10	0.20	0.30	0.40	0.50	0.60	0.70	0.80	0.90
0	0.4305	0.1678	0.0576	0.0168	0.0039	0.0006	0.0001	0.0000	0.0000
1	0.3826	0.3355	0.1976	0.0896	0.0312	0.0079	0.0013	0.0001	0.0000
2	0.1488	0.2936	0.2965	0.2090	0.1094	0.0413	0.0100	0.0011	0.0000
3	0.0331	0.1468	0.2541	0.2787	0.2188	0.1239	0.0467	0.0092	0.0004
4	0.0046	0.0459	0.1361	0.2322	0.2734	0.2322	0.1361	0.0459	0.0046
5	0.0004	0.0092	0.0467	0.1239	0.2188	0.2787	0.2541	0.1468	0.0331
6	0.0000	0.0011	0.0100	0.0413	0.1094	0.2090	0.2965	0.2936	0.1488
7	0.0000	0.0001	0.0013	0.0079	0.0312	0.0896	0.1976	0.3355	0.3826
8	0.0000	0.0000	0.0001	0.0006	0.0039	0.0168	0.0576	0.1678	0.4305
Total	1.0000	1.0000	1.0000	1.0000	1.0000	1.0000	1.0000	1.0000	1.0000

To use a binomial probability table,
1. locate the table or portion of the table with the required value of *n*;
2. locate the column headed by the required value of *p*;
3. locate the row for the required value of *x*;
4. read off the probability $P(x)$ at the intersection of the column with the required *p* and the row for the required *x*.

○ **EXAMPLE 8.3c**
In January 1991, with the introduction of the Goods and Services Tax, Canadian supermarkets had to adjust the prices of many non-food items. In one particular store it was determined that 40% of the adjusted prices were incorrect. Use Table 8.10 to calculate the probabilities that, of eight non-food items selected, further price corrections were necessary
a) for exactly six of the items;
b) for fewer than three of the items;
c) for between five and seven of the items;
d) for more than four of the items.

● **SOLUTION**
$n = 8$; $p = 0.40$.
The probabilities will be found in the column $p = 0.40$.
a) $P(6) = 0.041$
b) $P(< 3) = P(\leq 2) = P(0) + P(1) + P(2)$
$$= 0.0168 + 0.0896 + 0.2090 = 0.3154$$
c) $P(5 \text{ or } 6 \text{ or } 7) = P(5) + P(6) + P(7)$
$$= 0.1239 + 0.0413 + 0.0079 = 0.1731$$
d) $P(> 4) = P(\geq 5) = P(5) + P(6) + P(7) + P(8)$
$$= 0.1239 + 0.0413 + 0.0079 + 0.0006 = 0.1737$$

E. *Characteristics of the Binomial Distribution*

The main characteristics of the binomial distribution are its shape, mean and standard deviation.

1. The *shape* of the binomial distribution depends both on the number of trials n and the constant probability of success p. The graphs of three of the distributions for $n = 8$ in Figure 8.2 show what happens to the shape of a binomial distribution for different values of p.

   Note $n = 8$ is considered to be a small number of trials.

   a) When $P < 0.50$ and n is small, the distribution is *positively* skewed (see Figure 8.2, Diagram A).

FIGURE 8.2   Graphs of Binomial Distributions, $n = 8$

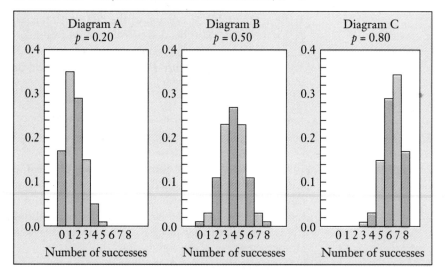

   b) When $P > 0.50$ and n is small, the distribution is *negatively* skewed (see Figure 8.2, Diagram C).
   c) When $p = 0.50$, the distribution is *symmetrical* for all values of n (see Figure 8.2, Diagram B).
2. The *mean* of the binomial distribution, $\mu = np$.
3. The *variance* of the binomial distribution $\sigma^2 = np(1 - p)$ and the *standard deviation* $\sigma = \sqrt{np(1 - p)}$.

○ **EXAMPLE 8.3d**
Determine the mean, variance and standard deviation for the probability distributions in Examples 8.3a, 8.3b, and 8.3c.

● **SOLUTION**
For Example 8.3a, $n = 8$, $p = 0.50$, $(1 - p) = 0.50$.

$$\mu = np = 8(0.50) = 4.00$$
$$\sigma^2 = np(1-p) = 8(0.50)(0.50) = 2.00$$
$$\sigma = \sqrt{np(1-p)} = \sqrt{2.00} = 1.4142$$

For Example 8.3b, $n = 8$, $p = 0.80$, $1 - p = 0.20$.

$$\mu = np = 8(0.80) = 6.40$$
$$\sigma^2 = np(1-p) = 8(0.80)(0.20) = 1.28$$
$$\sigma = \sqrt{np(1-p)} = \sqrt{1.28} = 1.1314$$

For Example 8.3c, $n = 8$, $p = 0.40$, $(1 - p) = 0.60$.

$$\mu = np = 8(0.40) = 3.20$$
$$\sigma^2 = np(1-p) = 8(0.40)(0.60) = 1.92$$
$$\sigma = \sqrt{np(1-p)} = \sqrt{1.92} = 1.3856$$

EXERCISE 8.3

1. If we repeat five tosses of one coin many times, what is the probability of four tails and one head appearing?

2. A board game uses a spinning wheel with five equal sections (shown below) to determine *payoffs*. If we repeat four spins of the wheel many times, what is the probability of "Stocks" showing once and "Bust" showing three times?

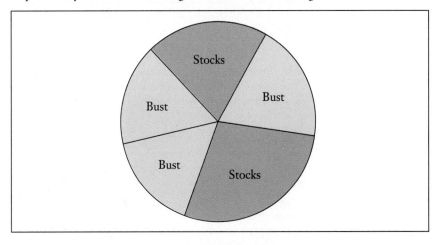

3. Auto glass companies now have a machine that can repair broken windshields with a success rate of 85%. If a random sample of six cars is taken from the customer records, determine the probability that
   a) exactly two windshields were successfully repaired;
   b) at least four windshields were successfully repaired.

4. A poll revealed that 12% of Canadians would vote for a new political party. If eight eligible voters are selected,
   a) what is the probability that exactly three will vote for a new party?
   b) what is the probability that no more than five will vote for a new political party?
   c) what is the probability that at least two will vote for a new political party?
5. Determine the mean and the standard deviation for Question 2.
6. Determine the mean and the standard deviation for Question 3.

REVIEW EXERCISE

1. An inventor is planning to go into business for himself. He uses the following estimates. Calculate the expected profits for his invention.

Event	Probability of event	Profit
Strong sales	0.04	$850 000
Fair sales	0.84	220 000
Poor sales	0.12	−75 000

2. The inventor in Question 1 uses the following estimates based on selling the rights to the product and collecting royalties. Calculate the expected profits.

Event	Probability of event	Profit
Strong sales	0.04	$500 000
Fair sales	0.84	100 000
Poor sales	0.12	5 000

3. In the past an automobile repair shop owner has noticed that sales increased as certain economic indicators decreased. Based on forecasts of these economic indicators and past observations, the owner has assigned probabilities to sales increases as follows:

Sales increase	Probability
5%	0.05
10%	0.15
15%	0.35
20%	0.30
25%	0.15

Calculate the expected average sales increase and the standard deviation.

4. When asked to quantify the effectiveness of a new promotional program based on their sales experience, four salespersons responded as follows:

Salesperson	A	B	C	D
Sales increase	$3000	$6000	$3000	$4000

   a) If all four persons gave their best estimates, what can be said about the nature of the probability estimate in each case?
   b) Compute the expected mean sales increase.
   c) Compute the standard deviation.

5. A manufacturer of photocopiers knows that the machines require regular service to avoid customer complaints. Despite this service, the failure rate of the machines is eight percent. In an office that is equipped with four copiers, determine the probability that
   a) none of the copiers will fail;
   b) at least one copier will fail.

6. In sensory research projects, two identical products are often given to participants for appraisal. If 10 persons are chosen, what is the probability that
   a) only one person will show a preference for product A, while nine will show a preference for product B?
   b) five persons will indicate a preference for product A, while the other five will indicate a preference for product B?

7. Revenue Canada estimates that five percent of Canadians "cheat" on their tax returns. If a random sample of 100 tax returns is chosen for auditing,
   a) what is the probability that exactly one taxpayer cheated?
   b) calculate the expected mean;
   c) calculate the standard deviation.

8. An accounts-receivable supervisor has determined that three percent of customers default on their payments. Currently there are 1200 credit customers with outstanding balances.
   a) Determine the expected mean number of bad debts.
   b) Compute the standard deviation.
   c) What is the probability that no more than five customers will default?

9. Given that $n = 30$, $p = 0.80$, use the binomial formula to determine
   a) $P(x > 28)$;
   b) $P(20 < x < 22)$;
   c) $P(x < 3)$.

10. Given that $n = 100$, $p = 0.25$, use the binomial formula to determine
   a) $P(x > 98)$;
   b) $P(78 < x < 80)$;
   c) $P(x < 4)$.

SELF-TEST

1. Given the following weather predictions, determine
   a) the mean expected rainfall;
   b) the standard deviation.

Rainfall (mm)	15	12	10	8	6	4	2
Probability	5%	15%	20%	30%	15%	10%	5%

2. The personnel department has found that 59% of college students accept job offers from the company. If 12 offers have been made to recent college graduates, what is the probability that
   a) three or fewer will accept?
   b) nine or more will accept?
   c) exactly seven will accept?

3. Compute the mean and the standard deviation for the distribution in Question 2.

Key Terms

Binomial distribution   216
Continuous probability distribution   210
Continuous random variable   210
Discrete probability distribution   209
Discrete random variable   209
Expected value   210
Probability distribution   208

Summary of Formulas

1. Discrete random variable:

a) Expected value (mean):

$$E(x) = \mu = \sum x \cdot P(x)$$

b) Variance:

$$V(x) = \sigma^2 = \sum (x - \mu)^2 \cdot P(x)$$
$$\text{or} \quad \sigma^2 = \sum x^2 \cdot P(x) - \mu^2$$

c) Standard deviation:

$$\sigma = \sqrt{\sigma^2}$$

2. Binomial probability distribution:

a) Binomial formula:
Probability of x successes in n trials:

$$P(x) = (_nC_x)(p^x)(1-p)^{n-x}$$

$$= \frac{n!}{(n-x)!\,x!}(p^x)(1-p)^{n-x}$$

b) Mean: $\mu = np$

c) Variance: $\sigma^2 = np(1-p)$

d) Standard deviation: $\sigma = \sqrt{\sigma^2} = \sqrt{np(1-p)}$

CHAPTER 9

The Normal Distribution

Introduction

Continuous variables are distinct from discrete variables in that they are not restricted to specific values and can assume an infinite number of values within a specified range. The computation of probabilities, expected values and standard deviations for continuous variables requires the use of integral calculus and is beyond the scope of this text. However, this does not prevent us from using the continuous probability distribution referred to as the normal distribution.

Objectives

Upon completion of this chapter you will be able to
1. understand the importance of the normal distribution in statistics;
2. define the characteristics of the normal curve;
3. understand the concept of the standardized normal curve;
4. compute and interpret z scores;
5. use a table of areas under the normal curve to solve problems involving the normal distribution;
6. use the normal distribution in appropriate situations as a replacement of the binomial distribution.

SECTION 9.1 ## Characteristics of the Normal Distribution

The **normal distribution** is the most important statistical probability distribution as far as practical applications are concerned. The three main reasons for its usefulness are as follows:
1. The distribution of much of the data collected in the physical and social sciences, and in business and industry is sufficiently close to that of the normal distribution to permit the utilization of the properties of the normal distribution in analyzing the actual distribution.

2. When the number of trials is large enough, the shape of some of the discrete probability distributions approximates a normal distribution. Specifically, even with n as small as 20, the shape of a binomial distribution becomes fairly symmetrical even when p is not very close to 0.50.
3. Because of certain properties of the sampling distribution of the means (see Chapter 10), the normal distribution provides the basis for statistical inference.

The graphical representation of the normal distribution (see Figure 9.1) is referred to as the **normal curve**. The following are some of the important characteristics of the normal curve:

FIGURE 9.1 **The Normal Curve**

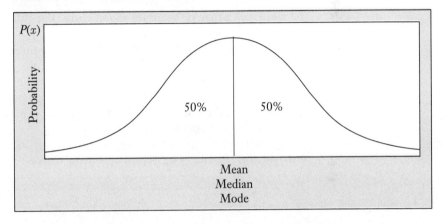

1. The curve is *bell-shaped* and has only one peak at the centre of the distribution.
2. The distribution is *symmetrical* about the vertical line drawn from the peak of the curve to the horizontal axis.
3. The measures of central tendency of the distribution — the *mean*, the *median* and the *mode* — are equal in value and are located at the peak of the normal curve.
4. The curve approaches the x axis gradually on either side of the mean but never touches the x axis. Theoretically, the tails of the distribution extend indefinitely in either direction.
5. Since the curve is completely symmetrical, the area to the left of the mean equals the area to the right of the mean, so that each side contains 50% of the total area under the curve.
6. The area under the curve represents probability. For any normal distribution, the probability is 50% that the continuous variable x will assume a value less than the mean and 50% that it will assume a value more than the mean.
7. The area between any two points under the curve represents the probability that the continuous variable x will assume some value within that interval.
8. The standard measure of central tendency of a normal distribution is its

mean μ. The standard measure of variability is the standard deviation σ. Any normal distribution is completely defined by the values of its mean μ and its standard deviation σ.

SECTION 9.2 ## Areas under the Normal Curve

The mean μ and the standard deviation σ completely define a normal curve. The standard deviation σ measures the extent to which the data under the normal curve deviate from the mean μ and is always measured from the mean μ.

Any normal curve has two *points of inflection*, that is, two points at which the slope of the curve becomes more horizontal than vertical, as shown in Figure 9.2.

If a vertical line is drawn from the point of inflection A to the x axis, the point of intersection of the vertical line and the x axis lies *one* standard deviation from the mean μ. The symbol used to designate this point is $+1\sigma$.

Because the values of the variable x to the right of the mean are *greater* than the mean, the symbol 1σ indicates that this point on the horizontal axis is located one standard deviation *above* the mean.

FIGURE 9.2 Points of Inflection of Normal Curve

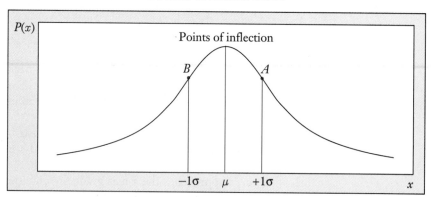

Since the normal curve is *symmetrical* around the mean, a vertical line drawn from the other point of inflection B is also one standard deviation from the mean and is designated by the symbol -1σ. Because values of the variable x to the left of the mean are *smaller* than the mean, the symbol -1σ indicates that this point on the horizontal axis is located one standard deviation *below* the mean.

For all normal curves, the area between μ and 1σ contains approximately 34.13% of the total area under the curve, as shown in Figure 9.3. Because of the symmetry of the normal curve, the area between μ and -1σ also contains approximately 34.13% of the total area.

FIGURE 9.3   Area between μ and ±1σ

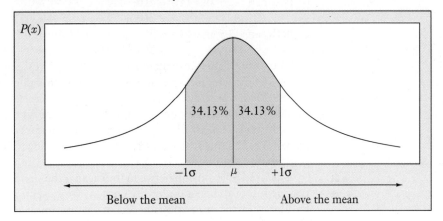

The horizontal axis can be marked by units each equal to one standard deviation. This creates a horizontal scale measured from the mean in terms of standard deviation units, as shown in Figure 9.4.

When verticals are drawn to the x axis at the points marked in standard deviation units, the total area under the normal curve is subdivided into smaller areas, each of which contains a certain proportion of the total area. The proportions are the same for all normal distributions.

The approximate proportions for each of these smaller areas are shown in Figure 9.4. As indicated, the proportion of the total area

between μ and 1σ = 0.3413;
between 1σ and 2σ = 0.1359;
between 2σ and 3σ = 0.0215.

FIGURE 9.4   Scale and Areas under Normal Curve

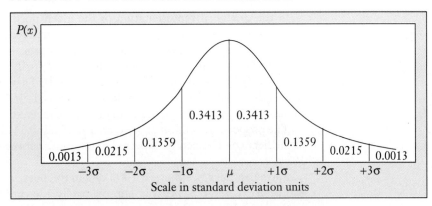

Because of the symmetry of the normal curve, the proportions are the same for the areas below the mean. From this, it follows that the proportion of the total area

a) between -1σ and $+1\sigma$ = 0.3413 + 0.3413 = 0.6826 = 68.26%;

b) between -2σ and $+2\sigma$ = 2(0.3413 + 0.1359)

$$= 2(0.4772) = 0.9544 = 95.44\%;$$

c) between -3σ and $+3\sigma$ = 2(0.3413 + 0.1359 + 0.0215)

$$= 2(0.4987) = 0.9974 = 99.74\%.$$

The three values calculated in (a), (b) and (c) contain slight rounding errors. The more accurate and frequently used values are 68.27%, 95.45% and 99.73%.

The mathematically derived normal curve never touches the horizontal axis. However, only 0.27% of the total area is in the two tails below -3σ and above $+3\sigma$. For this reason, the scale used for the normal distribution is usually restricted to the range -3σ to $+3\sigma$.

The fixed relationship between the mean μ the standard deviation σ and the proportion of the area under the curve is very useful in statistics because the frequency distribution of many sets of data resemble the normal distribution.

○ EXAMPLE 9.2a

Northern Taxi Inc. owns a fleet of 400 cars, which use an average of 12 L of fuel per 100 km with a standard deviation of 1.5 L of fuel per 100 km. Fuel consumption of the fleet of cars is normally distributed. Describe the data.

● SOLUTION

The mean, $\mu = 12$; the standard deviation, $\sigma = 1.5$; the number of observations, $N = 400$.

Knowing the μ and the σ of a set of data that is normally distributed is sufficient to describe the set of data and permits us to draw certain useful inferences.

In the graphical representation of the given normal distribution in Figure 9.5, place the value 12.0 at μ. The point marked 1σ on the horizontal scale is one standard deviation above the μ. Since $\mu = 12.0$ and $\sigma = 1.5$, the point 1σ is associated with a consumption of $12.0 + 1.5 = 13.5$.

Similarly, the point marked -1σ is one standard deviation below the mean; that is, the point -1σ is associated with ($12.0 - 1.5$) = 10.5. The remaining values can be marked in a similar manner, as shown in Figure 9.5.

The proportion of the area between μ and 1σ is 0.3413. This means 34.13% of the fleet vehicles can be expected to consume between 12.0 and 13.5 L per 100 km.

Furthermore, since there are 400 cars in the fleet, we can expect 400(0.3413) = 136.52, that is, approximately 136 cars, to consume between 12.0 and 13.5 L per 100 km.

The proportion between -2σ and $+2\sigma$ is 0.9545. This means that 95.45% of the vehicles can be expected to consume between 9.0 and 15.0 L per 100 km. Approximately 400(0.95945) = 382 cars consume between 9.0 and 15.0 L per 100 km.

The proportion of cars between -2σ and -3σ is 0.0215; that is, 2.15% of the cars, or approximately 9 vehicles, consume between 7.5 and 9.0 L per 100 km.

FIGURE 9.5   Normal Curve for Example 9.2a

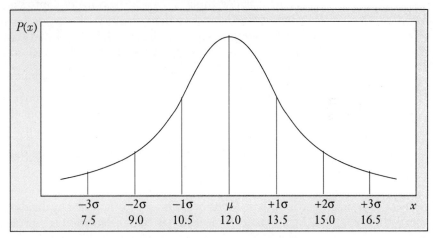

EXERCISE 9.2

1. For Example 9.2a determine the proportion of cars with a fuel consumption (per 100 km)
   a) between 10.5 L and 13.5 L;
   b) between 13.5 L and 16.5 L;
   c) between 9.0 L and 13.5 L;
   d) less than 10.5 L;
   e) more than 16.5 L.

2. A normal distribution of 640 observations has a mean of 1800 cm and a standard deviation of 90 cm.
   a) Draw a representative diagram for the distribution.
   b) Determine the proportion of observations
      i) between 1620 cm and 1980 cm;
      ii) between 1710 cm and 2070 cm;
      iii) above 1890 cm;
      iv) below 2070 cm.
   c) Determine the number of observations
      i) between 1710 cm and 1890 cm;
      ii) between 1530 cm and 1620 cm;
      iii) below 1890 cm;
      iv) above 2070 cm.

SECTION 9.3   The Standardized Normal Curve

A. Standard Normal Deviation — z Scores

The standard deviation σ describes the deviation of the x values from the mean μ. Any such deviation $(x-\mu)$ can be converted into standard units by dividing the deviation by the standard deviation.

For example, if $\mu = 12.0$, $\sigma = 1.5$, and a particular observation x has a value of 15.0, the deviation of x from μ, $(x - \mu) = (15.0 - 12.0) = 3.0$. This deviation is equivalent to $= \dfrac{3.0}{1.5} = +2.0$ standard deviation; that is, the observation $x = 15$ lies two standard deviations above the mean.

Similarly, an observation $x = 7.5$ deviates from the mean by $(x - \mu) = (7.5 - 12.0) = -4.50$. The number of standard deviations in the deviation $= \dfrac{-4.5}{1.5} = -3.0$ that is, the observation lies three standard deviations below the mean.

The location of any value x relative to the mean μ can be described in terms of standard deviations. This is accomplished by dividing the deviation $(x-\mu)$ by the standard deviation σ. This calculation is represented by the formula

$$z = \frac{\text{ACTUAL DEVIATION}}{\text{STANDARD DEVIATION}} = \frac{x - \mu}{\sigma}$$

where x is the value of an observation;
    μ is the population mean;
$(x - \mu)$ is the deviation of an observation x from the mean μ;
    σ is the population standard deviation;
    z is referred to as the z score and represents the number of standard deviations between a selected value x and the mean μ.

○ **EXAMPLE 9.3a**
Given that $\mu = 12.00$, $\sigma = 1.50$, determine and interpret the z score of
a) $x = 15.75$
b) $x = 11.25$
c) $x = 8.20$

● **SOLUTION**
a) For $x = 15.75$, $z = \dfrac{x - \mu}{\sigma} = \dfrac{15.75 - 12.00}{1.50} = \dfrac{3.75}{1.50} = +2.50$.
The z score $+2.5$ indicates that the value $x = 15.75$ lies 2.5 standard deviations above the mean.

b) For $x = 11.25$, $z = \dfrac{x - \mu}{\sigma} = \dfrac{11.25 - 12.00}{1.50} = \dfrac{-0.75}{1.50} = -0.50$.
The z score -0.50 indicates that the value $x = 11.25$ lies 0.50 standard deviations below the mean.

c) For $x = 8.20$, $z = \dfrac{x - \mu}{\sigma} = \dfrac{8.20 - 12.00}{1.50} = \dfrac{-3.80}{1.50} = -2.5333$.

The z score -2.5333 indicates that the value $x = 8.20$ lies 2.5333 standard deviations below the mean.

B. Table of Areas under the Normal Curve

Since a normal distribution is completely specified by the mean μ and the standard deviation σ, a *different* distribution results for each pair of values $\{\mu, \sigma\}$. However, every normal distribution can be transformed to the so-called **standard normal distribution** by changing the scale into a standard scale stated in terms of z scores. In this context, z scores are referred to as **standard normal deviates**.

This transformation is important because the areas under the standardized normal curve have been tabulated and can be applied to *any* normal distribution. Table 9.1 is a table of areas under the standardized normal curve, correct to four decimals, for values of z from 0.00 to 3.09.

The values listed in Table 9.1 (see also the inside front cover) are the probabilities that the random variable x will assume a value between the mean μ and a particular value of z. Entries for negative values of z are the same as for the corresponding positive values because of the symmetry of the normal curve and are not separately listed.

Table 9.1 can be used in two distinct ways:
1. to find areas to the right of, to the left of, or between z values;
2. to find z values for given areas.

C. Finding Areas for Given z Values

For a given z value, the associated table value can be found by looking in the left-hand column headed "z" and locating the row beginning with the first two digits (0.0 to 3.0) of the given z value. The desired table value is one of the 10 numbers listed in that row.

To select the correct table value, look across the top row in the table to locate the column headed by the last digit in the z value (0.00 to 0.09). The desired table value for the given z value is the number located at the point of intersection of the selected row and the selected column.

For example, to locate the table value for $z = 2.75$, locate the row beginning with 2.7 and the column headed by 0.05. The desired table value is 0.4970. This number represents the proportion of the total area between the mean μ and $z = 2.75$. It corresponds to the probability that the random variable x will assume a value in the interval between the μ and $z = 2.75$.

When using Table 9.1 keep the following points in mind:
1. The *total* area under the normal curve is 1.0000.
2. The mean μ (when $z = 0$) divides the total area into halves. The area to the left of μ *equals* the area to the right of μ; that is, each part equals 0.5000.
3. The table look-up for negative values of z is the same as for the corresponding positive values of z.

4. The table values represent the areas between μ and the given values of z. In many cases it will be necessary to add or subtract table values to obtain a specific area. For this reason it is highly recommended that you draw a diagram to identify the specific area under the curve.

TABLE 9.1   Areas under the Normal Curve

z	0.00	0.01	0.02	0.03	0.04	0.05	0.06	0.07	0.08	0.09
0.0	.0000	.0040	.0080	.0120	.0160	.0199	.0239	.0279	.0319	.0359
0.1	.0398	.0438	.0478	.0517	.0557	.0596	.0636	.0675	.0714	.0753
0.2	.0793	.0832	.0871	.0910	.0948	.0987	.1026	.1064	.1103	.1141
0.3	.1179	.1217	.1255	.1293	.1331	.1368	.1406	.1443	.1480	.1517
0.4	.1554	.1591	.1628	.1664	.1700	.1736	.1772	.1808	.1844	.1879
0.5	.1915	.1950	.1985	.2019	.2054	.2088	.2123	.2157	.2190	.2224
0.6	.2257	.2291	.2324	.2357	.2389	.2422	.2454	.2486	.2517	.2549
0.7	.2580	.2611	.2642	.2673	.2704	.2734	.2764	.2794	.2823	.2852
0.8	.2881	.2910	.2939	.2967	.2995	.3023	.3051	.3078	.3106	.3133
0.9	.3159	.3186	.3212	.3238	.3264	.3289	.3315	.3340	.3365	.3389
1.0	.3413	.3438	.3461	.3485	.3508	.3531	.3554	.3577	.3599	.3621
1.1	.3643	.3665	.3686	.3708	.3729	.3749	.3770	.3790	.3810	.3830
1.2	.3849	.3869	.3888	.3907	.3925	.3944	.3962	.3980	.3997	.4015
1.3	.4032	.4049	.4066	.4082	.4099	.4115	.4131	.4147	.4162	.4177
1.4	.4192	.4207	.4222	.4236	.4251	.4265	.4279	.4292	.4306	.4319
1.5	.4332	.4345	.4357	.4370	.4382	.4394	.4406	.4418	.4429	.4441
1.6	.4452	.4463	.4474	.4484	.4495	.4505	.4515	.4525	.4535	.4545
1.7	.4554	.4564	.4573	.4582	.4591	.4599	.4608	.4616	.4625	.4633
1.8	.4641	.4649	.4656	.4664	.4671	.4678	.4686	.4693	.4699	.4706
1.9	.4713	.4719	.4726	.4732	.4738	.4744	.4750	.4756	.4761	.4767
2.0	.4772	.4778	.4783	.4788	.4793	.4798	.4803	.4808	.4812	.4817
2.1	.4821	.4826	.4830	.4834	.4838	.4842	.4846	.4850	.4854	.4857
2.2	.4861	.4864	.4868	.4871	.4875	.4878	.4881	.4884	.4887	.4890
2.3	.4893	.4896	.4898	.4901	.4904	.4906	.4909	.4911	.4913	.4916
2.4	.4918	.4920	.4922	.4925	.4927	.4929	.4931	.4932	.4934	.4936
2.5	.4938	.4940	.4941	.4943	.4945	.4946	.4948	.4949	.4951	.4952
2.6	.4953	.4955	.4956	.4957	.4959	.4960	.4961	.4962	.4963	.4964
2.7	.4965	.4966	.4967	.4968	.4969	.4970	.4971	.4972	.4973	.4974
2.8	.4974	.4975	.4976	.4977	.4977	.4978	.4979	.4979	.4980	.4981
2.9	.4981	.4982	.4982	.4983	.4984	.4984	.4985	.4985	.4986	.4986
3.0	.4987	.4987	.4987	.4988	.4988	.4989	.4989	.4989	.4990	.4990

○ **EXAMPLE 9.3b**

Use Table 9.1 to determine the area under the normal curve
a) to the left of $z = 1.50$;
b) to the right of $z = -2.20$;
c) above $z = 2.50$;
d) below $z = -1.75$;
e) between $z = 1.25$ and $z = 2.96$;
f) between $z = -2.07$ and $z = -1.03$;
g) between $z = -2.33$ and $z = 1.64$;
h) below $z = -2.00$ or above $z = 2.00$.

● **SOLUTION**

a) The area between the mean μ and ($z = 1.50$) is 0.4332. The area below μ is 0.5000. The area to the left of ($z = 1.50$) is $0.5000 + 0.4332 = 0.9332$.

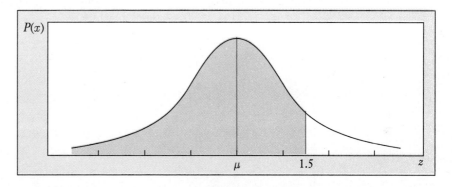

b) The area between μ and ($z = -2.20$) is 0.4861. The area to the right of μ = 0.5000. The area to the right of ($z = -2.20$) is $0.5000 + 0.4861 = 0.9861$.

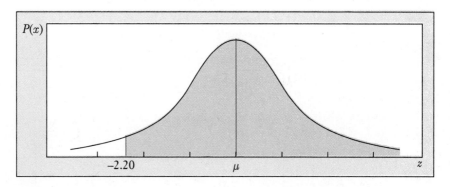

c) The area between μ and ($z = 2.50$) is 0.4938. The area above ($z = 2.50$) is $0.5000 - 0.4938 = 0.0062$.

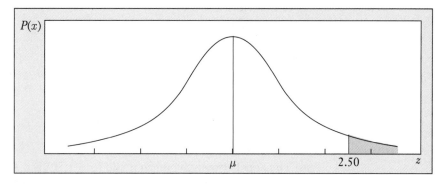

d) The area between μ and ($z = -1.75$) is 0.4599. The area below ($z = -1.75$) is $0.5000 - 0.4599 = 0.0401$.

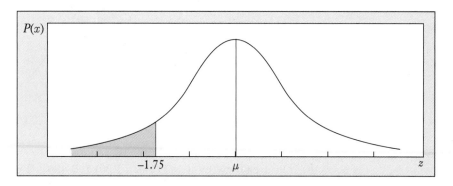

e) The area between μ and ($z = 2.96$) is 0.4985. The area between μ and ($z = 1.25$) is 0.3944. The area between ($z = 1.25$) and ($z = 2.96$) is $0.4985 - 0.3944 = 0.1041$.

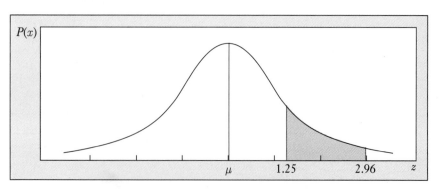

f) The area between μ and ($z = -2.07$) is 0.4808. The area between μ and ($z = -1.03$) is 0.3485. The area between ($z = -2.07$) and ($z = -1.03$) is 0.4808 − 0.3485 = 0.1323.

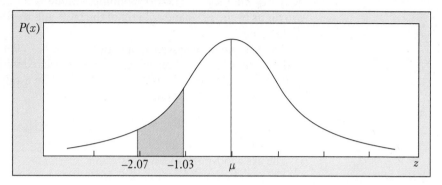

g) The area between μ and ($z = -2.33$) is 0.4901. The area between μ and ($z = 1.64$) is 0.4495. The area between ($z = -2.33$) and ($z = 1.64$) is 0.4901 + 0.4495 = 0.9396.

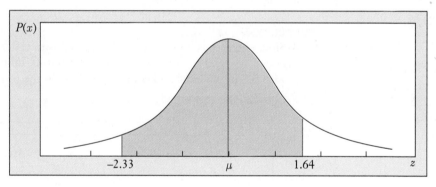

h) The area between μ and ($z = -2.00$) is 0.4772. The area below ($z = -2.00$) is 0.5000 − 0.4772 = 0.0228. The area between μ and ($z = 2.00$) is 0.4772. The area above ($z = 2.00$) is 0.5000 − 0.4772 = 0.0228. The total area below ($z = -2.00$) or above ($z = 2.00$) is 0.0228 + 0.228 = 0.0456.

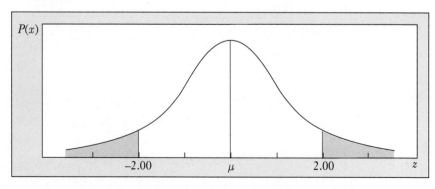

D. *Finding the z Value for a Given Area*

To determine the z value, locate the nearest value to the given area in the body of Table 9.1. The corresponding z value is found by locating the first two digits in the left-hand column of the table and the last digit in the top row of the table.

For example, if the given area between μ and z is 0.4384, the nearest value in the table is 0.4382. This table value is associated with 1.5 in the left-hand column of the table and 0.04 in the top row. The z value corresponding to a table value of 0.4384 is approximately 1.54.

○ **EXAMPLE 9.3c**

Determine the z value for each of the following.
a) The area above the mean μ is 0.2734.
b) The area below the mean μ is 0.3670.
c) The area above z is 0.10.
d) The area below z is 0.05.
e) The area below z is 0.95.
f) The area above z is 0.99.

● **SOLUTION**

a) Since the area is above the mean μ, z is positive. The area 0.2734, located between μ and the desired z value, is listed in Table 9.1 and is associated with 0.7 in the left-hand column and 0.05 in the top row.

Conclusion $z = +0.75$.

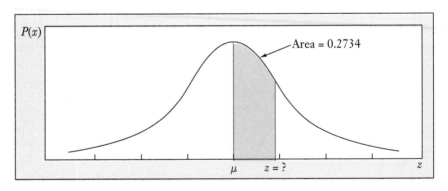

b) Since the area is below the μ, z is negative. The closest value in Table 9.1 to 0.3670 is 0.3665. This value is associated with a z of 1.11.

Conclusion $z = -1.11$.

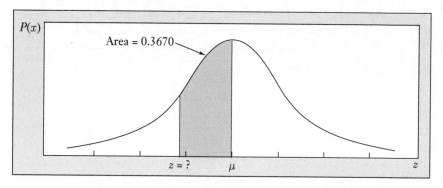

c) Since the given area is less than 0.50 and is above the desired z value, z must be positive. As the table values represent areas between μ and z, the area between μ and z is $0.5000 - 0.1000 = 0.4000$. The table value closest to 0.4000 is 0.3997 and is associated with a z value of 1.28.

Conclusion $z = +1.28$.

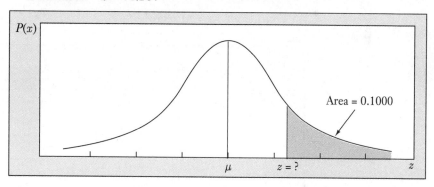

d) The given area is less than 0.5000 and below the desired z value. z is negative. The area between μ and z is $0.5000 - 0.0500 = 0.4500$. The table values closest to 0.4500 are 0.4495 and 0.4505 associated with the z values of 1.64 and 1.65.

Conclusion z is approximately -1.645.

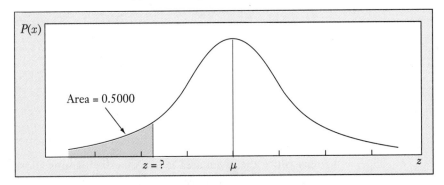

e) The given area is greater than 0.5000 and lies below the desired z value. The given area, made up of the area below the μ plus an area to the right of μ, is $0.9500 - 0.5000 = 0.4500$. z is positive.

Conclusion $z = +1.645$.

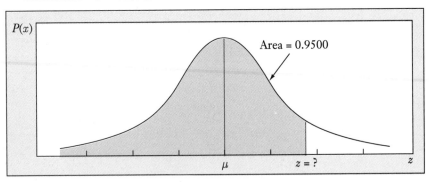

f) The given area is greater than 0.5000 and lies above the desired z value. The given area, made up of the area above μ plus an area to the left of μ, is $0.9900 - 0.5000 = 0.4900$. z is negative. The table value closest to 0.4900 is 0.4901 and is associated with a z value of 2.33.

Conclusion $z = -2.33$.

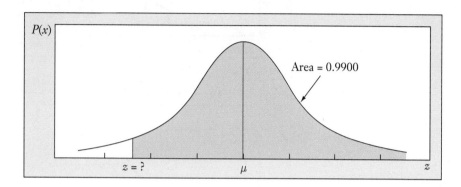

EXERCISE 9.3

1. In a normal distribution,
    a) the mean, median and mode are _____ in value;
    b) the symbol μ represents the _____ ;
    c) the symbol σ represents the _____ ;
    d) the area under the normal curve to the left of μ is _____ .

2. Using a table of areas under the normal curve, determine the area between μ and the following z values:
    a) $z = -1.30$ b) $z = 1.20$
    c) $z = 2.57$ d) $z = 1.0$
    e) $z = -0.55$ f) $z = -2.0$
    g) $z = -3.0$ h) $z = -1.0$

3. For a normal distribution with $\mu = 12.0$ and $\sigma = 3.00$, calculate z for the following values of x:
    a) $x = 9.0$ b) $x = 7.5$
    c) $x = 18.0$ d) $x = 12.3$
    e) $x = 20.5$ f) $x = 4.8$

4. Using a z-value table, determine the probability for each of the following:
    a) $-1.0 < z < 1.0$ b) $-2.0 < z < 1.0$
    c) $2.0 < z < 3.0$ d) $-2.0 < z < 2.0$
    e) $-3 < z < -2$ f) $-3 < z < 3$
    g) $-3 < z < -1$ h) $1 < z < 2$

5. Find the following shaded areas:

a)

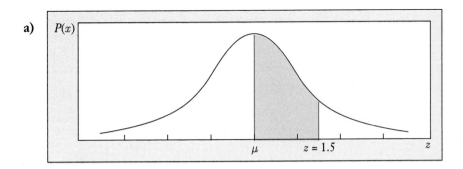

b)

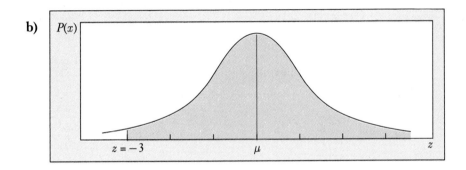

c)

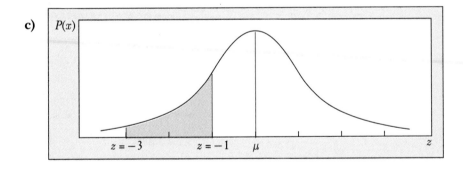

d)

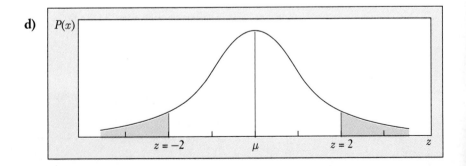

e)

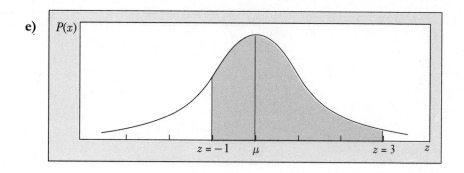

f)

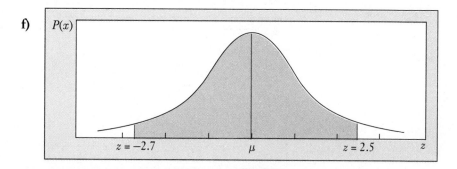

6. Find the z values associated with the following areas:

a)

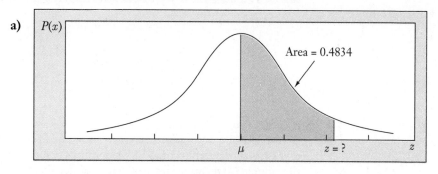

b)

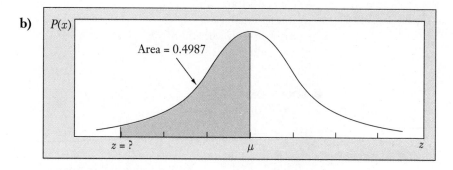

c)

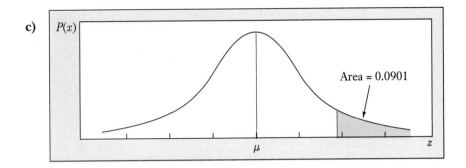

d)

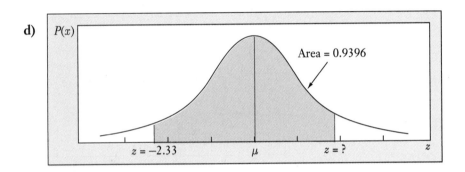

Applications of the Normal Curve

○ **EXAMPLE 9.4a**

A distributor's accounts receivable show an average balance of $200 with a standard deviation of $50. Assuming that the balances are normally distributed, determine the proportion of the accounts that
a) exceed $300;
b) are less than $50;
c) are between $50 and $150;
d) are between $125 and $350.

● **SOLUTION**

$\mu = 200; \sigma = 50$.

a) $z = \dfrac{x - \mu}{\sigma} = \dfrac{300 - 200}{50} = 2.00$. The area between μ and $(z = 2.00)$ is
0.4772. The required area $= 0.5000 - 0.4772 = 0.0228$.
2.28% of the accounts are expected to exceed $300.

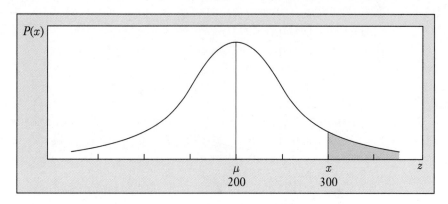

b) $z = \dfrac{50 - 200}{50} = -3.00$. The area between μ and z is 0.4987. The required
area $= 0.5000 - 0.4987 = 0.0013$.

0.13% of the accounts are expected to be less than $50.

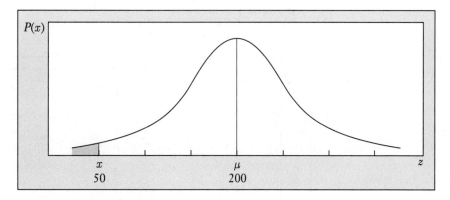

c) For x_1, $z = \dfrac{50 - 200}{50} = -3.00$. The area between μ and z is 0.4987. For
x_2, $z = \dfrac{150 - 200}{50} = -1.00$. The area between μ and z is 0.3413. The
required area $= 0.4987 - 0.3413 = 0.1574$.

15.74% of the accounts are expected to be between $50 and $150.

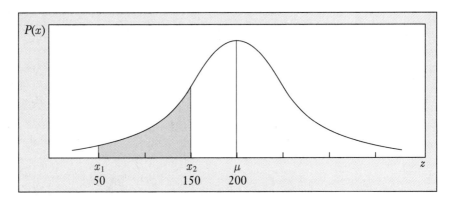

d) For x_1, $z = \dfrac{125 - 200}{50} = -1.50$. The area between μ and z is 0.4332. For x_2, $z = \dfrac{350 - 200}{50} = 3.00$. The area between μ and ($z = 3.00$) is 0.4987;

The required area $= 0.4332 + 0.4987 = 0.9319$.

93.19% of the accounts are expected to be between \$125 and \$350.

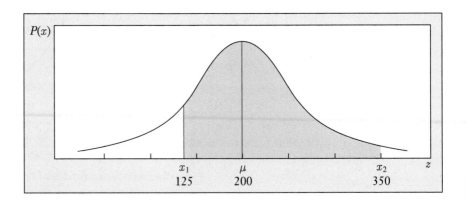

○ EXAMPLE 9.4b

The average price of a house in a certain area is \$500 000 with a standard deviation of \$50 000. Given that the distribution of the prices is normally distributed,

a) determine the proportion of houses that will sell for less than \$450 000;

b) compute the limits within which the middle 50% of the house prices will be found.

● SOLUTION

a) For $x = 450\ 000$, $z = \dfrac{450\ 000 - 500\ 000}{50\ 000} = -1.00$. The required area is $0.5000 - 0.3413 = 0.1587$.

15.87% of the houses in the area are expected to sell for less than \$450 000.

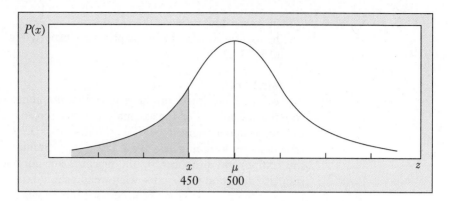

b) The middle 50% implies that 25% of the total area lies on either side of the mean μ. Let x_1 and x_2 represent the lower and upper limits respectively. The area between μ and x_2 is 0.2500. The closest value in Table 9.1 is 0.2486. The corresponding z value = 0.67.

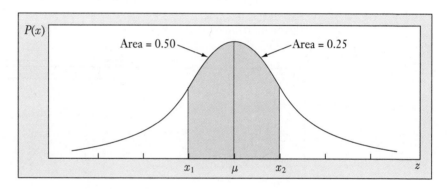

Substituting in the formula $z = \dfrac{x - \mu}{\sigma}$, we obtain for the upper limit x_2

$$0.67 = \frac{x - 500\ 000}{50\ 000}$$

$$0.67(50\ 000) = x - 500\ 000$$

$$33\ 500 = x - 500\ 000$$

$$x = 533\ 500$$

The upper limit is $533 500.

For the lower limit, $z = -0.67$. Substituting in the formula

$$-0.67 = \frac{x - 500\ 000}{50\ 000}$$

$$(-0.67)(50\ 000) = x - 500\ 000$$

$$-33\ 500 = x - 500\ 000$$

$$x = 466\ 500$$

The lower limit is $466 500.
The middle 50% of the house prices are expected to lie between $466 500 and $533 500.

○ **EXAMPLE 9.4c**

A plastics manufacturer produces injection mouldings of a component for a complex assembly. For the assembly to work properly the minimum weight of the component must be 500 g. Thus, it is important to keep rejects (components weighing less than 500 g) to a minimum. If the weights of the components produced by the machine are normally distributed with a standard deviation of 10 g, for what weight should the machine be set so that no more than five percent of the components produced will be rejected?

● **SOLUTION**

The desired weight is the mean weight μ.
The minimum acceptable weight $x = 500$ g.
The acceptable proportion of components with a weight less than 500 g = 0.05;
the proportion of components between μ and the minimum acceptable weight = 0.45.

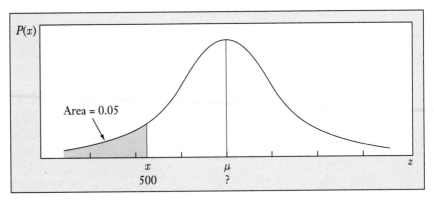

The closest values in Table 9.1 are 0.4495 and 0.4505.
The associated z values are 1.64 and 1.65. Since 0.45 is halfway between the two table values, a more precise value of $z = 1.645$.
Since $x = 500$ is located below the μ, $z = -1.645$.
The value of μ can be obtained by substituting in the formula

$$z = \frac{x - \mu}{\sigma}$$

$$-1.645 = \frac{500 - \mu}{10}$$

$$10(-1.645) = 500 - \mu$$

$$-16.45 = 500 - \mu$$

$$\mu = 516.45$$

The moulding machine should be set so that the mean weight of the components produced is 516.45 g.

○ **EXAMPLE 9.4d**

The plastics manufacturer has been told that if the machine is modified, the standard deviation can be reduced by 25% at a cost of $25 000. The cost of the plastic resin used in manufacturing the component is $2.00 per kg.

a) How many components would have to be produced to pay for the modification of the machine?

b) If the company manufactures two million components annually and requires a payback within two years, should management authorize the modification?

● **SOLUTION**

a) The new standard deviation = 75% of 10 = 7.5 g.
$z = -1.645$; $x = 500$.
The new mean setting point for the machine would be

$$-1.645 = \frac{500 - \mu}{7.5}$$

$$-1.645(7.5) = 500 - \mu$$

$$-12.3375 = 500 - \mu$$

$$\mu = 512.34$$

The average amount of resin saved per component = 516.45 − 512.34 = 4.11 g.

The cost of resin per gram $\dfrac{\$2.00}{1000}$ = $0.002.

The average saving per component = 4.11(0.002) = $0.00822.

The number of components required to pay back the cost of modification

$$= \frac{25\ 000}{0.00822} = 3\ 041\ 363.$$

b) The number of years required to produce 3 041 363 components

$$\frac{2\ 000\ 000}{3\ 041\ 363} = 1.52.$$

Since 1.52 years is within the company guidelines for a payback period, management should authorize the modification of the machine.

EXERCISE 9.4

1. A normal distribution of 2400 observations has a mean of $54.00 and a standard deviation of $6.00. Complete each of the following:

a) The value of -3σ is $_____.

b) The value of 3σ is $_____.

c) Since _____ percent of the observations lie between 3σ and -3σ, _____ observations lie between $_____ and $_____.

d) Since the value of 2σ is \$_____ and _____ percent of all observations lie below 2σ, _____ observations have a value less than \$_____ .

2. The lifespan of a particular brand of light bulb is normally distributed with a mean life of 400 hours and a standard deviation of 12 hours. Production for last week was 800 000 light bulbs. Use this information to answer the following questions.
   a) What is the z value of a light bulb having a life of 421 hours?
   b) How many light bulbs can be expected to have a life of less than 379 hours?
   c) What percent of the light bulbs should have a life less than 430 hours?
   d) The life of the most short-lived five percent of the light bulbs will likely be less than how many hours?
   e) What percent of the light bulbs can be expected to have a life between 385 hours and 415 hours?
   f) How many of the light bulbs can be expected to have a life between 409 hours and 427 hours?
   g) How many of the light bulbs can be expected to have a life of more than 382 hours?

3. For a normal distribution with $\mu = 3.5$ and $\sigma = 0.2$, find x if the area
   a) between μ and $z = 0.4980$;
   b) to the left of $z = 0.9505$;
   c) to the right of $z = 0.0041$;
   d) to the right of $z = 0.6808$;
   e) to the left of $z = 0.0778$;
   f) between $-z$ and $+z = 0.9544$.

4. Eight hundred students registered in a sports injury program. Of the registrants, 60 were invited for an interview based on the mark obtained in an entrance examination. Of those interviewed, 32 were admitted to the program. The examination marks were normally distributed with a mean of 85 and a standard deviation of 4.6.
   a) What mark did an applicant have to achieve to receive an interview?
   b) What percent of the applicants were rejected based on the entrance examination?
   c) What percent of all applicants were admitted to the program?

SECTION 9.5 Using the Normal Distribution to Approximate the Binomial Distribution

When dealing with the binomial distribution we noted that its shape is symmetrical for all values of n if $p = 0.50$. When $p \neq 0.50$ the shape of a binomial distribution becomes more and more symmetrical as n increases and gets closer and closer to a normal distribution. When $n > 30$ the binomial distribution approximates the normal distribution very closely unless p is either very large (close to 1.00) or very small (close to 0.00).

In general, the normal distribution can be used as a good approximation of a binomial distribution when np and $n(1-p)$ are both greater than *five*.

In addition to the above conditions for using the normal distribution to approximate the binomial distribution, allowance should be made for the discrete nature of the binomial distribution. This involves the use of the so-called *continuity correction factor*. In some cases this involves adding one-half of a unit to the x values in the normal distribution. In other cases, half a unit should be subtracted.

However, since in most practical applications the increased accuracy is not significant and becomes minimal for larger values of n, this correction factor will not be considered in this text in order to avoid the increased computational complexity in solving problems.

○ **EXAMPLE 9.5a**

A public-opinion poll found that 54% of women in a certain group read newspapers. Calculate the probability that in a group of 100 such women,
a) at least 50 read newspapers;
b) no more than 46 read newspapers;
c) between 45 and 65 read newspapers;
d) between 30 and 50 read newspapers;
e) exactly 54 women read newspapers.

● **SOLUTION**

Since the poll of any member in the group has just two possible outcomes, "read" or "not read," and the number of successes "read" is the result of counting, the distribution can be accepted to be binomial.

$n = 100$; $p = 0.54$; $(1-p) = 0.46$;
$np = 100(0.54) = 54$; $n(1-p) = 100(0.46) = 46$.

Since both np and $n(1-p)$ are greater than five, the normal distribution can be used to approximate the binomial distribution.

$$\mu = np = 100(0.54) = 54$$
$$\sigma^2 = np(1-p) = 100(0.54)(0.46) = 24.84$$
$$\sigma = \sqrt{24.84} = 4.9840$$

a) The requirement of "at least 50 readers" covers the range 50 to 100. Let $x = 50$. The value $x = 50$ is less than μ. The probability $P(\geq 50)$ is given by the area to the right of $x = 50$, as shown in the diagram.

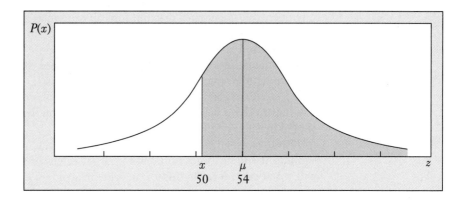

$$z = \frac{x - \mu}{\sigma} = \frac{50 - 54}{4.9840} = \frac{-4}{4.9840} = -0.80$$

The area between μ and (z = 0.80) is 0.2881. The required area is 0.2881 + 0.5000 = 0.7881. The probability that at least 50 of the women will read newspapers is approximately 79%.

b) "No more than 46" covers the range 0 to 46. Let x = 46. The value x = 46 is smaller than μ and lies to the left of μ. The probability $P(\leq 46)$ is given by the area to the left of x = 46 as shown in the diagram.

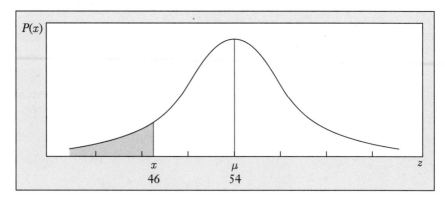

$$z = \frac{46 - 54}{4.9840} = \frac{-8}{4.9840} = -1.61$$

The area between μ and z = 1.61 is 0.4463. The required area is 0.5000 − 0.4463 = 0.0537. The probability that no more than 46 of the women read newspapers is approximately 5%.

c) "Between 45 and 65" covers the range 45 to 65. Let x_1 = 45 and x_2 = 65. The value x = 45 lies below μ, while the value x = 65 lies above μ. The probability $P(45 \leq x \leq 64)$ is given by the area between x_1 = 45 and x_2 = 65, as shown in the diagram.

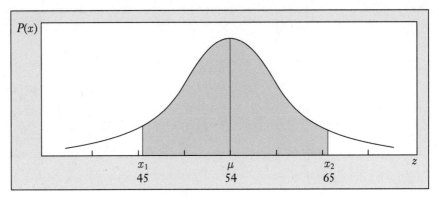

The z value for $x_1 = \dfrac{45 - 54}{4.9840} = \dfrac{-9}{4.9840} = 1.81$.

The z value for $x_2 = \dfrac{65 - 54}{4.9840} = \dfrac{11}{4.9880} = 2.21$.

The area between μ and x_1 is 0.4649. The area between μ and x_2 is 0.4864. The required area is $0.4649 + 0.4864 = 0.9513$. The probability that between 45 and 65 of the women read newspapers is approximately 95%.

d) "Between 30 and 50" covers the range 30 to 60. Let $x_1 = 30$ and $x_2 = 50$. Both values are smaller than μ and lie below μ. The probability $P(30 \leq x \leq 50)$ is given by the area between x_1 and x_2, as shown in the diagram.

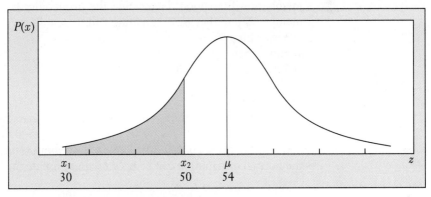

The z value for $x_1 = \dfrac{30 - 54}{4.9840} = \dfrac{-24}{4.9840} = -4.81$.

The z value for $x_2 = \dfrac{50 - 54}{4.9840} = \dfrac{-4}{4.9840} = -0.80$.

The z value -4.81 exceeds the range covered by Table 9.1. The greatest value in the table is 0.4990 for $z = 3.09$.

The area between μ and x_1 is 0.4999. The area between μ and x_2 is 0.2881. The required area is $0.4999 - 0.2881 = 0.2118$. The probability that between 30 and 50 women read newspapers is approximately 21%.

e) For "exactly 54" we need to create an interval. This is done by adding and subtracting 0.5 from 54. This means that $x_1 = 54.5$ and $x_2 = 53.5$.

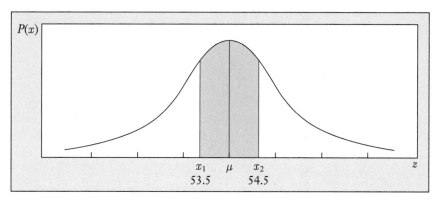

For x_1, $z = \dfrac{54.5 - 54}{4.9840} = 0.10$.

For x_2, $z = -0.10$.

The area between μ and x_1 is 0.0398, as is the area between μ and x_2. The required area $= 2(0.0398) = 0.0796$. The probability that exactly 54 of the women read newspapers is approximately 8%.

○ **EXAMPLE 9.5b**

A professional organization mails a salary survey questionnaire to its 2000 members every year and knows from experience that about 20% of the members will respond. Determine the probability that the response by the membership to the current mailing will be as follows:

a) fewer than 375;

b) more than 440;

c) between 375 and 440.

● **SOLUTION**

The given information is consistent with the characteristics of a binomial distribution for which $n = 2000$, $p = 0.20$, $1 - p = 0.80$. $np = 2000(0.20) = 400$ and $(1-p) = 2000(0.80) = 1600$. Since both values are greater than five, the normal distribution can be used to approximate the binomial distribution.

$$\mu = np = 2000(0.20) = 400;$$
$$\sigma^2 = np(1 - p) = 2000(0.20)(0.80) = 320;$$
$$\sigma = \sqrt{320} = 17.89.$$

a) Let $x = 375$.

$$z = \frac{375 - 400}{17.89} = \frac{-25}{17.89} = -1.40$$

The area between μ and x is 0.4192. The required area $= 0.5000 - 0.4192 = 0.0808$. The probability that fewer than 375 members will respond is approximately 8.08%.

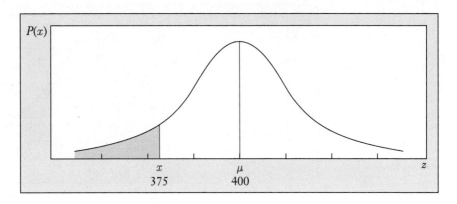

b) Let $x = 440$.

$$z = \frac{440 - 400}{17.89} = \frac{40}{17.89} = 2.24$$

The area between μ and x is 0.4875. The required area = 0.5000 − 0.4875 = 0.125. The probability that more than 440 members will respond is approximately 1.25%.

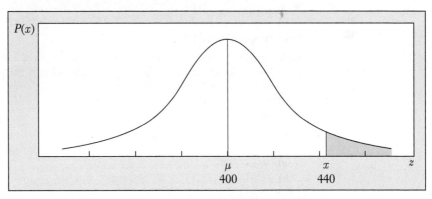

c) Let $x_1 = 375$ and $x_2 = 440$.

For x_1, $\quad z = \dfrac{375 - 400}{17.89} = -1.40$.

For x_2, $\quad z = \dfrac{440 - 400}{17.89} = 2.24$.

The area between μ and x_1 is 0.4192 and the area between μ and x_2 is 0.4875. The required area = (0.4192 + 0.4875) = 0.9067. The probability that between 375 and 440 members will respond is approximately 90.67%.

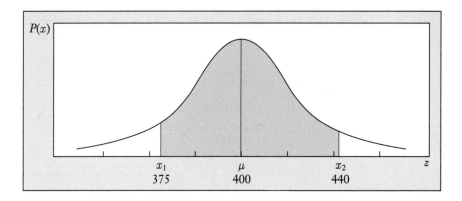

Note   The three problems cover the total area. The sum of the probabilities = $(0.0808 + 0.0125 + 0.9067) = 1.0000$.

EXERCISE 9.5

1. The senior sales agent for Randy's Real Estate Company closes a deal 65% of the time. If the manager believes that the agent will see 60 buyers next month, what is the probability that the agent's sales will be
   a) fewer than 30?
   b) more than 45?
   c) between 38 and 42?

2. A computer designed to understand spoken language is 60% accurate. If an 800-word letter is dictated to the computer, what is the probability that
   a) more than 500 words will be correct?
   b) between 450 and 520 words will be correct?
   c) fewer than 400 words will be correct?

3. Given that $n = 400$, $p = 80\%$, use the normal curve to calculate the following probabilities:
   a) $P(x \leq 300)$
   b) $P(x \geq 328)$
   c) $P(300 \leq x \leq 305)$

4. Given that $n = 3000$, $p = 10\%$, use the normal distribution to compute the following probabilities.
   a) $P(x \geq 260)$
   b) $P(320 \leq x \leq 350)$
   c) $P(x \leq 275)$

REVIEW EXERCISE

1. Look up the following values in a z table:
   a) $z = 0.65$

   b) $z = -1.82$
   c) $z = -3.04$
   d) $z = 2.33$

2. Tony was asked to look up the table value corresponding to $z = -2.01$. He said the value is 0.4821. Tina disagrees. If Tina is right, what z value did Tony look up?

3. Using a table of values under the normal curve, determine the areas for the following:
   a) $1.57 \leq z \leq 3.02$
   b) $z \leq -1.63$
   c) $-2.83 \leq z \leq -0.38$
   d) $z \geq 2.34$
   e) $-1.64 \leq z \leq 1.64$

4. What z values when doubled will also double the area (by using the table)?

5. For the following areas under the normal curve, determine the corresponding z values:
   a) the area to the right of z is 0.01;
   b) the area above z is 0.95;
   c) the area below z is 0.02;
   d) the area to the left of z is 0.90;
   e) the area between $-z$ and z is 0.95;
   f) the area between $-z$ and z is 0.98.

6. Determine the z value for the following probabilities;
   a) $P(\mu \leq x \leq z) = 0.4505$
   b) $P(-z \leq x \leq \mu) = 0.4901$
   c) $P(x \leq z) = 0.9207$
   d) $P(x \geq -z) = 0.9986$
   e) $P(x \leq -z) = 0.0322$
   f) $P(x \geq z) = 0.0401$

7. For a normal distribution, where $\mu = 70$ and $\sigma = 6$, calculate z for the following values of x:
   (a)73   **(b)** 69   **(c)** 60   **(d)** 85   **(e)** 90   **(f)** 52

8. Given $\sigma = 0.4$ and $z = -2.5$, determine the mean for $x = 6.8$.

9. The average family income in the city of Tortola is \$45 000, with a standard deviation of \$9000. Assuming that family income is normally distributed, determine the income level below which 80% of the families in Tortola live.

10. Sheridan College's Law and Enforcement Program received 1200 applications for 30 openings. The grades of the applicants were normally distributed with a mean of 78 and a standard deviation of 6. If applicants are chosen on the basis of grades, what will be the cutoff grade to fill the 30 openings?

11. Normally distributed observations such as a person's height or shoe size occur quite frequently in nature. Business people who are aware of this use it to their advantage. A purchasing agent for a large retailer buying 10 000 pairs of men's shoes uses the normal curve to decide on the order quantities for the various sizes. If men's average shoe size is 9 with a standard deviation of 1.5, what quantity should be ordered between sizes 7 and 11?

12. Amazed at the large choice of mutual funds, an investor used the published

five-year performance reports in selecting a mutual fund. The investor determined that the return on investment for 120 funds over the five-year period was normally distributed with a mean of 32% and a standard deviation of 8.2%. Further study of the funds indicated that funds in the top decile were high-risk investments. To narrow the choice, the investor decided to consider only those funds between the 80th and 90th percentile.

a) Determine the returns on investment that correspond to the 80th and the 90th percentile.

b) What was the number of funds from which the investor had to make a choice?

13. A survey of cars in a mall parking lot found that 105 of 420 cars were imports. If a random sample of 80 of the drivers is taken, what is the probability that 26 or fewer of the drivers have imported cars?

14. Tabulation of a strike vote showed that 55% of those voting cast their ballot in favour of strike action. If 50 voters are randomly selected, what is the probability that at least 24 of them voted in favour of strike action?

SELF-TEST

Questions **1** through **7** are based on the following sales invoice data:

	Company		
Statistical measure	AB	LM	ST
Arithmetic mean	480	525	400
Median	490	509	400
Mode	510	477	400
Standard deviation	40	50	32
Mean deviation	32	40	26
Quartile deviation	25	32	20

1. Which company's invoice amounts are normally distributed?

2. Estimate the range of the invoice amounts of company ST.

3. Above what sales amount do 75% of the sales invoices of company LM fall?

4. Outside what amounts do a total of 0.25% of the sales invoices of company ST fall?

5. Between what amounts do 50% of the sales invoices of company AB fall?

6. If the statistical measures for company ST are based on 600 invoices, how many show amounts between $350 and $450?

7. Determine the probability that a sales invoice of company ST will be for at least $495?

8. The distribution of order sizes for Macrae Company is normal, with a mean of $1500 and a standard deviation of $225.

a) What is the z value for an order of $1700?

    b) What proportion of the orders can be expected to be less than $1200?

    c) Out of 600 orders, how many are expected to fall between $1600 and $2000?

9. Phone calls to the customer service department of Nichol's Equipment Company concern warranty service 60% of the time. If 200 phone calls are taken, what is the probability

    a) that fewer than 100 phone calls were for warranty work?

    b) that fewer than 150 phone calls were for warranty work?

Key Terms

Normal curve   231
Normal distribution   230
Standard normal deviate   237
Standard normal distribution   237

Summary of Formulas

Standardized normal variable:

$$z = \frac{x - \mu}{\sigma}$$

CHAPTER 10

Sampling Distributions

Introduction

In most situations it is either physically or economically impractical to examine every item in a population. In order to obtain information about such populations a technique called sampling is often used.

Objectives

Upon completion of this chapter you will be able to
1. understand the importance of sampling and the main reasons for sampling;
2. distinguish between a population and a sample;
3. understand the relationship between the characteristics of a population (parameters) and of a sample (statistics) drawn from the population;
4. know the symbols used to differentiate between population parameter and sample statistics;
5. know the properties of the sampling distribution of the means and of the sampling distribution of proportions;
6. use the properties to solve problems.

SECTION 10.1 Sampling Considerations

A. The Importance of Sampling

Sampling involves selecting a portion of the population that is most representative of the characteristics of the population. The purpose of sampling is to provide sufficient information so that conclusions can be drawn about the characteristics of the population.

B. Reasons for Sampling

1. *Physical constraints.* It is often physically impossible to enumerate all items in a population.
2. *Time constraints.* Even if a census is physically possible, it is usually too time-consuming to be practical.
3. *Cost constraints.* It costs money to collect information. Obtaining information from a small portion of a population is less costly than taking a complete census of the population.
4. *Test constraints.* In many cases of industrial quality control, the item tested is destroyed during the test. In such cases, sampling is the only practical means of testing.

C. Sampling Methods

A variety of sampling methods are available. The different methods can be categorized under two main headings:
1. *Probability sampling.* In these methods, items are selected at random from the population and the probability of selecting the items is known.
2. *Non-probability sampling.* In these methods, items are selected from the population according to the judgment or purpose of the researcher.
   No one method of sampling is best. The appropriate method is determined by the nature of the population, the skill of the researcher and the economics of collecting data.
   A detailed examination of sampling surveys and sampling methods is beyond the scope of this text.

D. Sampling Concepts

1. *Population versus sample.* By **population** we mean the total of *all items* in the group of items in which we are interested. By **sample** we mean a *portion* of the population that has been selected for study.
   For example, if we are interested in the distribution of statistics marks for all 35 students registered in a statistics class, the population consists of the 35 statistics marks. If 10 marks are selected from the 35 marks, the 10 marks represent a sample.
2. *Finite population versus infinite population.* A **finite population** is a population that consists of a limited and specifically known number of items.
   An **infinite population** is a population for which the number of items is unlimited or for which the number of items is not specifically known. For practical purposes, a population consisting of a large number of items can be considered to be an infinite population.
   For example, a class known to consist of 35 students is a finite population, whereas the total number of trees in Canadian forests is considered to be an infinite population.
3. *Parameter versus statistic.* A **parameter** is a measure describing a *characteristic of a population*, such as the population mean and the population standard

deviation. A **statistic** is a measure describing a *characteristic of a sample*, such as the sample mean and the sample standard deviation.

For example, if the average statistics mark for a class of 35 students is known to be 70, the mean of 70 is a parameter. If the average mark for a sample of 10 students taken from the class is also 70, the mean of 70 is a statistic.

E. Summary of Distinctions between Population and Sample

In dealing with populations and samples we must keep in mind the difference in meaning of the terms parameter and statistic and the symbols used for both.

For population parameters we use Greek letter symbols such as μ, σ and π. For sample statistics we use Roman letters such as $\bar{x}$, s and p.

TABLE 10.1   Population versus Sample

Area of distinction	Population	Sample
Definition	Consists of all items in a group.	Consists of a selected portion of the group.
Characteristics	Are called parameters.	Are called statistics.
Symbols used	Greek letters	Roman letters
Mean	μ	$\bar{x}$
Standard deviation	σ	s
Proportion	π	p

F. Statistical Inference

If a sample is representative of the population, the sample results will reasonably mirror the population from which the sample was taken and can be used to estimate population characteristics.

The sample mean $\bar{x}$ is used to estimate the population mean μ. The sample standard deviation s is used to estimate the population standard deviation σ. The sample proportion p is used to estimate the population proportion π.

SECTION 10.2   The Sampling Distribution of the Means

A. Sampling Variation and the Sampling Distribution of the Means

When selecting a sample from a population, we can expect that the sample mean $\bar{x}$ will differ from the population mean μ. Furthermore, when selecting samples of a given size from a population, we can expect different samples to have different means. This variation in the possible sample means is referred to as **sampling variation**.

If all possible samples of a given size *n* are obtained from a population and their sample means are computed, the *listing of the computed means* forms a distribution referred to as the **sampling distribution of the means**.

B. Sampling Distribution of the Means Illustrated

○ **EXAMPLE 10.2a**

A group of four students had weekly study times as shown below:

Student	A	B	C	D
Hours per week	4	6	8	10

Construct the sampling distribution of the means for samples of size $n = 2$.

● **SOLUTION**

The sampling distribution of the means is obtained by listing all possible samples of size $n = 2$, and computing the mean of each sample as shown in Table 10.2 below.

The listing of the values obtained as the means of all the possible samples represents the sampling distribution of the means, illustrated in Figure 10.1.

TABLE 10.2 Sampling Distribution of the Means

Possible samples of size, $n = 2$		Calculation of sample means	Sampling distribution of the means $n = 2$
Students	Hours		
A, B	4, 6	$\dfrac{4 + 6}{2}$	5
A, C	4, 8	$\dfrac{4 + 8}{2}$	6
A, D	4, 10	$\dfrac{4 + 10}{2}$	7
B, C	6, 8	$\dfrac{6 + 8}{2}$	7
B, D	6, 10	$\dfrac{6 + 10}{2}$	8
C, D	8, 10	$\dfrac{8 + 10}{2}$	9

FIGURE 10.1   Graph of the Sampling Distribution of the Means

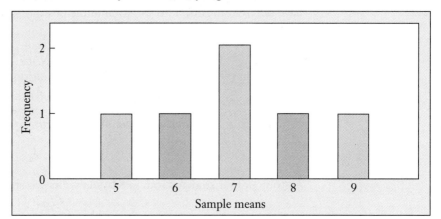

C. The Mean of the Sampling Distribution of the Means

The *mean of the sampling distribution of the means*, denoted by the symbol $\mu_{\bar{x}}$, is given by

$$\mu_{\bar{x}} = \frac{\text{SUM OF ALL POSSIBLE SAMPLE MEANS FOR THE GIVEN SAMPLE SIZE}}{\text{NUMBER OF POSSIBLE SAMPLES OF SIZE } n}$$

In our case the number of samples = 6 and their means are as listed in Table 10.2.

$$\mu_{\bar{x}} = \frac{5 + 6 + 7 + 7 + 8 + 9}{6} = \frac{42}{6} = 7$$

A useful relationship between the mean of the sampling distribution of the means $\mu_{\bar{x}}$ and the population mean μ can be obtained by comparing $\mu_{\bar{x}}$ and μ:

The population mean, $\mu = \dfrac{4 + 6 + 8 + 10}{4} = \dfrac{28}{4} = 7$.

The implication is that $\mu_{\bar{x}} = \mu$; that is, the mean of the sampling distribution of the means equals the population mean.

D. Standard Deviation of the Sampling Distribution of the Means

While $\mu_{\bar{x}} = \mu$, the individual sample means $\bar{x}$ differ from $\mu_{\bar{x}}$ and therefore from μ. To determine the extent to which a sample mean $\bar{x}$ can be expected to differ from the population mean μ, a measure of dispersion known as the *standard deviation of the sampling distribution of the means* can be computed.

This standard deviation is denoted by $\sigma_{\bar{x}}$ and is commonly referred to as the **standard error**. The computation of the standard error is shown in Table 10.3.

TABLE 10.3 **Computation of the Standard Deviation of the Sampling Distribution of the Means**

Sample	Sample mean $\bar{x}$	Deviation of the sample mean from the population mean, $(\bar{x} - \mu)$	Squared deviation $(\bar{x} - \mu)^2$
A, B	5	$5 - 7 = -2$	4
A, C	6	$6 - 7 = -1$	1
A, D	7	$7 - 7 = 0$	0
B, C	7	$7 - 7 = 0$	0
B, D	8	$8 - 7 = 1$	1
C, D	9	$9 - 7 = 2$	4
		$\sum(\bar{x} - \mu)^2 = 10$	

$$\text{VARIANCE} = \frac{\sum(\bar{x} - \mu)^2}{\text{NUMBER OF SAMPLES}} = \frac{10}{6} = \frac{5}{3}$$

The standard error, $\sigma_{\bar{x}} = \sqrt{\dfrac{5}{3}}$

E. *Relationship between the Standard Error $\sigma_{\bar{x}}$ and the Population Standard Deviation σ*

The standard deviation of the sampling distribution of the means — that is, the standard error — can be obtained without the computations shown in Table 10.3 from the population standard deviation σ by means of the following formulas:

$$\sigma_{\bar{x}} = \frac{\sigma}{\sqrt{n}}, \text{ for infinite populations (N unknown)}$$

and

$$\sigma_{\bar{x}} = \frac{\sigma}{\sqrt{n}}\sqrt{\frac{N - n}{N - 1}}, \text{ for finite populations}$$

where N is the size of the population;

n is the sample size;

$\sqrt{\dfrac{N - n}{N - 1}}$ is called the **finite correction factor**.

In our example, the variance σ^2 and the population standard deviation σ are readily obtainable by computing the individual squared deviations, as shown in Table 10.4.

TABLE 10.4   Computation of Variance

Student	Deviation $(x - \mu)$	Squared deviation $(x - \mu)^2$
A	$4 - 7 = -3$	9
B	$6 - 7 = -1$	1
C	$8 - 7 = 1$	1
D	$10 - 7 = 3$	9
		$\sum(x - \mu)^2 = 20$

$$\sigma^2 = \frac{\sum(x - \mu)^2}{N} = \frac{20}{4} = 5; \quad \sigma = \sqrt{5}$$

Since N is known, the population is finite. The appropriate formula is

$$\sigma_{\bar{x}} = \frac{\sigma}{\sqrt{n}}\sqrt{\frac{N - n}{N - 1}}, \quad \text{where} \quad \sigma = \sqrt{5}, \quad N = 4, \quad n = 2.$$

$$\sigma_{\bar{x}} = \frac{\sqrt{5}}{\sqrt{2}}\sqrt{\frac{4 - 2}{4 - 1}} = \frac{\sqrt{5}}{\sqrt{2}}\frac{\sqrt{2}}{\sqrt{3}} = \frac{(\sqrt{5})(\sqrt{2})}{(\sqrt{2})(\sqrt{3})} = \frac{\sqrt{5}}{\sqrt{3}} = \sqrt{\frac{5}{3}}$$

EXERCISE 10.2

1. The test scores for a group of students are given below:

Student	A	B	C	D	E
Test score	4	5	5	7	9

   a) Determine the population mean and the population standard deviation.
   b) Construct the sampling distribution of the means for samples of size, $n = 3$.
   c) Compute the mean of the sampling distribution of the means and establish that $\mu_{\bar{x}} = \mu$.
   d) Compute the standard error from the basic data.
   e) Use the appropriate formula to verify the result in (d).

2. The hourly earnings of a group of employees are listed below:

Employee	A	B	C	D	E	F
Hourly earnings	$8	$9	$6	$8	$9	$5

   a) Compute the population mean and the population standard deviation.
   b) Construct the sampling distribution of the means for samples of size, $n = 3$.
   c) Determine $\mu_{\bar{x}}$ and establish that $\mu_{\bar{x}} = \mu$.
   d) Compute $\sigma_{\bar{x}}$ from the basic data.
   e) Use the appropriate formula to verify the result in (d).

SECTION 10.3 ## Properties of the Sampling Distribution of the Means

The sampling distribution of the means has certain properties that are useful in drawing conclusions about populations from sample information. These properties are summarized below.

1. The mean of the sampling distribution of the means is equal to the population mean, $\mu_{\bar{x}} = \mu$.
2. The standard deviation of the sampling distribution of the means (referred to as the standard error) is given by

$$\sigma_{\bar{x}} = \frac{\sigma}{\sqrt{n}} \qquad \text{for infinite populations}$$

and

$$\sigma_{\bar{x}} = \frac{\sigma}{\sqrt{n}} \sqrt{\frac{N-n}{N-1}} \qquad \text{for finite populations.}$$

3. If the sample size $n > 30$, the sampling distribution of the means will approximate the normal distribution.
4. If the population is normally distributed, the sampling distribution of the means will be normal, regardless of sample size.

The following useful implications are based on properties 3 and 4:

a) There is a 68.26% chance that any sample mean will fall between $\mu \pm 1\sigma_{\bar{x}}$.
b) There is a 95.45% chance that any sample mean will fall between $\mu \pm 2\sigma_{\bar{x}}$.
c) There is a 99.73% chance that any sample mean will fall between $\mu \pm 3\sigma_{\bar{x}}$.

○ **EXAMPLE 10.3a**
Anne's Catering Service has selected a random sample of 100 orders. If it is known that the mean for all orders is $120 and the standard deviation is $25,

a) what is the mean of the sampling distribution of the means for sample size 100?
b) what is the standard deviation of the sampling distribution of the means?
c) what is the likelihood that the sample mean will fall between $\mu \pm 1\sigma_{\bar{x}}$?
d) what are the chances that the sample mean will fall between 115 and 125?
e) within what range of values does the sample mean have a 98% chance of falling?
f) what is the probability that the sample mean will fall within $4 of the true mean?

● **SOLUTION**
$n = 100; \mu = 120; \sigma = 25$.

a) The mean of the sampling distribution of the means $\mu_{\bar{x}} = \mu = 120$ (Property 1).
b) Since N is not known, the population is considered to be infinite. The standard deviation of the sampling distribution of the means (standard error)

$$\sigma_{\bar{x}} = \frac{\sigma}{\sqrt{n}} = \frac{25}{\sqrt{100}} = \frac{25}{10} = 2.5 \qquad \text{(Property 2)}$$

c) Since $n > 30$, the sampling distribution of the means will approximate the normal distribution. The likelihood that the sample mean will fall between $\mu \pm 1\sigma_{\bar{x}} = 68.26\%$ (Property 3).

d) For $x = 115$,

$$z = \frac{115 - 120}{2.5} = \frac{-5}{2.5} = -2.0$$

For $x = 125$,

$$z = \frac{125 - 120}{2.5} = \frac{5}{2.5} = +2.0$$

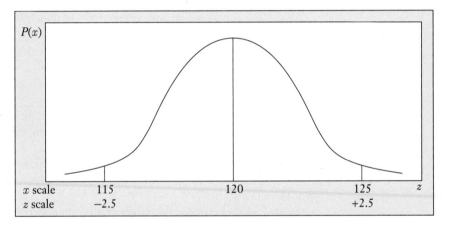

| x scale | 115 | 120 | 125 | z |
| z scale | −2.5 | | +2.5 | |

The range 115 to 125 represents the interval $\mu \pm 2\sigma_{\bar{x}}$. The likelihood that the sample mean will fall within this interval is 95.45%.

e) Let x_1 and x_2 represent the lower limit and the upper limit of the range respectively. For a normal distribution, the area between x_1 and x_2 is symmetrical about μ. The proportion of the area between μ and x_1 is 0.4900, as is the area between μ and x_2.

From Table 9.2, the z value associated with an area of 0.4900 is 2.33.

Substituting in the formula $z = \dfrac{\bar{x} - \mu}{\sigma_{\bar{x}}}$, we can compute the values for x_1 and x_2.

For x_1, $\quad -2.33 = \dfrac{x_1 - 120}{2.50}$ $\qquad$ For x_2, $\quad 2.33 = \dfrac{x_2 - 120}{2.50}$

$$-5.825 = x_1 - 120 \qquad\qquad 5.825 = x_2 - 120$$

$$x_1 = 114.175 \qquad\qquad\qquad x_2 = 125.875$$

For a 98% chance, the range will be \$114.175 to \$125.875.

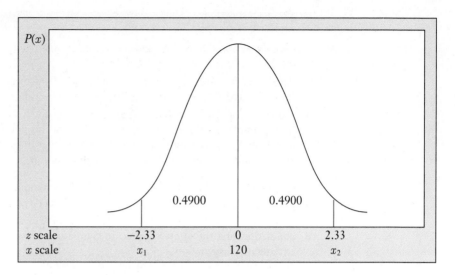

f) To fall within \$4 of the true mean, the sample mean should fall into the range \$116 to \$124.

For $x = 116$,

$$z = \frac{116 - 120}{25} = \frac{-4}{25} = -1.6$$

For $x = 124$,

$$z = \frac{124 - 120}{25} = \frac{4}{25} = 1.6$$

The area associated with a z value of 1.6 is 0.4452.

The area between 116 and 124 is $2(0.4452) = 0.8902$.

The sample mean has an 89.02% chance of falling within \$4 of the true mean of \$120.

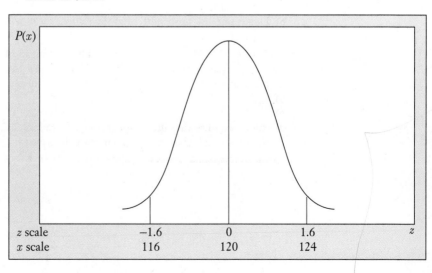

EXERCISE 10.3

1. Given that $\mu = 800$, $\sigma = 54$ and $n = 64$, determine
   (a) $\sigma_{\bar{x}}$ (b) $\mu + 3\sigma_{\bar{x}}$ (c) $\mu - 3\sigma_{\bar{x}}$

2. Given that $\mu = 250$, $\sigma = 24$, $n = 36$ and $N = 600$, determine
   (a) $\sigma_{\bar{x}}$ (b) $\mu \pm \sigma_{\bar{x}}$ (c) $\mu \pm 2\sigma_{\bar{x}}$

3. A normal population of 1500 has a mean of 400 and a standard deviation of 35. If all possible samples of size 25 are drawn from this population, and the sample means are computed and arranged in the form of a frequency distribution,
   a) determine the mean of the frequency distribution;
   b) compute the standard deviation of the frequency distribution;
   c) comment on the shape of the frequency distribution.

4. A large population is known to have a mean of $11 500 and a standard deviation of $2750. If all possible samples of size 225 are drawn from this population, and the sample means are computed and arranged in the form of a frequency distribution,
   a) what is the mean of the frequency distribution?
   b) what is the standard deviation of the frequency distribution?
   c) what is the shape of the frequency distribution?

5. Given that $\mu = 7.5$, $\sigma_{\bar{x}} = 0.4$ and $n = 49$, determine
   (a) $\mu_{\bar{x}}$ (b) σ

6. Given that $\mu_{\bar{x}} = 1.000$, $\sigma_{\bar{x}} = 0.004$ and $\sigma = 0.06$, determine
   (a) μ (b) n

7. A sample of 64 is taken from a large population. The population has a mean of $200 and a standard deviation of $20.
   a) Determine the mean of the sampling distribution of the sample means.
   b) Compute the standard error.
   c) Identify the value that indicates that the sampling distribution of the means is approximately normal.
   d) Determine the range of values within which the sample mean has a 95.44% chance of falling.
   e) Determine the probability that the sample mean will fall within $5 of the population mean.

8. A normally distributed population of 900 has a mean of 0.25 mm and a standard deviation of 0.06 mm. If a random sample of 16 is chosen, determine
   a) the shape of the sampling distribution of the means;
   b) $\mu_{\bar{x}}$;
   c) $\sigma_{\bar{x}}$;
   d) the probability that the sample mean will be less than 0.22 mm;
   e) the probability that the sample mean will exceed 0.27 mm.

SECTION 10.4 The Sampling Distribution of Proportions

A. Introduction

There are many situations in business and industry in which we are interested in the number of items in a population that have a particular characteristic. Consider the following examples:

1. The registrar's office of a college needs to know how many students are female or how many are male. The characteristic of interest is "female" or "male."

2. A candidate for a political office wants to know how many votes she or he is likely to get in an election. The characteristic of interest is "number of votes."

3. A manufacturer of computer chips wants to know how many chips in a production run will be usable. The characteristic of interest is "number of usable chips."

In such situations the variable to be analyzed is the proportion of responses that have a particular characteristic. This proportion is determined by

$$\frac{\text{NUMBER OF ITEMS THAT HAVE THE CHARACTERISTIC}}{\text{TOTAL NUMBER OF ITEMS}}$$

The computation of the proportion for all possible samples of a given size results in the construction of the **sampling distribution of proportions**.

The sampling distribution of proportions is a binomial distribution since it deals with situations in which there are only two possible outcomes and the distribution is a result of counting. Provided the sample size is sufficiently large ($n > 30$), we can use the properties of the normal distribution.

The sampling distribution of proportions can be used to consider the relationship between the population proportion and the possible values that a sample proportion may assume.

For purposes of analysis, the symbol π (read "pie") is used to denote the population proportion. The sample proportion is denoted by p

where $p = \dfrac{x}{n}$

x = the number of items in a sample having a specific characteristic;
n = the total number of items in the sample.

For a sample of 30 students, consisting of 18 female and 12 males, the sample proportion having the characteristic "female,"

$$p(\text{female}) = \frac{18}{30} = 0.60 = 60\%.$$

The sample proportion having the characteristic "male,"

$$p(\text{male}) = \frac{12}{30} = 0.40 = 40\%.$$

B. Properties of the Sampling Distribution of Proportions

1. The mean of the sampling distribution of proportions, denoted by μ_p, is equal to the population proportion:

$$\mu_p = \pi$$

2. The standard deviation of the sampling distribution of proportions, referred to as the **standard error of proportions** and denoted by σ_p, is given by

$$\sigma_p = \sqrt{\frac{\pi(1-\pi)}{n}} \qquad \text{for infinite populations}$$

and

$$\sigma_p = \sqrt{\frac{\pi(1-\pi)}{n}}\sqrt{\frac{N-n}{N-1}} \qquad \text{for finite populations.}$$

3. For $n > 30$ and provided that $n\pi$ and $n(1-\pi)$ are both greater than 5, the sampling distribution of proportions approximates the normal distribution.

   On the basis of Property 3, chances are
   a) 68.26% that a sample proportion p lies between $\pi \pm 1\sigma_p$;
   b) 95.45% that a sample proportion p lies between $\pi \pm 2\sigma_p$;
   c) 99.73% that a sample proportion p lies between $\pi \pm 3\sigma_p$.

Note   The use of the normal distribution as an approximation to the binomial distribution is not appropriate for small values of n.

○ **EXAMPLE 10.4a**
A sample of 100 items is taken randomly from a production run of 2500 items. Normally, 90% of a production run meets quality standards and is acceptable. Determine
a) the mean of the sampling distribution of proportions;
b) the standard error of proportions;
c) the range of values within which the sample proportion has a 99.73% chance of falling;
d) the chances that the sample proportion will fall within five percentage points of the population proportion.

● **SOLUTION**
The characteristic considered is "acceptable."
π (acceptable) $= 0.90$; $n = 100$; $N = 2500$.
a) The mean of the sampling distribution of proportions, $\mu_p = \pi = 0.90$.
b) Since N is known, the population is a finite population. The standard error of proportions σ_p is given by

$$\sigma_p = \sqrt{\frac{\pi(1-\pi)}{n}}\sqrt{\frac{N-n}{N-1}}$$

$$= \sqrt{\frac{0.90(0.10)}{100}}\sqrt{\frac{2500-100}{2500-1}}$$

$$= \sqrt{\frac{0.09}{100}}\sqrt{\frac{2400}{2499}}$$

$$= \sqrt{0.0009}\sqrt{0.9603842}$$

$$= (0.03)(0.9799919)$$

$$= 0.0294$$

c) The proportion 0.9973 defines the interval $\pi \pm 3\sigma_p$; that is,
   $0.90 \pm 3(0.0294) = 0.90 \pm 0.0882$.
   The upper limit of the range $= 0.90 + 0.0882 = 0.9882$;
   the lower limit of the range $= 0.90 - 0.0882 = 0.8118$.
   The sample proportion of acceptable production has a 99.73% chance of
   falling into the range 0.8118 to 0.9882, that is, 81.18% to 98.82%.
d) The interval to be considered is 0.90 ± 0.05, that is, 0.85 to 0.95.

$$n\pi = 100(0.90) = 90 > 5$$
$$n(1-\pi) = 100(0.10) = 10 > 5$$

Since $n > 30$ and both $n\pi$ and $n(1-\pi)$ are greater than 5, the properties
of the normal curve can be used.

$$\text{For } p = 0.85, \; z = \frac{0.85 - 0.90}{0.0294} = \frac{-0.05}{0.0294} = -1.70;$$

$$\text{for } p = 0.95, \; z = \frac{0.95 - 0.90}{0.0294} = \frac{0.05}{0.0294} = 1.70.$$

The area between π and $p = 0.85$ is 0.4554;
the area between π and $p = 0.95$ is 0.4554;
the combined area is $2(0.4554) = 0.9108$.

There is a 91.08% chance that the sample proportion will fall within five
percentage points of the population proportion.

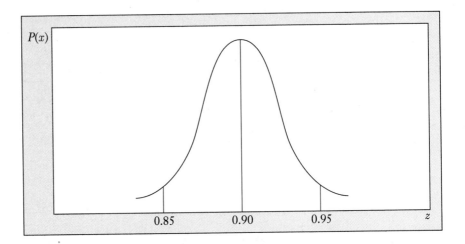

EXERCISE 10.4

1. Ward One contains 12 000 constituents. In a recent plebiscite 9600 constituents voted in favour of a "Green Plan" for the city. If 1000 voters were randomly chosen, what is the probability that the sample proportion in favour of the plan will be between 0.78 and 0.82?

2. The Lead Pipe Company has made a shipment of 400 lengths of pipe. Ten lengths were returned as they did not meet specifications. Historically, the rejection rate has been two percent. Determine the chances of selecting a sample of 400 pipes containing 10 or more rejects.

3. Given that $\sigma_p = \sqrt{\dfrac{0.85(0.15)}{40}} \sqrt{\dfrac{350-40}{350-1}}$, determine
   a) the size of the population;
   b) the sample size;
   c) the population proportion.

4. What is the maximum value that $\pi(1 - \pi)$ can take?

5. Compute the standard error for each of the following:
   a) $\pi = 0.60$, $n = 36$;
   b) $\pi = 0.75$, $n = 40$, $N = 500$;
   c) $\pi = 0.35$, $n = 100$, $N = 1000$;
   d) $\pi = 0.05$, $n = 81$, $N = 250$.

6. Determine the interval
   a) $\pi \pm 1.65\,\sigma_p$ given $\pi = 0.42$, $n = 48$, $N = 360$;
   b) $\pi \pm 2.50\,\sigma_p$ given $\pi = 0.92$, $n = 144$.

7. The college division of Prentice-Hall wants to determine the proportion of unsaleable books for a large production run. In the past the proportion of unsaleable books has been 7%. Compute the chances that the sample proportion for a sample of size 225 will fall within three percentage points of the population proportion.

8. A sample of 144 students is taken from a population of 2000 high-school students to estimate the proportion of students who can be expected to register at a college. Assuming that the true population proportion is 35%, determine the probability that the sample proportion will be no less than 30%.

REVIEW EXERCISE

1. Compute the upper and lower limits for the following intervals:
   a) $\mu \pm 2\sigma_{\bar{x}}$, given $\sigma = 0.22$, $n = 121$, $\mu = 0.75$;
   b) $\mu \pm 3\sigma_p$, given $\sigma = 1.8$, $n = 50$, $N = 800$, $\mu = 28$.

2. Compute the interval limits such that the given percent of the means of samples of the given size fall between the limits.
   a) 95% of all samples of size 100 for $\mu = 35$, $\sigma = 4$;
   b) 90% of all samples of size 225 for $\mu = 1200$, $\sigma = 36$.

3. a) For a distribution with a $\mu = 31.2$, $\sigma = 3.7$, $N = 800$, compute the upper limit so that one percent of the means of samples of size 50 fall above the upper limit.
   b) For a population having a mean of 356 and a standard deviation of 33, determine the lower interval limit so that 98% of all samples of size 121 have a mean greater than the lower limit.

4. Compute the limits for the following intervals:
   a) $\mu \pm 2.5\sigma_{\bar{x}}$, given that $\mu = 16\ 000$, $\sigma = 4200$, $N = 500$ and $n = 40$;
   b) $\mu \pm 1.8\sigma_{\bar{x}}$, given that $\mu = 4500$, $\sigma = 375$, $n = 64$.

5. In a normal distribution with a mean of 71.2 and a standard deviation of 6.8, a sample of 38 items is drawn. Determine the probability that the mean of the sample will be less than 68.

6. A normal distribution has a $\mu = 1.69$ and $\sigma = 0.18$. From this distribution a sample of 16 observations is randomly selected.
   a) Comment on the shape of the sampling distribution.
   b) What are the chances that the sample mean falls between 1.56 and 1.78?

7. The Canada goose has an average weight of 2.4 kg with a standard deviation of 0.61 kg. For a flock of 36 Canada geese flying overhead,
   a) determine the probability that some of the geese weigh more than 4 kg;
   b) compute the probability that the mean weight of the flock falls between 2.1 and 2.7 kg.

8. A gas-station owner is planning to add more gas pumps. The owner is certain that gasoline sales are normally distributed with a mean of 43.8 L and a standard deviation of 3.5 L.
   a) Will the owner add gas pumps if the criteria for adding pumps require that 10% of gas sales are more than 48.7 L?
   b) If a random sample of 16 customers is taken, what is the probability that the average gas sale is less than 41.5 L?

9. It was determined that 160 graduating high-school students had each watched an average of 15 000 hours of television in their lives with a standard deviation

of 4650 hours. For a random sample of 40 graduates,

a) compute the standard error;

b) determine the interval in which 90% of the sample means can be expected to fall;

c) calculate the probability that the sample selected will have a mean of less than 13 700 hours.

10. From a normally distributed large population with a mean of 0.25 mm and a standard deviation of 0.03 mm, a random sample of nine observations is chosen.

a) Comment on the shape of the sampling distribution of the means.

b) Compute $\sigma_{\bar{x}}$,

c) What proportion of the population is expected to measure less than 0.24 mm?

d) What proportion of the population is expected to measure between 0.28 mm and 0.31 mm?

e) What is the probability that the sample mean falls between 0.23 mm and 0.27 mm?

f) What are the chances that the sample mean will be less than 0.26 mm?

11. At D.F.R. Investments, average sales for the 124 sales representatives are $3 200 000 with a standard deviation of $150 000. For a random sample of 36 sales representatives,

a) what is the standard deviation of the sampling distribution of the means?

b) how many sales reps have sales of less than $2 840 000?

c) what proportion of the sales staff have sales between $2 750 000 and $2 960 000?

d) what is the probability that the sample has a mean of more than $3 250 000?

e) what are the chances that the sample mean falls between $3 160 000 and $3 180 000?

12. The average price of houses sold in a city in April was $215 000 with a standard deviation of $28 770. If a sample of 36 of the 128 houses sold had a mean of $206 000, will a 95% interval estimate include this sample?

13. The average balance in a local bank's chequing accounts was $4700 with a standard deviation of $820. A random sample of 70 chequing accounts showed a mean balance of $4810. If the bank has 1100 chequing accounts, what is the chance of selecting a sample of 70 accounts with a mean greater than $4810?

14. A distributor of VCRs believes that the average age of videocassette recorders in his target market is 4.8 years with a standard deviation of 1.1 years. Determine the probability that the mean age of a sample of 75 is less than 5 years.

15. In Canada, 47.3% of the population is Catholic, while 41.2% is Protestant. In a random sample of 200 persons, 88 were Catholic and 86 were Protestant. What is the probability of selecting a sample of 200 persons

a) with fewer than 86 Protestants?

b) with fewer than 86 Catholics?

16. A robot that sorts glass from containers holding recycled items is 96% correct. If a sample of 500 recycled items is selected, what is the probability that more than 6% of the items will be sorted incorrectly?

17. Over a 365-day period a weather forecast is correct 68% of the time. If you

are planning a 21-day holiday during a forecasted sunny period, what is the probability that you will have at least 15 days of sunshine?

18. An internal audit showed that 3% of the company's 820 accounts payable were processed incorrectly. If 150 accounts are reviewed at random, what is the probability that fewer than 6 accounts payable were processed incorrectly?

19. Of 1070 students registered in a business program, 78% came directly from high school. For a sample of 82 students, determine the interval that contains the sample proportion 98% of the time.

20. Given that the population proportion is 0.32 and the sample size is 48, determine the upper and lower limit of the interval into which the sample proportion will fall with a probability of 90%.

21. In the stockbrokerage industry, 4% of cold calls (selling to individuals by telephone) result in a transaction. A minimum of 30 transactions a month is required for a stockbroker to maintain employment with the brokerage firm. If a new employee averages 25 cold calls a day for the 20 working days in the first month of employment, what is the probability of continued employment?

22. Match the correct definition to the symbol shown in the first column.

    a) N i) population proportion
    b) $\bar{x}$ ii) value of an observation
    c) μ_p iii) sample size
    d) σ iv) sample mean
    e) $\sigma_{\bar{x}}$ v) sample proportion
    f) π vi) population mean
    g) n vii) mean of the sampling distribution of the means
    h) $\mu_{\bar{x}}$ viii) mean of the sampling distribution of proportions
    i) x ix) population size
    j) p x) population standard deviation
    k) μ xi) standard error of proportions
    l) σ_p xii) standard error of the mean

SELF-TEST

1. Write the symbol that represents each of the following:
   a) the mean of the sampling distribution of the means;
   b) the standard error of proportions;
   c) the mean of the sampling distribution of proportions;
   d) the standard deviation of the sampling distribution of the means.

2. The accountant at Saw-Me Lumber Company knows that the company's suppliers are paid on the average within 34.2 days with a standard deviation of 3.1 days. If a random sample of 60 paid invoices is pulled, what is the probability that the sample average will be between 33 and 34 days?

3. A normally distributed population representing the heart rate of resting persons has a mean of 62.5 beats per minute and a standard deviation of 7.6 beats per minute.

a) Determine the standard error for a sample of 24.

b) Calculate the interval within which the sample mean has an 88% chance of falling.

4. The 1992 salaries of the employees of a national accounting firm averaged $45 280 with a standard deviation of $5790. Revenue Canada selected 45 out of the 389 employees to audit their pension plan contributions.

a) What was the salary of an employee in the top five percent of the firm's payroll?

b) What is the upper limit of the interval above which the sample mean has a five percent chance of falling?

5. The 1991 census showed that an urban community in New Brunswick consisted of 5129 households. The distribution of children per household was normal, with an average of 2.6 and a standard deviation of 0.69.

a) How many households had between two and three children?

b) What is the probability that a household chosen will have between two and three children?

c) If a sample of 9 households is drawn, what is the probability that the sample mean will fall between 2 and 3?

d) Comment on the difference between questions **(b)** and **(c)**.

6. During the last five summers, Josie sold yogurt on the main street of Grand Bend, and every summer 29% of sales were strawberry-flavoured products. For a random sample of 110 customers, what is the probability that between 25 and 35 selected a strawberry-flavoured product?

7. Peter runs a food concession at the carnival every Labour Day weekend. This year 68% of 2127 customers ordered a cola drink. A random selection of 250 customers were given free hats courtesy of Peter. What is the probability that more than 150 people wearing a free hat purchased a cola?

8. What sample size is required to maintain a standard error of no more than 1.26 for a large population having a standard deviation of 10.5?

Key Terms

Summary of Formulas

1. Sampling distribution of the means:

a) Mean:

$$\mu_{\bar{x}} = \mu$$

b) Standard error:

i) $\sigma_{\bar{x}} = \dfrac{\sigma}{\sqrt{n}}$, when the population size (N) is infinite or unknown.

ii) $\sigma_{\bar{x}} = \dfrac{\sigma}{\sqrt{n}} \cdot \sqrt{\dfrac{N-n}{N-1}}$, when the population size (N) is known.

2. Sampling distribution of proportions:

a) Mean:

$$\mu_p = \pi$$

b) Standard error:

i) $\sigma_p = \sqrt{\dfrac{\pi(1-\pi)}{n}}$, when the population size (N) is infinite or unknown.

ii) $\sigma_p = \sqrt{\dfrac{\pi(1-\pi)}{n}} \cdot \sqrt{\dfrac{N-n}{N-1}}$, when the population size (N) is known.

3. Sample proportion:

$$p = \dfrac{x}{n}$$

CHAPTER 11

Interval Estimation

Introduction

The value of a sample mean or sample proportion is used to develop an interval within which we can expect the population mean or population proportion to lie. The width of this interval depends on how sure we want to be about the estimate of the population parameter. The more confident we want to be, the wider the interval becomes.

Objectives

Upon completion of this chapter you will be able to
1. define the terms point estimate, interval estimate, confidence level (confidence coefficient, degree of confidence), confidence limits, and confidence interval;
2. construct confidence intervals about a sample mean at specified confidence levels for a large sample ($n > 30$) when
    a) the standard deviation of the population is known, and
    b) the standard deviation of the population is not known;
3. construct confidence intervals about a sample proportion for specified confidence levels for large samples ($n > 30$);
4. determine the sample size required for a specified level of confidence and a specified maximum allowable error when estimating a population mean or a population proportion;
5. construct confidence intervals about a sample mean at specified confidence levels for small samples ($n < 30$).

SECTION 11.1 Estimating Concepts

A. Estimators

The characteristics of a sample, referred to as *statistics*, are used to estimate the characteristics of a population, referred to as *parameters*. The statistics $\bar{x}$, p and s are used to estimate the corresponding *parameters* μ, π and σ. A statistic used in this way is called an **estimator**.

To be accepted as a good estimator, a statistic must have three properties: it must be *unbiased*, *efficient* and *consistent*.

An **unbiased estimator** is one whose expected value (long-run average) is equal to the parameter being estimated. On this basis, the sample mean $\bar{x}$ is an unbiased estimator of the population mean μ and the sample proportion p is an unbiased estimator of the population proportion π.

The sample standard deviation s is obtained as the square root of the sample variance. When calculating the population variance, we used the formula $\sigma^2 = \dfrac{\Sigma(x - \mu)^2}{N}$. The corresponding formula for the sample variance is modified by dividing by $(n-1)$ in order to use the sample standard deviation as an unbiased estimator of the population standard deviation.

This means that, provided we use the formula $s^2 = \dfrac{\Sigma(x - \bar{x})^2}{n - 1}$, the sample variance s^2 is an unbiased estimator of σ^2 and the sample standard deviation $s = \sqrt{s^2}$ is an unbiased estimator of the population standard deviation σ.

Note To avoid confusion, all references to sample variance and sample standard deviation in the remaining sections of this chapter and the remaining chapters of this text are in terms of the unbiased estimators s^2 and s.

An *efficient* estimator is one that has a relatively small variance. This implies that if a large number of samples are drawn from a population the computed values of the estimator are close to one another.

A *consistent* estimator is one in which the difference between the estimator and the parameter becomes smaller as the sample size increases.

B. Point Estimation

When a sample statistic is used to estimate the corresponding population parameter, the result is called a **point estimate**, and the process of estimating with a single number is referred to as **point estimation**.

○ **EXAMPLE 11.1a**

The production department of a beverage factory fills bottles with various fruit juices and soft drinks on a contract basis. One of the production lines is scheduled to fill 280-mL cans with soda pop. A random sample taken from the run provided the following data in terms of the number of millilitres in the sampled cans:

| 284 | 278 | 285 | 294 | 277 |
| 270 | 265 | 282 | 276 | 279 |

Based on the information gathered from the sample,
a) obtain the best estimate of the average fill for the production run;
b) compute the associated standard deviation.

● **SOLUTION**

Computation details for finding the mean $\bar{x}$ and the standard deviation s are shown in Table 11.1.

TABLE 11.1   Computational Details for Finding $\bar{x}$ and s

Observed sample value x	Deviation from sample mean $(x - \bar{x})$	Squared deviation $(x - \bar{x})^2$	Squared sample value x^2
284	$284 - 279 =$ 5	25	80 656
278	$278 - 279 =$ -1	1	77 284
285	$285 - 279 =$ 6	36	81 225
294	$294 - 279 =$ 15	225	86 436
277	$277 - 279 =$ -2	4	76 729
270	$270 - 279 =$ -9	81	72 900
265	$265 - 279 = -14$	196	70 225
282	$282 - 279 =$ 3	9	79 524
276	$276 - 279 =$ -3	9	76 176
279	$279 - 279 =$ 0	0	77 841
$\sum x = 2790$	$\sum(x - \bar{x}) =$ 0	$\sum(x - \bar{x})^2 = 586$	$\sum x^2 = 778\ 996$

a) The sample mean,

$$\bar{x} = \frac{\sum x}{n} = \frac{2790}{10} = 279.$$

Based on the sample result, the best estimate of the population mean $\mu = 279$. This estimate of μ is a point estimate.

b) The unbiased sample variance,

$$s^2 = \frac{\sum(x - \bar{x})^2}{n - 1} = \frac{586}{9} = 65.11.$$

The unbiased sample standard deviation,

$$s = \sqrt{s^2} = \sqrt{65.11} = 8.069.$$

Alternately, the sample variance can be determined by means of the short-cut formula:

$$s^2 = \frac{n\sum x^2 - (\sum x)^2}{n(n - 1)}$$

where $n =$ the number of observations in the sample;

x = the value of an individual observation in the sample.

$$s^2 = \frac{10(780\ 116) - (2792)^2}{(10)(9)} = \frac{7\ 801\ 160 - 7\ 795\ 264}{90} = \frac{5896}{90} = 65.5111$$

$$s = \sqrt{65.5111} = 8.09389.$$

○ **EXAMPLE 11.1b**

In a sample of 40 business administration students, 22 were female. What is the estimated proportion of female students in the business administration program?

● **SOLUTION**

The characteristic considered is "female." The sample proportion of females is

$$p(\text{female}) = \frac{x}{n} = \frac{\text{Number of females in the sample}}{\text{Total number of students in the sample}}$$

$$= \frac{22}{40} = 0.55 = 55\%$$

Since p is an unbiased estimator of π, an unbiased point estimate of the proportion of females in the entire business administration program is 55%.

C. *Interval Estimation*

Because of *sampling variation* we know that different samples will have different sample means or sample proportions. We also know that the mean or proportion of any one sample is unlikely to have the same value as the population mean or population proportion. Thus, a point estimate of the population mean or population proportion can be expected to differ from the true population mean or true population proportion, but we have no knowledge of how large the difference may be and we have no way of assessing its size.

To overcome this weakness of point estimates, we construct an interval around the sample mean $\bar{x}$ or the sample proportion p in such a way that we can be "confident" that the "true" population mean μ or the "true" population proportion π lies somewhere within this interval. This process of estimating the value of a population parameter within a range of values is called **interval estimation**.

SECTION 11.2 Interval Estimation around the Mean When σ Is Known

A. Construction of Confidence Intervals

Estimation problems in which the population standard deviation is considered to be known are usually associated with situations for which a considerable amount of historical data are available, such as repetitive manufacturing processes.

○ **EXAMPLE 11.2a**

(Refer to Example 11.1a.) Assume that it is known that when the filling machine is working properly, the quantity of soda pop in the cans is normally distributed with a mean of 280 mL and a standard deviation of 5 mL. In addition to the first sample with a mean of 279, four more samples of size 10 drawn during the production run had means of 282, 274, 277 and 284 respectively.
a) Construct the interval $\mu \pm 1.96\,\sigma_{\bar{x}}$.
b) Construct the interval $\bar{x} \pm 1.96\,\sigma_{\bar{x}}$ around each of the five sample means.
c) Draw a graph of the sampling distribution of the means and show graphically the relationship among μ, the five sample means and the intervals.

● **SOLUTION**

a) $\mu = 280 \,;\, \sigma = 5\,$.

$$\sigma_{\bar{x}} = \frac{\sigma}{\sqrt{n}} = \frac{5}{\sqrt{10}} = \frac{5}{3.16228} = 1.58$$

The interval
$$\mu \pm 1.96\,\sigma_{\bar{x}} = 280 \pm 1.96(1.58) = 280 \pm 3.10$$
$$= 276.90 \text{ to } 283.10$$

b) The desired interval $\bar{x} \pm 1.96\,\sigma_{\bar{x}} = \bar{x} \pm (1.96)(1.58) = \bar{x} \pm 3.10\,$.
For the sample with $\bar{x} = 279\,$, the interval is $279 \pm 3.10\,$,
that is, from 275.90 to 282.10.
For the sample with $\bar{x} = 282\,$, the interval is 278.90 to 285.10.
For the sample with $\bar{x} = 274\,$, the interval is 270.90 to 277.10.
For the sample with $\bar{x} = 277\,$, the interval is 273.90 to 280.10.
For the sample with $\bar{x} = 284\,$, the interval is 280.90 to 287.10.

c) The relationship among μ, the five sample means and the intervals is shown in Figure 11.1.

FIGURE 11.1 **Relationship between the Intervals Containing μ and $\bar{x}$**

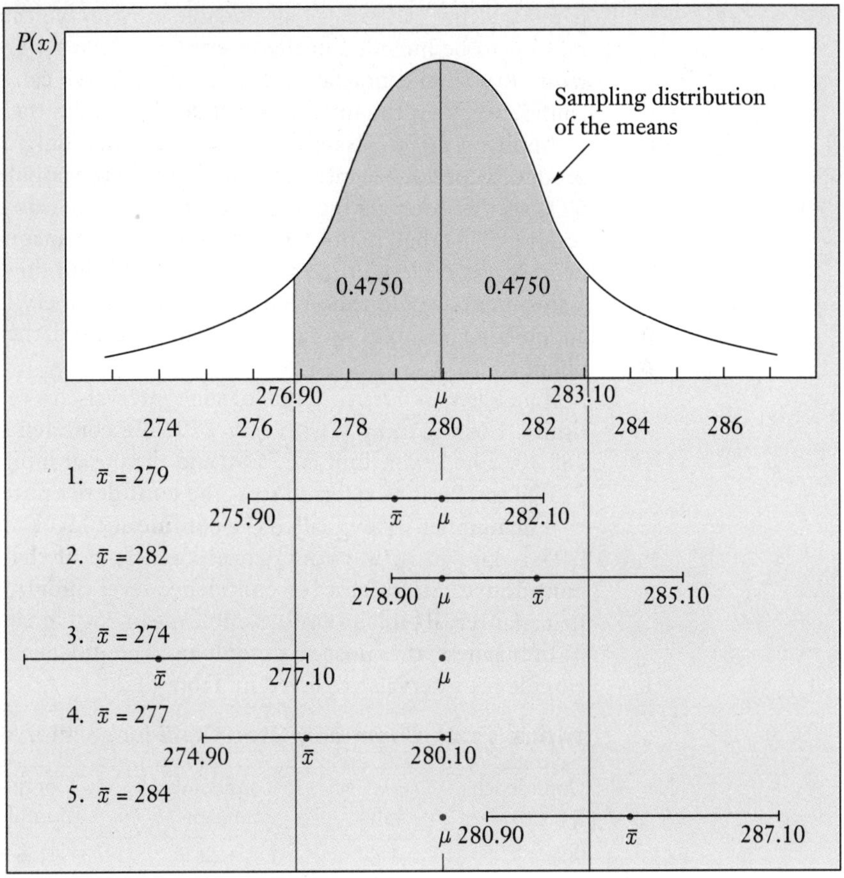

Note For Sample 1 the mean $\bar{x} = 279$ falls into the range 276.90 to 283.10; the population mean μ falls into the interval $279 \pm 1.96\sigma_{\bar{x}}$.

For samples 2 and 4 the means $\bar{x} = 282$ and $\bar{x} = 277$ also fall into the range 276.90 to 283.10; the population mean μ falls into the intervals $282 \pm 1.96\sigma_{\bar{x}}$ and $277 \pm 1.96\sigma_{\bar{x}}$.

For samples 3 and 5 the means $\bar{x} = 275$ and $\bar{x} = 284$ fall outside the range 276.90 and 283.10; the population mean μ is not included in the intervals $275 \pm 1.96\sigma_{\bar{x}}$ and $284 \pm 1.96\sigma_{\bar{x}}$.

B. Confidence Intervals and Confidence Levels

Since the area under the normal curve between μ and $z = 1.96$ is 0.4750, the area included in the inverval $\mu \pm 1.96\sigma_{\bar{x}}$ is 0.9500, that is, 95% of the total area. This means that we can expect that for $\mu = 280$ mL the means of 95% of all samples of size 10 fall between 276.90 mL and 283.10 mL.

Conversely, as indicated by Figure 11.1, the interval $\bar{x} \pm 1.96\sigma_{\bar{x}}$ includes the population mean, provided that $\bar{x}$ falls between 276.90 mL and 283.10 mL. It follows that for 95% of all possible samples we can expect the population mean μ to be included in the interval $\bar{x} \pm 1.96\,\sigma_{\bar{x}}$. For the remaining 5% of samples μ should not fall into this interval. We can say that we have "95% confidence" that the interval $\bar{x} \pm 1.96\,\sigma_{\bar{x}}$ includes the population mean μ.

Applying this to Example 11.2a, we see that only three of the five samples produced confidence intervals that included the population mean; that is, only 60% of the intervals included μ. This is so far removed from the expected result of 95% that, in practice, the production manager would suspect that the process was no longer in control and would shut down the line immediately. Adjustments would then be made to the machinery before starting up again and further samples would be taken to ensure the correct fill was being obtained.

The *upper* and *lower* limits of such intervals are referred to as **confidence limits**. For the sample with $\bar{x} = 279$, the confidence limits are 275.90 and 282.10. The lower limit is 275.90 and the upper limit is 282.10. The interval 275.90 to 282.10 is referred to as the **confidence interval**.

The number 95% is called the **confidence level** and its decimal equivalent 0.95 is known as the confidence coefficient of the interval estimate. The **confidence coefficient** (or confidence level) indicates the *degree of certainty* that an interval built around a sample mean $\bar{x}$ will include the population mean μ. In business, the most commonly used confidence levels and the associated confidence intervals are listed in Table 11.2.

TABLE 11.2 Commonly Used Confidence Intervals

Confidence level	z value	Confidence interval	Confidence intervals for Example 11.2a, $\sigma_{\bar{x}} = 1.58$
90%	1.64	$\bar{x} \pm 1.64\,\sigma_{\bar{x}}$	$\bar{x} \pm 1.64(1.58) = \bar{x} \pm 2.59$
95%	1.96	$\bar{x} \pm 1.96\,\sigma_{\bar{x}}$	$\bar{x} \pm 1.96(1.58) = \bar{x} \pm 3.10$
98%	2.33	$\bar{x} \pm 2.33\,\sigma_{\bar{x}}$	$\bar{x} \pm 2.33(1.58) = \bar{x} \pm 3.68$
99%	2.58	$\bar{x} \pm 2.58\,\sigma_{\bar{x}}$	$\bar{x} \pm 2.58(1.58) = \bar{x} \pm 4.08$

Note The higher the confidence level, the wider the confidence interval. As the interval widens, the precision of the estimate decreases. If the interval becomes too wide, the estimate may have no practical value.

Confidence levels also indicate the chance of being wrong. Being 90% confident means that there is a 10% chance of error. A 95% confidence level implies a 5% chance of error. Conversely, a 2% chance of error suggests a 98% degree of confidence. *The confidence coefficient plus the chance of error always add up to 1.*

The width of the confidence interval and the precision of the interval estimate depend on the standard deviation and the choice of confidence level. In general, when σ is known, the confidence interval is given by

$$\bar{x} \pm z\sigma_{\bar{x}}$$

where $\bar{x}$ = the sample mean (point estimate);

$\quad z$ = the value associated with a given level of confidence;

$\quad \sigma_{\bar{x}}$ = the standard error of the mean found by

$$\sigma_{\bar{x}} = \frac{\sigma}{\sqrt{n}} \text{ for infinite populations, or}$$

$$\sigma_{\bar{x}} = \frac{\sigma}{\sqrt{n}} \sqrt{\frac{N-n}{N-1}} \text{ for finite populations.}$$

○ EXAMPLE 11.2b

Assuming that the mean fill of a sample of 25 cans taken from the bottling line in Example 11.2a was 278, determine the confidence intervals for confidence levels of
a) 90%   b) 95%   c) 98%.

● SOLUTION

$\bar{x} = 278 ; n = 25 ; \sigma = 5 ; N$ is unknown.

$$\sigma_{\bar{x}} = \frac{\sigma}{\sqrt{n}} = \frac{5}{\sqrt{25}} = \frac{5}{5} = 1$$

a) For 90% confidence, $z = 1.64$;
   the confidence interval $\bar{x} \pm z\sigma_{\bar{x}} = 278 \pm 1.64(1)$.
   The upper limit of the interval = 278 + 1.64 = 279.64;
   the lower limit of the interval = 278 − 1.64 = 276.36.
   The 90% confidence interval covers the range 276.36 to 279.64.
b) For 95% confidence, $z = 1.96$;
   the confidence interval = $278 \pm 1.96(1)$.
   The 95% confidence interval covers the range 276.04 to 279.96.
c) For 98% confidence, $z = 2.33$.
   The confidence interval = $278 \pm 2.33(1)$.
   The 98% confidence interval covers the range 275.67 to 280.33.

Note   If, as stated, the true population mean $\mu = 280$, the 90% confidence interval does not contain μ. With regard to the 95% confidence interval, μ is very close to the upper limit but still outside the interval. μ falls within the 98% confidence interval.

○ EXAMPLE 11.2c

Compute the 90% confidence interval for a sample of 36 items with a mean of 1.20 taken from a population of 600 items. It is known that the population standard deviation is 0.24.

● SOLUTION

$N = 600 ; n = 36 ; \bar{x} = 1.20 ; \sigma = 0.24$; for 90% confidence, $z = 1.64$.
   Since σ is known, the confidence interval is given by $\bar{x} \pm z\sigma_{\bar{x}}$.
   Since N is known, the population is finite and the standard error is given by

$$\sigma_{\bar{x}} = \frac{\sigma}{\sqrt{n}}\sqrt{\frac{N-n}{N-1}} = \frac{0.24}{\sqrt{36}}\sqrt{\frac{600-36}{600-1}}$$

$$= \frac{0.24}{6}\sqrt{\frac{564}{599}} = 0.04\sqrt{0.9414693}$$

$$= 0.04(0.9703499) = 0.0388.$$

The 90% confidence interval is $1.20 \pm 1.64(0.0388) = 1.20 \pm 0.064$. The interval in which the true population mean is located with 90% certainty is the range 1.136 to 1.264.

EXERCISE 11.2

1. A single number used to estimate a population parameter is referred to as _____.

   A range of values used to estimate a population parameter is referred to as _____.

   A statistic used to estimate a parameter is referred to as _____, while a specific value of a statistic is referred to as _____.

2. As the confidence level increases, the interval estimate becomes _____, while the precision of the estimate _____.

   The confidence interval when σ is known is given by the expression _____.

3. The following information is available for a normally distributed infinite population:

   $\bar{x} = \$150$; $\sigma = \$35$; $n = 25$; confidence coefficient = 0.95. *1.96*

   a) Compute the upper limit of the interval estimate.

   b) Comment on the shape of the sampling distribution of the means.

4. Given $\sigma = 1000$, $\bar{x} = 870\ 000$, $N = 900$ and $n = 100$,

   a) determine the confidence interval for a degree of confidence of
      i) 0.80, ii) 0.90, iii) 0.98;

   b) state what happens to the size of the confidence interval as the degree of confidence increases.

5. From a sample of 40 patients, Dr. D. Zees would like to estimate the average amount charged per visit per patient. Assuming $\mu = \$35$ and $\sigma = \$9$, what are the chances that the sample mean will have a value within $1 of the true mean?

6. Last month John sent 850 packages by courier. He would like to know the average cost of sending a package. A random selection of 35 charges averaged $10.50. Assuming $\sigma = \$1.22$,

   a) compute the standard error;

   b) construct the 95% confidence interval around the sample mean.

7. A sample of 200 is selected from a population whose mean is known to be 50 with a standard deviation of 9.9. Determine the probability that the sample mean falls within two units of the population mean.

8. The Steel Pipe Company has received a shipment of 900 lengths of pipe. The company's quality-control engineer wants to estimate the average diameter of

the pipes to see if the shipment meets minimum standards. Historically, the standard deviation of the diameters of pipe shipments has been 5 mm. A sample of 45 pipes had a mean diameter of 150 mm. Compute the confidence interval with the chance of a 1% error.

Interval Estimation around the Mean for Large Samples ($n > 30$) when σ Is Unknown.

As stated earlier, estimation problems in which the population standard deviation is considered to be known are usually associated with situations for which a considerable amount of historical data are available. In most situations, however, σ is not known. In addition, we likely know little or nothing about the shape of the distribution from which the sample is drawn.

The sampling distribution of the means of any distribution approaches a normal distribution for samples of size $n > 30$. This permits the use of z values to construct confidence intervals when σ is unknown in the same way as when σ is known, provided the sample size n is greater than 30.

When σ is unknown and $n > 30$, the confidence interval in which μ can be found with a given degree of confidence is given by

$$\overline{x} \pm z s_{\overline{x}}$$

where $\overline{x}$ = the sample mean;

z = the value associated with a given confidence level;

$s_{\overline{x}}$ = the standard error found by

$$s_{\overline{x}} = \frac{s}{\sqrt{n}} \text{ for infinite populations, or}$$

$$s_{\overline{x}} = \frac{s}{\sqrt{n}} \sqrt{\frac{N-n}{N-1}} \text{ for finite populations;}$$

s = the sample standard deviation;

n = the sample size;

N = the population size.

○ **EXAMPLE 11.3a**

The Stagger Inn Tavern wants to estimate the average dollar purchase per customer. A sample of 81 customers spent an average of $4.00 with a standard deviation of $0.63. Estimate the true average expenditure per customer with 95% confidence.

● **SOLUTION**

$n = 81$; $\overline{x} = 4.00$; $s = 0.63$;

the z value associated with a 95% confidence level is 1.96;

since N is unknown, the population is considered infinite;

$$s_{\bar{x}} = \frac{s}{\sqrt{n}} = \frac{0.63}{\sqrt{81}} = \frac{0.63}{9} = 0.07;$$

since σ is unknown, the interval estimate is given by

$$\bar{x} \pm z s_{\bar{x}} = 4.00 \pm 1.96(0.07) = 4.00 \pm 0.1372.$$

The upper limit of the interval = $4.00 + 0.1372 = 4.1372$;
the lower limit of the interval = $4.00 - 0.1372 = 3.8628$.

The confidence interval in which the true average expenditure per customer can be found with 95% certainty is \$3.86 to \$4.14 approximately.

○ **EXAMPLE 11.3b**

The average height of a sample of 36 students out of 1600 students registered in a high school is 170 cm with a standard deviation of 10 cm. Construct a confidence interval with a 1% chance of error.

● **SOLUTION**

$N = 1600$; $n = 36$; $\bar{x} = 170$; $s = 10$;
for a 1% chance of error, the confidence level is 99%;
the z value for the 99% confidence level = 2.58;
since N is known, the population is finite;

$$s_{\bar{x}} = \frac{s}{\sqrt{n}}\sqrt{\frac{N-n}{N-1}} = \frac{10}{\sqrt{36}}\sqrt{\frac{1600-36}{1600-1}} = \frac{10}{6}\sqrt{\frac{1564}{1599}}$$

$$= 1.6667\sqrt{0.97811132} = 1.6667(0.9890) = 1.6484.$$

Since σ is unknown and $n > 30$, the confidence interval around the mean is given by

$$\bar{x} \pm z s_{\bar{x}} = 170 \pm 2.58(1.6484) = 170 \pm 4.25.$$

The upper limit of the interval = $170 + 4.25 = 174.25$;
the lower limit = $170 - 4.25 = 165.75$.

The confidence interval in which the true average height of the group of students can be found with 99% confidence is 165.75 cm to 174.25 cm.

○ **EXAMPLE 11.3c**

The scores of 32 students on a diagnostic test drawn from 1200 students registered in the first year of a college program were as follows:

14	9	12	23	11	15	7	14
6	13	18	19	21	17	9	10
16	18	15	20	8	13	17	24
11	16	14	13	18	15	12	16

Construct the 80% confidence interval in which the true average test score for the 1200 students is located.

● **SOLUTION**
$N = 1200; n = 32$

TABLE 11.3 Calculation of the Mean and Standard Deviation of the Sample

x	x^2	x	x^2	x	x^2	x	x^2
14	196	6	36	16	256	11	121
9	81	13	169	18	324	16	256
12	144	18	324	15	225	14	196
23	529	19	361	20	400	13	169
11	121	21	441	8	64	18	324
15	225	17	289	13	169	15	225
7	49	9	81	17	289	12	144
14	196	10	100	24	576	16	256
105	1541	113	1801	131	2303	115	1691

$$\sum x = 105 + 113 + 131 + 115 = 464$$
$$\sum x^2 = 1541 + 1801 + 2303 + 1691 = 7336;$$

$$\overline{x} = \frac{\sum x}{n} = \frac{464}{32} = 16.50;$$

$$s^2 = \frac{n \sum x^2 - (\sum x)^2}{n(n-1)} = \frac{32(7336) - (464)^2}{32(31)} = \frac{234\,752 - 215\,296}{992}$$

$$= \frac{19\,456}{992} = 19.612903.$$

$$s = \sqrt{s^2} = \sqrt{19.612903} = 4.428646;$$

$$s_{\overline{x}} = \frac{s}{\sqrt{n}} \sqrt{\frac{N-n}{N-1}} = \frac{4.428646}{\sqrt{32}} \sqrt{\frac{1200-32}{1200-1}}$$

$$= \frac{4.428646}{5.656854} \sqrt{\frac{1168}{1199}} = 0.782881 \sqrt{0.974145}$$

$$= 0.782881(0.986988) = 0.7727.$$

For an 80% confidence level, the areas between μ and $\pm z$ must each equal 40% or 0.4000. The corresponding z values $= \pm 1.28$. Since σ is unknown and $n > 30$, the 80% confidence interval $= 14.50 \pm 1.28(0.7727) = 14.50 \pm 0.989$, that is, from 13.51 to 15.49.

EXERCISE 11.3

1. A sample of the billings of 361 doctors revealed that their mean annual income was $210 000 with a standard deviation of $34 000. Compute the 90% confidence limits.

2. A credit manager has chosen 225 accounts out of 16 000 and found that the average payment period was 45 days with a standard deviation of 4 days. Determine the confidence interval within which the average payment period for all accounts can be expected to be found with a 2% chance of error.

3. For the confidence interval $41 \pm 2.33 \left(\dfrac{2}{\sqrt{36}} \right) \sqrt{\dfrac{2000 - 36}{2000 - 1}}$, determine

   (a) N (b) n (c) $\bar{x}$ (d) s (e) $s_{\bar{x}}$ (f) the confidence level.

4. For the confidence interval $120 \pm 1.65 \left(\dfrac{13}{\sqrt{49}} \right)$, determine

   a) the sample size;
   b) the sample standard deviation;
   c) the sample mean;
   d) the confidence level;
   e) the standard error;
   f) the upper confidence limit.

5. The Dew Drop Pub wants to estimate the average number of litres of beer sold per day. A sampling of 48 business days produced a daily average of 125 L with a standard deviation of 16 L. Compute the confidence limits at the 90% confidence level.

6. Dr. Ayad, dentist, would like to know the average number of fillings of each of his 2160 patients. A random sample of the files of 90 patients showed an average of 5 fillings per patient with a standard deviation of 1.2 fillings. Determine the confidence interval for the 95% confidence level.

7. The Cheer Up Answering Service has collected the following data about the number of calls handled per day:

355	410	416	423	435	396	409
372	473	442	427	419	438	453
466	412	415	433	442	387	403
438	342	383	404	373	449	398
408	427	437	453	422	417	435
409	430	444	433	428	437	416
410	429	442	397	412	423	425

Construct the 85% confidence interval.

8. Construct the 80% confidence interval for the following sample results taken from a normal distribution:

1985	636	2422	1103
3888	3962	2939	3106
2291	4909	4982	3621
1179	3665	3643	4319

SECTION 11.4 Interval Estimation of Proportions for Large Samples ($n > 30$) when π Is Unknown

Since the sample proportion p is an unbiased estimator of π, it can be used as a point estimate of π. When both np and $n(1-p)$ are greater than 5, we can use the normal distribution. It is generally accepted that this occurs when the sample size is greater than thirty ($n > 30$). The confidence interval in which π is then expected to lie is given by

$$p \pm zs_p$$

where $s_p = \sqrt{\dfrac{p(1-p)}{n}}$ for infinite populations

and

$$s_p = \sqrt{\dfrac{p(1-p)}{n}}\sqrt{\dfrac{N-n}{N-1}} \text{ for finite populations.}$$

○ **EXAMPLE 11.4a**

The manager of a credit union wants to estimate the proportion of members whose biweekly salaries are deposited by electronic transfer. A random sample of 250 depositors indicated that 95 were on direct deposit. Compute the 95% confidence interval of the true proportion of direct deposit accounts.

● **SOLUTION**

The characteristic considered is "on direct deposit."

The sample proportion, $p = \dfrac{x}{n} = \dfrac{95}{250} = 0.38 = 38\%$.

$$s_p = \sqrt{\dfrac{p(1-p)}{n}} = \sqrt{\dfrac{(0.38)(0.62)}{250}} = \sqrt{0.000924} = 0.0307$$

$$np = (250)(0.38) = 95 \quad \text{and} \quad n(1-p) = 250(0.62) = 155$$

Since both np and $n(1-p)$ are greater than 5, we can use the normal distribution.

For 95% confidence, $z = 1.96$.

The confidence interval $= p \pm zs_p = 0.38 \pm 1.96(0.0307) = 0.38 \pm 0.060$, that is, 0.32 to 0.44.

The manager can be 95% confident that between 32% and 44% of the members are on direct deposit.

○ **EXAMPLE 11.4b**

The quality-control engineer of a parts manufacturer has taken a random sample of 60 from a production run of 900 parts and found 9 defective parts.
a) What is a point estimate of the proportion of defective parts?
b) What is the interval estimate of the proportion of non-defective parts with a 2% chance of error?

● **SOLUTION**

a) The characteristic considered is "defective."

The proportion of defective parts = $\dfrac{9}{60}$ = 0.15 = 15%.

A point estimate of the proportion of defective parts is 15%.

b) The characteristic considered is "non-defective." The number of non-defective parts in the sample = 60 − 9 = 51.

The proportion of non-defective parts, $p = \dfrac{51}{60}$ = 0.85 = 85%.

$N = 900$; $n = 60$.

$$s_p = \sqrt{\frac{p(1-p)}{n}} \sqrt{\frac{N-n}{N-1}} = \sqrt{\frac{(0.85)(0.15)}{60}} \sqrt{\frac{900-60}{900-1}}$$

$$= \sqrt{0.002125} \sqrt{0.9343715} = (0.0461)(0.9666) = 0.0446$$

$$np = 60(0.85) = 51 \quad \text{and} \quad n(1-p) = 60(0.15) = 9$$

Since both np and $n(1-p)$ are greater than 5, we can use the normal distribution.

For a 2% chance of error, $z = 2.33$.

The confidence interval = $p \pm zs_p = 0.85 \pm 2.33(0.0446)$
$$= 0.85 \pm 0.104,$$

that is, 0.746 to 0.954.

The interval in which the true proportion of non-defective parts can be found with a 2% chance of error is between 74.6% and 95.4%.

EXERCISE 11.4

1. Given that $p = 0.32$, $n = 40$, and $N = 210$, compute the 90% confidence interval.

2. In a sample, 9 out of 52 observations are successes. Construct the confidence interval for the true proportion of successes in the population from which the sample was drawn with a 1% chance of error.

3. Only card-carrying college instructors can vote on union issues. Of the 320 instructors at a college in Saskatchewan a sample of 90 showed that 54 were card-carrying union members. Determine the interval estimate of the proportion of instructors at the college who can vote on union issues with a 5% chance of error.

4. A pollster working for an incumbent member of Parliament obtained the following information on an issue:

For, 16;   Against, 18;   Undecided, 6.

Using the poll, calculate the 98% confidence limits for the true proportion of voters in favour of the issue.

5. A production department experiences difficulties in the manufacture of part number 910512. A random sample of 400 parts revealed that 112 parts were defective.

   a) What is a point estimate of the proportion of non-defective parts?

b) Regarding the defective parts, what is the standard error of proportion?

c) What is the interval estimate for the population proportion of defective parts using a confidence coefficient of 0.90?

6. A survey of 240 college students indicated that 96 preferred coffee to tea.

   a) What is the point estimate of the proportion of students who prefer coffee?

   b) Compute the standard error of proportion.

   c) What is the interval estimate of the population proportion of students who prefer tea, with a 20% chance of error?

SECTION 11.5 Determination of Sample Size

A. The Maximum Error of an Interval Estimate

For the confidence intervals $\bar{x} \pm z\sigma_{\bar{x}}$, $\bar{x} \pm zs_{\bar{x}}$ and $p \pm zs_p$, the terms $z\sigma_{\bar{x}}$, $zs_{\bar{x}}$ and zs_p establish the maximum distance from the sample mean $\bar{x}$ or the sample proportion p in which the population parameters μ or π can be found with a given degree of certainty. These three terms describe the **maximum error** of an interval estimate on either side of μ or π. (See the diagram below.)

For example, in the interval $\bar{x} \pm 4$ the maximum distance from $\bar{x}$ to the upper and lower limits of the class interval is 4. This means that for the given chance of error, 4 is the maximum error that can be made when using the sample mean $\bar{x}$ to estimate the population mean μ.

B. Finding the Sample Size for Estimating a Population Mean

The process of interval estimation requires the use of samples. In most practical business situations it is first necessary to determine the size of sample that will generate the data from which an estimate of the population mean μ will be obtained.

To determine the proper sample size, three factors must be considered:

1. the confidence level for the interval estimate;
2. the size of the maximum allowable error;
3. the standard deviation of the population from which the sample is taken.

The first two factors are influenced by the nature of the research and must be specified by the researcher.

The third factor that must be considered in determining the sample size is the population standard deviation. In some situations it is possible to use the standard deviation obtained for existing data as an estimate of σ. If no such data are available, a preliminary sample must be taken to compute s as an estimate of σ.

The two formulas

$$\text{MAXIMUM ERROR} = zs_{\bar{x}} \quad \text{and} \quad s_{\bar{x}} = \frac{s}{\sqrt{n}}$$

can then be used to calculate the sample size n.

For example, if the specified maximum error is 4, the confidence level is set at 95% (in which case $z = 1.96$), and the sample standard deviation is 15.00, the required sample size can be found by first substituting in the formula

$$\text{MAXIMUM ERROR} = zs_{\bar{x}}$$
$$4.00 = 1.96\,s_{\bar{x}}$$
$$s_{\bar{x}} = \frac{4.00}{1.96} = 2.0408$$

Now substituting in the second formula

$$s_{\bar{x}} = \frac{s}{\sqrt{n}}$$

we obtain

$$2.0408 = \frac{15.00}{\sqrt{n}}$$
$$\sqrt{n} = \frac{15.00}{2.0408} = 7.35$$
$$n = 7.35^2 = 54.0225 \approx 55$$

The two formulas can be combined into one to yield the formula

$$\text{MAXIMUM ERROR} = z\left(\frac{s}{\sqrt{n}}\right)$$

from which we obtain

$$\sqrt{n} = \frac{zs}{\text{MAXIMUM ERROR}}$$
$$n = \left(\frac{zs}{\text{MAXIMUM ERROR}}\right)^2$$

In our example,

$$n = \left(\frac{1.96(15.00)}{4.00}\right)^2 = (7.35)^2 = 54.0225 \approx 55$$

○ EXAMPLE 11.5a

The loans manager of a credit union wants to estimate the average outstanding loan balance with 99% confidence and a maximum error of $200. To obtain information about the standard deviation a sample of 20 loans was randomly selected. This sample showed an average loan balance of $8000 with a standard deviation of $1200. What sample size is required to estimate the true average balance with 99% confidence and a maximum error of $200?

● **SOLUTION**

For 99% confidence, $z = 2.58$; the specified maximum error $= 200$; the standard deviation $s = 1200$.

$$n = \left(\frac{zs}{\text{MAXIMUM ERROR}}\right)^2 = \left(\frac{2.58(1200)}{200}\right)^2 = (15.48)^2 = 239.6304$$

The required sample size is 240.

C. Finding the Sample Size for Estimating a Proportion

To determine the sample size required to estimate a proportion we again need information about three factors:
1. the desired level of confidence;
2. the maximum allowable error (in this case stated as a percent);
3. the population proportion π.

Again, the values for the first two of these factors must be specified by the researcher. The third factor, the population proportion π, is the value that we wish to estimate at the specified level of confidence and the specified maximum error. For determining the sample size we can again resort to historical data available for situations closely resembling that being studied, or we can take a small sample to get a rough estimate of π.

However, when determining the sample size for estimating proportions we can determine the maximum sample size that will satisfy the specifications for the confidence level and the maximum error by utilizing the *maximum error term* in the confidence interval.

The two applicable formulas,

$$\text{MAXIMUM ERROR} = zs_p \quad \text{and} \quad s_p = \sqrt{\frac{p(1-p)}{n}}$$

can be combined into the one formula to calculate n directly:

$$n = \frac{z^2(p)(1-p)}{(\text{MAXIMUM ERROR})^2}$$

Since the value $p(1-p)$ becomes a *maximum* when $p = 0.5$, that is, $p(1-p) = (0.5)(0.5) = 0.25$,

$$\text{MAXIMUM VALUE OF } n = \frac{0.25(z^2)}{(\text{MAXIMUM ERROR})^2}$$

This approach to computing the sample size ensures that the sample size is not too small, and can be used when it is not possible to obtain an estimate of the true population proportion. However, if the actual population proportion is either much larger or much smaller than 50%, the sample size may be much larger than needed. This, in turn, may lead to unnecessary costs.

○ EXAMPLE 11.5b

Suppose the manager of the credit union in Example 11.4a wanted to estimate the proportion of members whose biweekly salaries are deposited by electronic transfer within five percentage points of the true population proportion with 95% confidence.

a) Determine the sample size, assuming that the manager has no idea what the proportion might be and does not want to spend time taking a sample.

b) Determine the sample size, assuming that the manager uses as an estimate of π the sample proportion of 38% obtained from the random sample of 250 depositors.

● SOLUTION

For 95% confidence, $z = 1.96$; the maximum error $= 5\% = 0.05$.
For (a), by setting $p = 0.5$,

$$n\,(\text{maximum}) = \frac{0.25(z^2)}{(\text{MAXIMUM ERROR})^2} = \frac{0.25(1.96)^2}{(0.05)^2}$$

$$= \frac{0.25(3.8416)}{0.0025} = 384.16$$

The manager would sample 385 accounts.
For (b), $p = 0.38$.

$$n = \frac{z^2(p)(1-p)}{(\text{MAXIMUM ERROR})^2} = \frac{(1.96)^2(0.38)(0.62)}{(0.05)^2}$$

$$= \frac{3.8416(0.2356)}{0.0025} = 362.03$$

The manager should sample 363 accounts.

Note When dealing with calculations of sample size, always round up to the next whole number.

EXERCISE 11.5

1. Compute the sample size, given the following information:
   a) $\sigma_{\bar{x}} = 8.2$; $\sigma = 319.8$
   b) $s_{\bar{x}} = 0.13$; $s = 2.08$
   c) $\sigma_{\bar{x}} = 29.5$; $\sigma = 383.5$; $N = 1000$
   d) allowable error $= 12$; $z = 1.96$, $\sigma = 144$
   e) allowable error $= 0.05$; $z = 2.58$; $p = 0.70$

2. A population has a mean of $680 and a standard deviation of $25. Determine the sample size so that the sample mean will be in the interval $670 to $690 at a confidence level of

a) 98%;   **b)** 90%.

3. A local car dealer believes sales are in part determined by the relationship between the retail price of the automobiles and current average family income. The dealership has hired a market research firm to determine average family income in the area. Existing data indicate that the standard deviation is $3600. How large a sample should be taken to determine current average family income with 90% confidence and a maximum allowable error of $200?

4. Users of a large computer system have been complaining about the slow response time to their database inquiries. Management has decided to study the problem by randomly sampling the terminals. Project guidelines require 95% confidence in the results and a maximum allowable error of 0.5 seconds. A small pilot study found a mean response time of 6.8 seconds, with a standard deviation of 1.5 seconds. How large a sample should be taken?

5. All new employees of Hard Sell Inc. are subject to a performance review to determine whether they should be offered a permanent job. An important consideration in assessing performance for sales staff is their success rate for closing deals. Management wants to be 98% confident with a maximum error of 2% in estimating an employee's true ability to close deals.
   a) Compute the sample size required without any estimate of an employee's success rate.
   b) Compute the sample size knowing that during a particular time period deals were closed 30% of the time.

6. A cable company wants to know the proportion of homes in a suburban area still using TV antennas. In other similar areas 37% of the homes use TV antennas. Determine the sample size if the company wants to know the true proportion with a maximum error of 0.025 at the 90% confidence level.

SECTION 11.6 Interval Estimation around the Mean for Small Samples ($n \leq 30$) when σ Is Unknown

A. The t Distribution

When the size of the sample is small ($n \leq 30$), the use of the normal distribution for the construction of confidence intervals is no longer appropriate. In this case we use student's **t distribution**.

The shape of the t distribution is similar to that of the normal distribution in that it is bell-shaped, symmetrical and continuous. However, the t distribution is more widely dispersed and flatter, and its shape depends on the sample size.

The areas under the standardized t distribution curve depend on a factor known as **degrees of freedom**. For the t distribution, the degrees of freedom are defined to be one less than the sample size; that is the degrees of freedom, d.f. $= (n - 1)$.

As the number of degrees of freedom increases with increasing sample size, the shape of the t distribution approaches the normal distribution more and more closely, and when $n > 30$ the normal distribution can be used.

B.   The Concept of "Degrees of Freedom"

To understand the concept of degrees of freedom, consider the following example.

To estimate μ we need to know $\bar{x}$. The sample mean $\bar{x}$ is computed using the formula $\bar{x} = \dfrac{\sum x}{n}$.

For a given value of $\bar{x}$, $(n-1)$ of the observation making up $\sum x$ can have any value, but the last observation must have a value that, when added to the other $(n-1)$ values, equals n times the mean.

For example, if $n = 6$ and $\bar{x} = 7$, $\sum x = 6(7) = 42$. The first five values can be chosen in any way, such as 6, 4, 9, 10 and 8. The last value must equal $42 - (6 + 4 + 9 + 10 + 8) = 42 - 37 = 5$. If the first five observations have the values 3, 7, 8, 9 and 11, the last value $= 42 - (3 + 7 + 8 + 9 + 11) = 42 - 38 = 4$.

This means that for a given mean $\bar{x}$ and $n = 6$, the first five values are free to vary, but the last value has no such freedom. The computation of $\bar{x}$ is subject to $(6 - 1) = 5 = (n - 1)$ degrees of freedom.

C.   Determination of the Confidence Interval

When σ is unknown and $n \leq 30$, the confidence interval is given by

$$\bar{x} \pm t_{n-1}\, s_{\bar{x}}$$

where $\bar{x}$ = the sample mean;

   n = the sample size;

   N = the population size;

   $(n - 1)$ = degrees of freedom;

   t_{n-1} is found in a table of t values;

   $s_{\bar{x}} = \dfrac{s}{\sqrt{n}}$ for infinite populations, or

   $s_{\bar{x}} = \dfrac{s}{\sqrt{n}} \sqrt{\dfrac{N - n}{N - 1}}$ for finite populations.

D. Using the t Table

A table of values of the t distribution for degrees of freedom from 1 to 30 is provided in Table 11.4. The figures in the body of the table are the t values associated with a specified number of degrees of freedom for the areas in the tails of the curve representing a t distribution. Depending on the design of a t table, the table look-up can be based either on the area in one tail or the combined area in both tails (see diagrams A and B below).

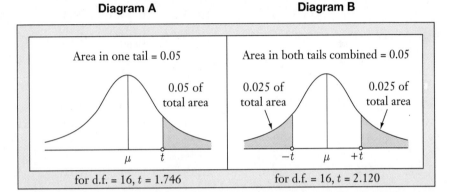

| **Diagram A** | **Diagram B** |

for d.f. = 16, t = 1.746   for d.f. = 16, t = 2.120

The columns in a t table are headed by selected proportions of the area in one tail or both tails combined. The most common proportions selected for one-tail tables are 0.10, 0.05, 0.025, 0.01 and 0.005 (see Diagram A).

The corresponding proportions for the combined area are 0.20, 0.10, 0.05, 0.02 and 0.01 (see Diagram B). The proportions for the combined area represent the chance of error when constructing confidence intervals and correspond to the confidence coefficients 0.80, 0.90, 0.95, 0.98 and 0.99 respectively.

Table 11.4 is designed for looking up t values for the *combined* area in the two tails (see Diagram B).

○ **EXAMPLE 11.6a**

Find the t values for the following:
a) Sample size n = 10 at a confidence level of 90%.
b) Sample size n = 6 at a confidence level of 99%.
c) Sample size n = 24 with a 20% chance of error.
d) Sample size n = 13 with a 2% chance of error.

TABLE 11.4 *t* Values for Areas in Both Tails Combined

Degrees of freedom	Area in both tails combined — proportion of total area				
	0.20	0.10	0.05	0.02	0.01
1	3.078	6.314	12.706	31.821	63.657
2	1.886	2.920	4.303	6.965	9.925
3	1.638	2.353	3.182	4.541	5.841
4	1.533	2.132	2.776	3.747	4.604
5	1.476	2.015	2.571	3.365	4.032
6	1.440	1.943	2.447	3.143	3.707
7	1.415	1.895	2.365	2.998	3.499
8	1.397	1.860	2.306	2.896	3.355
9	1.383	1.833	2.262	2.821	3.250
10	1.372	1.812	2.228	2.764	3.169
11	1.363	1.796	2.201	2.718	3.106
12	1.356	1.782	2.179	2.681	3.055
13	1.350	1.771	2.160	2.650	3.012
14	1.345	1.761	2.145	2.624	2.977
15	1.341	1.753	2.131	2.602	2.947
16	1.337	1.746	2.120	2.583	2.921
17	1.333	1.740	2.110	2.567	2.898
18	1.330	1.734	2.101	2.552	2.878
19	1.328	1.729	2.093	2.539	2.861
20	1.325	1.725	2.086	2.528	2.845
21	1.323	1.721	2.080	2.518	2.831
22	1.321	1.717	2.074	2.508	2.819
23	1.319	1.714	2.069	2.500	2.807
24	1.318	1.711	2.064	2.492	2.797
25	1.316	1.708	2.060	2.485	2.787
26	1.315	1.706	2.056	2.479	2.779
27	1.314	1.703	2.052	2.473	2.771
28	1.313	1.701	2.048	2.467	2.763
29	1.311	1.699	2.045	2.462	2.756
30	1.310	1.697	2.042	2.457	2.750
z value	1.28	1.64	1.96	2.33	2.58

Note When $n > 30$ the *t* values become close enough to the corresponding *z* values so that the normal distribution can be used.

● **SOLUTION**

a) The number of degrees of freedom,

$$d.f. = (n-1) = (10-1) = 9.$$

For a confidence level of 90% the chance of error is 10%. The combined area in the two tails is 0.10; the area in each tail is 0.05. In Table 11.4 locate the row headed by 9 and the column headed by 0.10. The desired t value = 1.833.

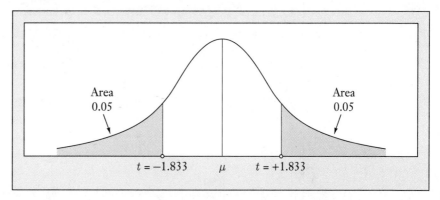

b) For a 99% confidence level the combined area in the tails is 0.01 (0.005 in each tail). For $n = 6$, d.f. = 5. The t value = 4.032.

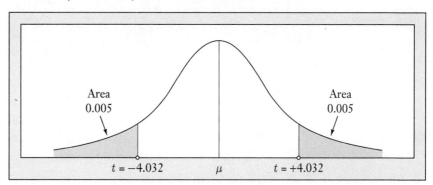

c) For a 20% chance of error the combined area in the tails is 0.20 (0.10 in each tail). For $n = 24$, d.f. = 23. The t value = 1.392.

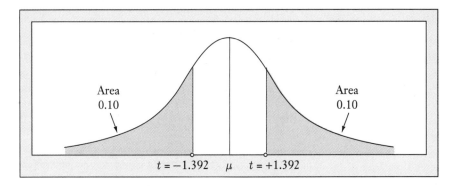

d) For a 2% chance of error the combined area in the tails is 0.02 (0.01 in each tail). For $n = 13$, d.f. $= 12$. The t value $= 2.681$.

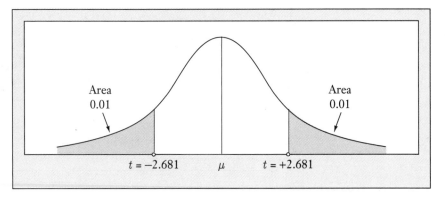

E. *Constructing Confidence Intervals Using t Values*

When σ is unknown, the confidence interval in which the mean can be found for small samples ($n \leq 30$) with a given degree of confidence is given by

$$\overline{x} \pm t_{n-1} s_{\overline{x}}$$

where $\overline{x}$ = the mean of the sample;

t_{n-1} = the value found in a t table for $(n-1)$ degrees of freedom for a specified chance of error = $(1 -$ the confidence coefficient);

$s_{\overline{x}}$ = the standard error.

○ EXAMPLE 11.6b

A sample of size 25 has a mean of 150 and a standard deviation of 35. Construct a 99% confidence interval.

● SOLUTION

$n = 25$; $\overline{x} = 150$; $s = 35$; N is unknown; the population is infinite.

The standard error for an infinite population

$$s_{\overline{x}} = \frac{s}{\sqrt{n}} = \frac{35}{\sqrt{25}} = \frac{35}{5} = 7$$

Since $n < 30$, the use of the t table is appropriate.
For 99% confidence the chance of error is 1%.
For $n = 25$, d.f. = 24.
From Table 11.4 the t value for 24 degrees of freedom and a chance of error of $0.01 = 2.797$.
The confidence interval is given by
$\overline{x} \pm t_{n-1} s_{\overline{x}} = 150 \pm 2.797(7) = 150 \pm 19.579$, that is, 130.421 to 169.579.

○ EXAMPLE 11.6c

To reduce costs, the management of a delivery service is considering the use of propane or natural gas for the company's fleet of 256 panel trucks as an alternative to gasoline. To assist in the determination of cost savings, the fleet manager took a random sample of five trucks and measured their current gasoline consumption in kilometres per litre as follows:

$$9.0 \quad 10.8 \quad 8.4 \quad 10.3 \quad 11.7$$

Estimate the true gasoline consumption at the 95% confidence level.

● SOLUTION

Since σ is unknown and $n < 30$, the interval in which μ will be found is $\overline{x} \pm t_{n-1} s_{\overline{x}}$.
The chance of error (area in both tails combined) = 0.05 and d.f. = $(n-1)$ = $(5-1) = 4$.
The t value for d.f. = 4 and a chance of error of $0.05 = 2.776$.
The sample mean $\overline{x}$ and the unbiased sample standard deviation s can be computed from the sample data.

x	9.0	10.8	8.4	10.3	11.7	$\sum x = 50.2$
x^2	81.00	116.64	70.56	106.09	136.89	$\sum x^2 = 511.18$

$$\overline{x} = \frac{\sum x}{n} = \frac{50.2}{5} = 10.04$$

$$s^2 = \frac{n\sum x^2 - (\sum x)^2}{n(n-1)} = \frac{5(511.18) - (50.2)^2}{5(4)}$$

$$= \frac{2555.90 - 2520.04}{20} = \frac{35.86}{20} = 1.793$$

$$s = \sqrt{1.793} = 1.3390.$$

Since $N = 256$, the population is finite.

$$s_{\bar{x}} = \frac{s}{\sqrt{n}}\sqrt{\frac{N-n}{N-1}} = \frac{1.3390}{\sqrt{5}}\sqrt{\frac{256-5}{256-1}} = \frac{1.3390}{2.2361}\sqrt{\frac{251}{255}}$$

$$= 0.5988\sqrt{0.984314} = 0.5988(0.9921) = 0.5941$$

The confidence interval $= 10.04 \pm 2.776(0.5941) = 10.04 \pm 1.65$.
The interval in which μ is expected to be found is 8.39 to 11.69.
The management of the delivery service can be 95% confident that the true average gasoline consumption by its fleet of trucks is somewhere between 8.39 and 11.69 km/L.

○ **EXAMPLE 11.6d**
Assume that the manager of the delivery service took another sample of size 30. This sample yielded a mean of 9.95 km/L with a standard deviation of 1.35 km/L. Construct a 95% confidence interval using
a) the t tables; b) the z tables.

● **SOLUTION**
$N = 256$; $n = 30$; $\bar{x} = 9.95$; $s = 1.35$.

$$s_{\bar{x}} = \frac{1.35}{\sqrt{30}}\sqrt{\frac{256-30}{256-1}} = \frac{1.35}{5.4772}\sqrt{\frac{226}{255}} = 0.2465\sqrt{0.8863}$$

$$= 0.2465(0.9414) = 0.2320$$

a) For 95% confidence the chance of error $= 0.05$; d.f. $= 29$; the corresponding t value $= 2.045$.
   The confidence interval $= \bar{x} \pm t_{n-1}s_{\bar{x}} = 9.95 \pm 2.045(0.2320)$
   $= 9.95 \pm 0.474$.
   The range of values in which μ is expected to be found is 9.476 to 10.424 km/L.
b) For 95% confidence, $z = 1.96$.
   The confidence interval $= \bar{x} \pm zs_{\bar{x}} = 9.95 \pm 1.96(0.2320) = 9.95 \pm 0.455$.
   The range of values in which μ is expected to be found is 9.495 to 10.405 km/L.

Note Since the t value is larger than the z value, the t value gives a wider confidence interval.

EXERCISE 11.6

1. Determine the t value for each of the following:
   a) $n = 25$; the chance of error is 1%.
   b) $n = 20$; the confidence coefficient is 0.80.
   c) $n = 5$; the area in the two tails combined is 0.10.
   d) $n = 11$; the area in the right tail is 0.01.
   e) $n = 30$; the area in each tail is 0.025.

2. A sample of 12 items has a mean of 7.3 and a standard deviation of 2.4. Compute the 98% confidence limits for the population mean.

3. Steelco's fastener division uses statistical process control in the production of head bolts for the Big Three automakers. This requires the operators to take a random sample of eight bolts at randomly chosen intervals. One such sample had a mean of 8.00 mm and a standard deviation of 0.02 mm. Determine the confidence interval using a confidence coefficient of 0.95.

4. In an effort to improve cash flow, the supervisor of an accounts receivable department decided to randomly check 20 of the company's 500 accounts on a daily basis. Yesterday's sample mean was $10 800 with a standard deviation of $1350. Estimate the true mean with a 10% chance of error.

5. The owner of a restaurant wants to estimate the daily average consumption of beef. A sample of seven days revealed the following daily usage in kilograms:

    17.5   14.3   20.8   16.5   18.0   16.8   21.4

    Construct the confidence interval for a 20% chance of error.

6. A survey of employees in comparable positions in different companies yielded the following hourly wages:

$12.60	$9.60	$10.40	$11.00	$9.20
$10.90	$11.00	$9.80	$12.00	$10.50

Estimate the average hourly wage with 90% confidence if the total number of employees having comparable positions is 800.

REVIEW EXERCISE

1. Calculate the confidence interval for each of the following:
   a) $\sigma = 25$, $\bar{x} = 300$, $n = 49$, confidence coefficient = 0.90;
   b) $\sigma = 9$, $\bar{x} = 36$, $n = 16$, $N = 330$, chance of error = 0.01;
   c) $s = 2$, $\bar{x} = 1300$, $n = 144$, $N = 4800$, confidence level = 98%;
   d) $s = 0.80$, $\bar{x} = 16.00$, $n = 16$, chance of error = 0.05.

2. Determine the confidence interval for each of the following:
   a) $\sigma = 3$, $\bar{x} = 20$, $n = 81$, $N = 1090$, chance of error = 0.20;
   b) $s = 0.40$, $\bar{x} = 4.60$, $n = 24$, $N = 730$, confidence level = 99%;
   c) $\sigma = 11$, $\bar{x} = 640$, $n = 25$, confidence coefficient = 0.95;
   d) $s = 40$, $\bar{x} = 1600$, $n = 36$, chance of error = 0.10.

3. From the equation $z\sigma_{\bar{x}} = 2.33 \left(\dfrac{67.20}{\sqrt{64}} \right)$, determine
   a) the sample size;
   b) the confidence level;
   c) the standard deviation;
   d) the standard error;
   e) the maximum allowable error.

4. From the equation $t_{n-1} s_{\bar{x}} = 2.120 \left(\dfrac{14.00}{\sqrt{17}} \right) \sqrt{\dfrac{583}{599}}$, determine

   a) the population size;
   b) the confidence level;
   c) the standard deviation;
   d) the standard error;
   e) the maximum error.

5. A sample of 56 is drawn from a population that has a standard deviation of 3.4 g. If the sample mean is 375 g, construct the confidence interval that contains the true mean with 90% certainty.

6. Given a population of 4400 with a standard deviation of $3280, compute the upper and lower limits of an 85% confidence interval for a sample of 150 with a sample mean of $21 850.

7. J.L. Condie Publishers wants to know the time it takes from finding an author to write a textbook to publishing it. A random sample of 64 textbooks had a mean time of 20.8 months with a standard deviation of 3.6 months. Using a confidence coefficient of 0.90, determine the time interval in which the true mean can be expected to be found.

8. The owner of a Tom Norton Donut Shop is reviewing staffing needs for the morning shift based on the length of lineup. If the lineup is six or more customers, another employee will be added to the morning shift. A record of the number of customers in the line taken at 45 different occasions over a period of one week showed an average lineup of 4.2 customers with a standard deviation of 0.8. Construct the 90% confidence interval to help the owner in making the staffing decision.

9. A truckload of fresh lobsters from the east coast awaits unloading at the food terminal. The buyer wants only market-size lobsters (0.50 kg or over). To check on the weight specification, the buyer takes a random sample of 32 lobsters. The total weight of the sample was 20 kg and the standard deviation was 0.21 kg. The buyer wants only a 2% chance of error.

   a) Compute the upper and lower limits of the confidence interval.
   b) Based on the confidence interval, should the buyer accept or reject the shipment?
   c) Recalculate the confidence interval if the shipment consists of 200 lobsters.

10. Aglobulin's chamber of commerce has 793 members. In a confidential survey of 60 members it was learned that the members had an average income of $83 427 with a standard deviation of $5382. Compute the confidence interval at the 99% level.

11. Friday's trading volume on the TSE was 23 412 083 shares. A random sample of 50 stocks had a mean trading price of $16.42 and a standard deviation of $4.89. Calculate the interval limits for the average trading price at a confidence coefficient of 0.95.

12. ABC Distributors owns a fleet of 180 cars. A random sample of 36 cars showed their average mileage for the year to be 22 580 km with a standard deviation of 2915 km. Determine the true average mileage with 98% certainty.

13. Look up the t value for each of the following:
   a) $n = 20$, the combined area = 0.10;

b) $n = 15$, the area in one tail = 0.10;
c) $n = 9$, the chance of error = 0.05;
d) $n = 26$, the confidence coefficient = 0.99.

14. The t values for each of the following are not listed in the t tables included in the text. Look up the two closest values in the t table that are on either side of the required value.
a) $n = 18$, the area in the right tail = 0.04;
b) $n = 28$, the combined area = 0.15;
c) $n = 7$, the confidence level = 96%;
d) $n = 14$, the chance of error = 0.08.

15. A psychologist doing research into dreams has determined that the 1380 cases reviewed showed a standard deviation of 1.8 for the number of dreams per night. If a sample of 12 from the cases reviewed had a mean of 4.6 dreams per night, construct a confidence interval around the sample mean with a 2% chance of error.

16. For a population with a standard deviation of 13, construct an 80% confidence interval around the sample mean of 110 for a sample of size 9.

17. A pizzeria offers its product free if it is not served within 12 minutes after the order has been placed. Before starting this promotion, the owner had taken a small sample to obtain an estimate of the true average time elapsed between placing the order and serving the customer. Elapsed time, in minutes, for the sample was as follows:

9.3	10.0	9.8	10.1	9.5	10.3
9.1	10.5	9.9	10.5	10.1	9.7

a) Calculate the sample mean and the unbiased standard deviation.
b) Construct a 99% confidence interval.

18. Quick Clean Window Services currently charges residential customers according to the number and size of windows being cleaned. The problem with this approach is that it requires the owner to visit the customer and prepare a quote. To determine if the need for a quote could be eliminated by charging a flat fee regardless of the number of windows to be cleaned, the owner decided to research the average time spent by a work crew for cleaning the windows of a number of residences. The study showed the following times (in hours):

     3.2   3.8   3.5   3.9   4.0   3.1   3.4   3.5

a) Determine the sample mean and the standard error.
b) Determine the upper and lower limits for a 0.98 confidence coefficient.
c) Should the owner go with the flat-fee approach?

19. During a 60-lap Formula One race in Mont Tremblant, Faster Foster's pit crew randomly sampled 10 laps and found the mean speed per lap to be 158 km/h with a standard deviation of 12.6 km/h. Assuming that the lap speed is normally distributed, compute the confidence interval in which the true average lap speed can be expected to be found with a 5% chance of error.

20. Robin's Bar and Grill uses an ice-cream dispensing machine that is precalibrated and can be set to dispense small, medium or large ice-cream cones. A random sample of 24 small ice-cream cones weighed an average of 86 g with a standard deviation of 6.1 g. Compute the 98% confidence interval for the mean weight of a small cone filled by the machine.

21. The province of Alberta wants to estimate the proportion of women employed in the public sector. A random sample of 150 public-service employees contained 51 women. Construct a 90% confidence interval for the true proportion of women in Alberta's public sector.

22. A survey of 250 newlyweds found that 73% of them had had premarital sex. Construct a 90% confidence interval for the true proportion of newlyweds that had had premarital sex.

23. Laura's Catering Service provides a bar and bartender for catering parties. Last year, 156 parties needed a bar. A random sample of 39 of these parties showed that rye was ordered for 58% of the parties. Compute the 95% confidence interval in which the true proportion of parties for which rye was ordered can be found.

24. Out of 40 male students selected from a high-school population of 376 male students, 28 watched the TV program "The Young and the Restless." Construct the 98% confidence interval.

25. Calculate the approximate sample size required given the following information:
maximum allowable error = 1000
confidence coefficient   = 0.90
sample standard deviation = 12 195

26. Compute the approximate sample size for the following:
maximum error   = 0.02
chance of error   = 0.02
sample proportion = 0.80

27. The true proportion for a population is 0.55. There is 90% certainty that a sample proportion will fall between 0.53 and 0.57. What sample size was used to compute the confidence limits?

28. The true mean for a population is known to be 6350 kg. It is also known that there is a 99% chance that a sample mean will fall between 6410 kg and 6290 kg. A random sample taken from the population had a standard deviation of 421 kg. What sample size was used to determine the confidence interval?

29. The research department of Gordon Products Limited has estimated the total cost of a survey of 300 households to be $34 500. The project manager believes the total survey cost to be prohibitive and wants to know if the total cost can be reduced by using a smaller sample without changing the maximum allowable error. The research department indicated that a smaller sample could be used by reducing the proposed 95% confidence level but warned that this would result in an increase in the sampling error.
a) Calculate the sample size if the confidence level is dropped to 90%.
b) Determine the increase in the sampling error and the cost savings by reducing the confidence level to 90%.

30. A freelance writer would like to submit a business article on executive salaries including an interval estimate for the true average executive salary. The publisher is very interested in the idea, provided that the writer meets the following requirements for publication:
i) the maximum error in the estimate must not exceed $2500;
ii) the estimate must have no more than a 5% chance of error.

To determine the appropriate sample size, the writer took a small random survey of executives and obtained the following data ($000):

| 85 | 82 | 79 | 98 | 105 | 89 | 95 | 100 |

Compute the sample size that will meet the publisher's conditions.

31. Algonquin Airlines would like know the proportion of passengers who require special assistance. The company wants to select a sample that provides 95% confidence that the proportion of special-needs passengers is estimated within two percentage points of the true proportion. How large a sample is required?

32. A city's urban planning department is looking into rerouting trucks around the downtown core. Joseph Lahda, the senior planner, has proposed that a traffic study be done to establish the proportion of trucks using downtown streets. The study is to provide 98% confidence in the results, with a maximum allowable error of 0.015. Previous studies on file suggest that passenger car traffic accounts for 83% of traffic volume. How large a sample should be used for the study?

SELF-TEST

1. The variation in the number of persons employed by the businesses of a small community is indicated by the standard deviation of 4 persons. A random sample of 32 businesses showed the average number of persons employed to be 10.
   a) Construct a confidence interval in which the true average number of employees can be found with 95% certainty.
   b) Construct the corresponding interval if the number of businesses in the community is 210.

2. An import-export company has randomly selected 35 days and determined that the average daily cost of postage is $32.85, with a standard deviation of $2.15. The sample also indicated that 65% of the company's mail is international.
   a) Construct a confidence interval for the mean daily cost of postage with a chance of error of no more than 2%.
   b) Construct an 85% confidence interval for the proportion of domestic mail.

3. Peter sorts mail in the local post office and was speculating about the mean number of pieces of mail he put into each post-office box. His daughter Alexandra, a business student, suggested that he randomly choose 55 boxes and provide her with the count of pieces of mail in each box. For the sample provided by her father the next day, Alexandra found the mean to be 6.8 pieces of mail per box, with a standard deviation of 2.7 pieces. Construct a 99% confidence interval for the average number of pieces of mail per post-office box.

4. Cutrate Lumber Company would like to know the proportion of boards rejected per lift. A sample of 60 lifts showed that 9.8% of the boards in each lift were non-saleable. Construct a 90% confidence interval for the true proportion of rejects.

5. A small brewery monitors the alcohol content of its beer by taking 6 samples

from every vat brewed. The following are the percent alcohol content data for the samples taken from vat No. 010527-M:

$$5.3 \quad 5.0 \quad 5.1 \quad 5.4 \quad 5.3 \quad 5.3$$

a) Calculate the mean and standard deviation for the alcohol content of the samples taken.

b) Construct a confidence interval for the mean alcohol content with a 2% chance of error.

6. A major chain of food stores wants to determine the proportion of customers who buy on impulse. The company decided to use a sample large enough to estimate the proportion of impulse buyers within 2.5 percentage points of the true proportion with 95% certainty. How large should the sample be?

7. Wedgewood Golf Products plans to retail golf tees in a most unusual manner. For a flat price, customers put one hand into a large jar and take out as many tees as they can grab. To determine the flat price, the manager wants to estimate the average number of tees per handful with a maximum error of one tee and a 5% chance of error. For the small number of store employees who took part, the standard deviation was 3 tees. What is the required size for the sample?

Key Terms

Summary of Formulas

1. Sample mean: $\bar{x} = \dfrac{\sum x}{n}$

2. Sample variance: $s^2 = \dfrac{\sum (x - \bar{x})^2}{n - 1}$

or $s^2 = \dfrac{n\sum x^2 - (\sum x)^2}{n(n - 1)}$

3. Sample standard deviation: $s = \sqrt{s^2}$

4. Confidence intervals for mean:

  a) When σ is known:

    i) $\bar{x} \pm z\sigma_{\bar{x}} = \bar{x} \pm z\dfrac{\sigma}{\sqrt{n}}$ when N is infinite or unknown.

    ii) $\quad\quad = \bar{x} \pm z\dfrac{\sigma}{\sqrt{n}}\sqrt{\dfrac{N-n}{N-1}}$ when N is known.

  b) When σ is not known and the sample size is large $(n > 30)$:

    i) $\bar{x} \pm zs_{\bar{x}} = \bar{x} \pm z\dfrac{s}{\sqrt{n}}$ when N is infinite or unknown.

    ii) $\quad\quad = \bar{x} \pm z\dfrac{s}{\sqrt{n}}\sqrt{\dfrac{N-n}{N-1}}$ when N is known.

  c) When σ is not known and the sample size is small $(n \leq 30)$:

    i) $\bar{x} \pm \sigma_{n-1}\, s_{\bar{x}} = \bar{x} \pm \sigma_{n-1}\dfrac{s}{\sqrt{n}}$ when N is infinite or unknown.

    ii) $\quad\quad = \bar{x} \pm \sigma_{n-1}\dfrac{s}{\sqrt{n}}\sqrt{\dfrac{N-n}{N-1}}$ when N is known.

5. Confidence interval for a proportion when π is unknown $(n > 30)$:

  i) $p \pm zs_p = p \pm z\sqrt{\dfrac{p(1-p)}{n}}$ when N is infinite or unknown.

  ii) $\quad\quad = p \pm z\sqrt{\dfrac{p(1-p)}{n}}\sqrt{\dfrac{N-n}{N-1}}$ when N is known.

6. Sample size for mean:

$$n = \left(\frac{zs}{\text{MAXIMUM ERROR}}\right)^2$$

7. Sample size for a proportion:

  i) $n = \dfrac{s^2(p)(1-p)}{(\text{MAXIMUM ERROR})^2}$

  ii) MAXIMUM VALUE OF $n = \dfrac{0.25(s^2)}{(\text{MAXIMUM ERROR})^2}$

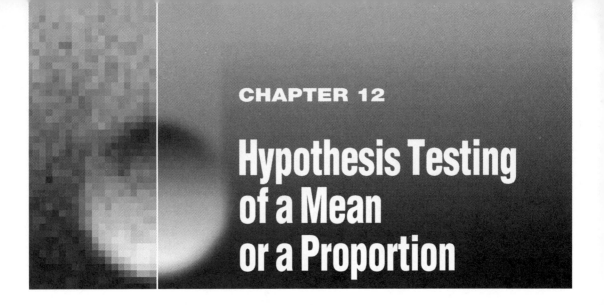

CHAPTER 12

Hypothesis Testing of a Mean or a Proportion

Introduction

When estimating the population mean or the population proportion from sample information, our main concern was determining the interval within which μ or π could be expected to be found for a chosen confidence level. In this chapter we will deal with hypothesis testing as another way of drawing conclusions about the population mean μ or the population proportion π from the sample statistics $\bar{x}$ or p.

Objectives

Upon completion of this chapter you will be able to
1. understand the concept of hypothesis testing;
2. set up null and alternate hypotheses for given problem situations in both descriptive and symbolic terms;
3. distinguish between one-tail and two-tail tests;
4. understand the concept of significance level and the importance of choosing the significance level for a test;
5. determine the critical value of the test statistic for a test;
6. state the decision rule for a test;
7. compute the sample test statistic and interpret the result;
8. understand the concept of Type I and Type II error;
9. determine the actual significance level (p value) of a test.

SECTION 12.1 ## The Hypothesis Testing Procedure

The basic idea behind hypothesis testing is to formulate a hypothesis and then decide, on the basis of sample evidence, whether to accept the hypothesis as a reasonable description of the situation or to reject it as an unreasonable description.

The procedure for carrying out a statistical hypothesis test is as follows:
1. Formulate the null hypothesis and the alternate hypothesis.
2. Identify the type of test to be used.
3. Select the level of significance.
4. Identify the appropriate statistic to be used.
5. Determine the critical region for the test statistic.
6. State the decision rule.
7. Take a sample and compute the value of the sample test statistic.
8. Apply the appropriate decision rule and interpret the result.

In this procedure, steps 1 to 6 must precede the taking of the sample and must not be changed when the value of the sample test statistic has been determined. The reason for doing this is to prevent the creation of hypotheses to fit the sample result.

SECTION 12.2 Basic Concepts

A. Null Hypothesis and Alternate Hypothesis

In many business situations we start with a belief (hunch) that the population mean μ or the population proportion π is *greater than*, *less than* or *not equal to* some specific value.

For example, we may believe that
1. a filling machine that is supposed to fill packages with 500 g of some product is incorrectly set — that is, the contents of the packages do not weigh 500 g;
2. the concentration of some pollutant in the atmosphere is greater than 10 ppm (parts per million);
3. the yield of a new production process is greater than the yield of the current process;
4. the time required to perform a task after a training program is less than the time required to perform the task before the training program;
5. the proportion of defective items in a consignment is greater than 5%.

More often than not, we cannot test the truth of such statements directly. However, problems of this nature can be approached indirectly by assuming that the logical alternative of this belief is not reasonable.

In our five examples we set up the following statements as the logical alternatives to our beliefs:
1. Our belief is that the machine is not filling the packages with 500 g. The logical alternative of this belief is that the machine is correctly set — that is, the net weight in the package equals 500 g.
2. The concentration of the pollutant in the atmosphere equals 10 ppm.
3. The yield of the new production process equals the current yield.
4. The time required to perform the task after the training program equals the time required before the training program.
5. The proportion of defective items in the consignment equals 5%.

In the context of hypothesis testing any such logical alternative is referred to as the **null hypothesis**. Denoted by H_0, the null hypothesis is a statement about some specific value of μ or π that we hope to be able to reject at the completion of the testing procedure.

The statement that we hope to demonstrate to be true is referred to as the **alternate hypothesis**. Denoted by H_a, the alternate hypothesis is a statement about a range of values for μ or π.

For belief 1, the hypotheses, stated in *descriptive* terms, are

H_0 : The net weight of the packages equals 500 g.
H_a : The net weight of the packages is not equal to 500 g.

Stated in symbolic terms,

$H_0 : \mu = 500$ g.
$H_a : \mu \neq 500$ g.

Note that the focus of the test of hypothesis is on the null hypothesis. Upon taking a sample, we use the relevant sample statistic ($\bar{x}$ or p) to demonstrate whether it is reasonable to *accept* the null hypothesis as being true or to *reject* it as being false.

If the value of the sample statistic is such as to indicate that the null hypothesis H_0 is unlikely to be true, then the alternate hypothesis H_a is considered to be a more reasonable statement about the true state of affairs.

B. Types of Tests — One-Tail Tests versus Two-Tail Tests

The range of values specified by the alternate hypothesis determines the type of test. An alternate hypothesis that specifies a complete range of values either greater than or less than the specific value stated in the null hypothesis defines a one-sided, or *one-tail test*.

The direction of the symbol for "greater than" ($>$) or "less than" ($<$) can be used to identify the test as being a right-tail test or a left-tail test. Alternate hypotheses of the type $\mu > 500$ involve a right-tail test (the symbol $>$ points to the right) while alternate hypotheses of the type $\mu < 500$ involve a left-tail test (the symbol $<$ points to the left).

Alternate hypotheses of the type $\mu \neq 500$ imply that μ may take a complete range of values below 500 as well as above 500. This type of test is called a two-sided, or two-tail test.

The null hypotheses, alternate hypotheses and types of tests for the five situations described above are summarized in Table 12.1.

C. Significance Level

The purpose of hypothesis testing is to determine what values of the sample test statistic will lead us to decide whether to reject the null hypothesis as false or to accept it as true.

To do so we need to establish the probability level at which we are prepared to reject the null hypothesis as being a very unlikely statement about the true

TABLE 12.1   **Summary of Hypotheses and Types of Test**

Example	Hypothesis and type of test	Meaning of test
1	$H_0 : \mu = 500$ g $H_a : \mu \neq 500$ g Two-tail test	The symbol $\neq$ in H_a implies that we expect the true population mean to be either more than or less than 500 g.
2	$H_0 : \mu = 10$ ppm $H_a : \mu > 10$ ppm Right-tail test	The symbol $>$ in H_a implies that we expect the true population mean to be greater than 10 ppm.
3	$H_0 : \mu = y$ kg/batch $H_a : \mu > y$ kg/batch Right-tail test	The symbol $>$ in H_a implies that we expect the true population mean to be greater than y kg per batch.
4	$H_0 : \mu = t$ min $H_a : \mu < t$ min Two-tail test	The symbol $<$ in H_a implies that we expect the true population mean to be smaller than t minutes.
5	$H_0 : \pi = 0.5$ $H_a : \pi > 0.5$ Right-tail test	The symbol $>$ in H_a implies that we expect the true population proportion to be more than 0.05.

population parameter and therefore willing to accept the alternate hypothesis as being a more likely statement about it.

The probability chosen is called the **significance level** of the test. Denoted by the Greek letter α (read "alpha"), the significance level represents the *chance of error*, that is, the risk that the null hypothesis will be rejected when it is in fact true.

In business situations the significance levels are often set by company policy. The most frequently chosen values of α are 0.10, 0.05, 0.02 and 0.01. These selections are referred to as 10%, 5%, 2% and 1% levels of significance.

D.   *Test Statistic to Be Used*

As we are only dealing with testing a population mean or a population proportion, the test statistic of concern will involve either the normal distribution or the t distribution. This means that, depending on the testing situation, the statistic to be used will be either a z value or a t value.

The choice of distribution in a particular test depends on whether or not the population distribution is known to be normal and on the sample size. If the population is known to be *normal*, the test statistic to be used will be a z value, regardless of the size of the sample. If the population is not known to be normal, the test statistic will be a z value for large samples ($n > 30$) and a t value for small samples ($n \leq 30$).

When the test statistic is a z value, Table 12.2 can be used to select the appropriate value for the chosen significance level and the type of test to be used.

TABLE 12.2   z Values for Hypothesis Testing

Significance level α	z value	
	One-tail test	Two-tail test
0.01	2.33	2.58
0.02	2.055	2.33
0.05	1.645	1.96
0.10	1.28	1.645

When the test statistic is a t value, use Table 11.4.

E. Critical Values and Critical Regions

Referred to as **critical values**, the test statistics divide the total area under the curve representing the specific sampling distribution into two regions: the **acceptance region** and the **rejection region** (see figures 12.1, 12.2 and 12.3).

For a two-tail test the significance level establishes the combined area in the tails of a standardized sampling distribution below the negative value of the associated test statistic z or t and above their positive values.

A two-tail test is appropriate for a hypothesis test involving a "not equal to" ($\neq$) situation (see Figure 12.1).

FIGURE 12.1   Critical Regions for Two-Tail Tests

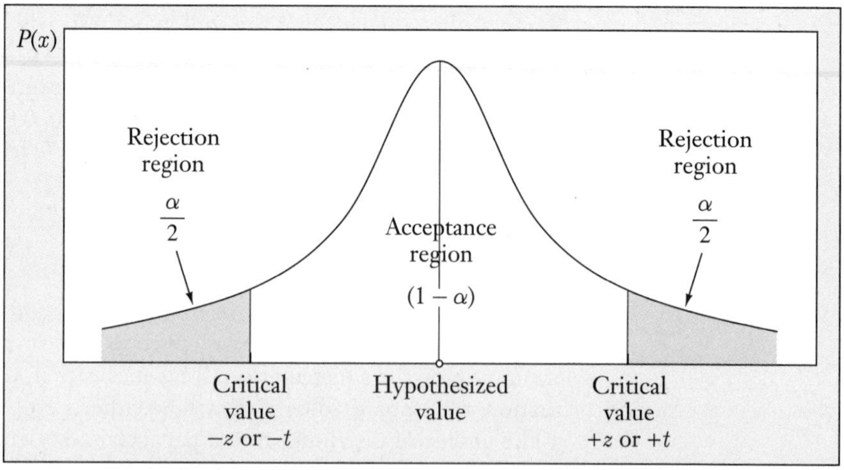

For a right-tail test the chosen significance level determines the area in the tail to the right of the positive value of the test statistic. The critical value of the test statistic divides the total area under the curve representing the sampling distribution into the acceptance region and the rejection region.

A right-tail test is appropriate for a hypothesis test involving a "greater than" ($>$) situation (see Figure 12.2).

FIGURE 12.2 **Critical Regions for Right-Tail Tests**

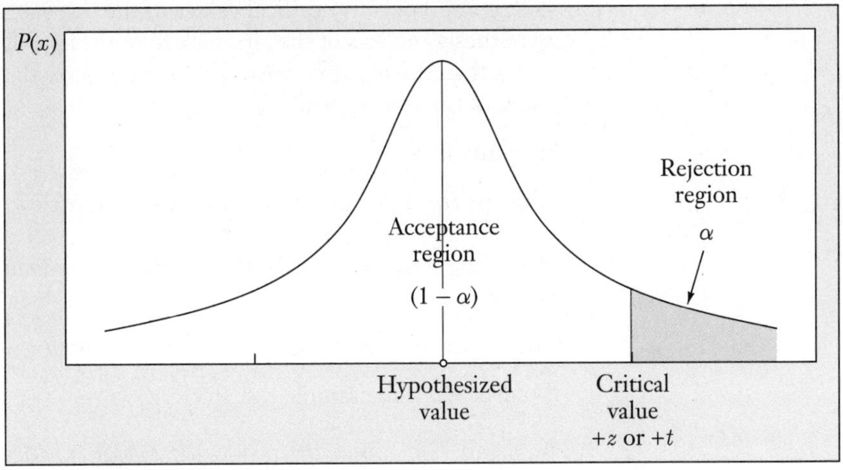

For a left-tail test, α represents the area in the tail below the negative value of the test statistic. This critical value of the test statistic divides the total area under the curve representing the sampling distribution into the acceptance region and rejection region.

FIGURE 12.3 **Critical Regions for Left-Tail Tests**

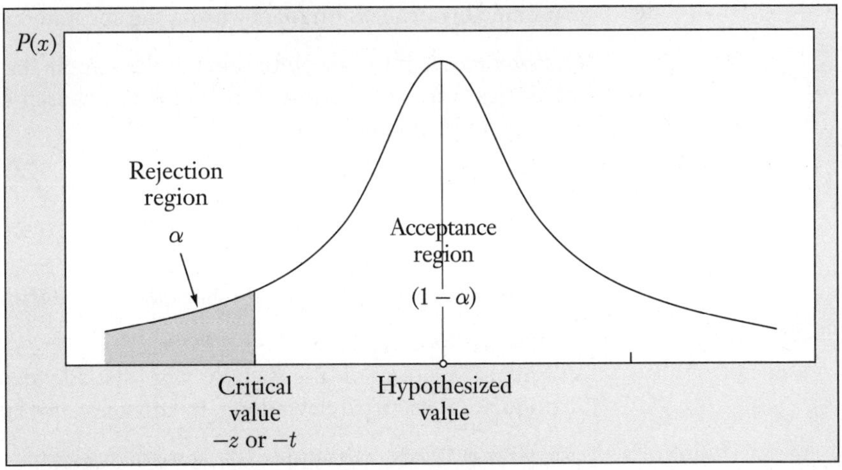

A left-tail test is appropriate for a hypothesis test involving a "less-than" (<) situation (see Figure 12.3).

F. *Stating the Decision Rule*

The decision rule outlines the conditions under which the null hypothesis will be accepted or rejected.

For a two-tail test the decision rule takes the following form:

"Accept the null hypothesis if the sample test statistic falls between the positive and negative critical values of the test statistic, or reject the null hypothesis and accept the alternate hypothesis if the sample test statistic is less than the negative critical value or greater than the positive critical value of the test statistic."

In symbolic terms:

"Accept H_0 if $(-z$ or $-t\,) <$ sample test statistic $< (+z$ or $+t\,)$

or

Reject H_0 and accept H_a if the sample test statistic $< (-z$ or $-t\,)$ or $> (+z$ or $+t\,)$."

For a right-tail test the decision rule is of the form

"Accept H_0 if the sample test statistic $< (+z$ or $+t\,)$

or

Reject H_0 and accept H_a if the sample test statistic $> (+z$ or $+t\,)$."

For a left-tail test the decision rule is of the form

"Accept H_0 if the sample test statistic $> (-z$ or $-t)$

or

Reject H_0 and accept H_a if the sample test statistic $< (-z$ or $-t\,)$."

The specific ways of stating the decision rule for each type of test can be combined in one decision rule by using the absolute values of the test statistics.

Decision rule   If the absolute value of the sample test statistic is greater than the critical value of the test statistic for the chosen significance level, reject H_0, otherwise do not.

In symbolic terms,

If |sample test statistic| > |critical value of z or t| reject H_0, otherwise do not.

G. *Computing the Value of the Sample Test Statistic*

1. *When testing a mean.*

   The calculation of the sample test statistic depends on whether the population standard deviation σ is known or not known.

   a) If σ is *known*, the sample test statistic

   $$z = \frac{\overline{x} - \mu}{\sigma_{\overline{x}}}$$

   where $\overline{x}$ = the sample mean;
   μ = the hypothesized population mean;
   σ = the population standard deviation;
   n = the sample size;
   $\sigma_{\overline{x}}$ = the standard error of the mean.

b) If σ is *not known*, the sample test statistic

$$z(\text{or } t) = \frac{\overline{x} - \mu}{s_{\overline{x}}}$$

where $\overline{x}$ = the sample mean;
μ = the hypothesized mean;
s = the sample standard deviation;
n = the sample size;
$s_{\overline{x}}$ = the estimated standard error of the mean.

2. *When testing a proportion.*
Provided $n > 30$ and $n\pi(1 - \pi) > 3$, the sample test statistic for testing a proportion is found by

$$z = \frac{p - \pi}{\sigma_p}$$

where p = the sample proportion;
π = the hypothesized population proportion;
n = the sample size;
σ_p = the standard error of proportion.

EXERCISE 12.2

1. A hypothesis test is to be performed for each of the following statements:
   a) A fuel additive is said to improve performance by at least 10%.
   b) Company A's dot matrix printer ribbons will average one million characters.
   c) Vitamin XYZ reduces cholesterol levels.
   For each of the statements,
   i) write a reasonable null and alternate hypothesis in descriptive terms;
   ii) state the hypotheses in (i) in symbolic form;
   iii) indicate the type of test to be used.

2. A hypothesis test is to be performed for each of the following statements:
   a) The average person has an IQ of 100.
   b) 70% of cola drinkers prefer Coke to Pepsi.
   c) Washing the hulls of boats in dry dock will reduce the zebra mussel population in the Great Lakes.
   For each of the statements,
   i) write a reasonable null and alternate hypothesis in descriptive terms;
   ii) state the hypotheses in (i) in symbolic form;
   iii) indicate the type of test to be used.

SECTION 12.3 Hypothesis Testing of a Mean

The choice of test statistic in a particular test depends on
1. whether the distribution of the population is known to be normal or not;

2. whether the sample size is large $(n > 30)$ or small $(n \leq 30)$ (when we do not know that the population distribution is normal).

Based on the above factors, we will examine three cases.

Case 1   The population distribution is known to be normal.

When the population is normal, the sampling distribution of the means will also be normal, regardless of the sample size. The use of the z statistic is appropriate in this case.

If σ is known, $z = \dfrac{\overline{x} - \mu}{\sigma_{\overline{x}}}$. If σ is unknown, $z = \dfrac{\overline{x} - \mu}{s_{\overline{x}}}$.

Case 2   A large sample $(n > 30)$ is taken from a population distribution whose shape is unknown and σ is not known.

When the sample size is greater than 30, the sampling distribution of the means will approximate the normal distribution. The use of the z statistic is appropriate in this case.

$$z = \frac{\overline{x} - \mu}{s_{\overline{x}}}$$

Case 3   A small sample $(n \leq 30)$ is taken from a population whose shape is not known and σ is not known.

When the shape of the sampling distribution is not known, the appropriate sampling distribution of the means is the t distribution and the appropriate test statistic is the t value.

$$t = \frac{\overline{x} - \mu}{s_{\overline{x}}}$$

Care must be taken in using the t tables depending on whether the test is a two-tail test or a one-tail test.

The t table supplied in this text (Table 11.4) gives t values for the combined area in both tails. For a *two-tail* test the table proportions 0.20, 0.10, 0.05, 0.02 and 0.01 correspond *directly* to the significance levels {20%, 10%, 5%, 2%, 1%}. For example, for a two-tail test at 5% level of significance, the appropriate t value can be located in the column headed by 0.05. However, in the case of a *one-tail* test, the level of significance must be *doubled* to locate the proper column. For a one-tail test at the 5% level of significance, the appropriate t value can be located in the column headed by $2(0.05) = 0.10$.

○ **EXAMPLE 12.3a**

The population is normal with an assumed mean of 500 and a standard deviation of 35. A sample of 25 items had a mean of 483. Conduct a two-tail test with a significance level of 0.01.

● **SOLUTION**

$\mu = 500$; $\overline{x} = 483$; $n = 35$; $\sigma = 35$; $\alpha = 0.01$.

Step 1   Statement of hypotheses.
Null hypothesis, $H_0 : \mu = 500$
Alternate hypothesis, $H_a : \mu \neq 500$

Step 2   Type of test.
Two-tail test is specified.

Step 3   Significance level.
$\alpha = 0.01$

Step 4   Since the population is known to be normal, the z statistic is appropriate regardless of the sample size.

Step 5   For a significance level of 1% and a two-tail test, the critical value of the test statistic $z = 2.58$.

Step 6   Decision rule.
Reject H_0 and accept H_a if the absolute value of the sample test statistic is greater than 2.58.

Step 7   Compute the sample test statistic.
Since σ is known,

$$\sigma_{\bar{x}} = \frac{\sigma}{\sqrt{n}} = \frac{35}{\sqrt{25}} = \frac{35}{5} = 7.00$$

FIGURE 12.4   Critical Regions for Two-Tail Test

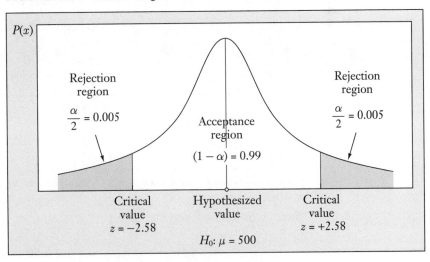

$$z = \frac{\bar{x} - \mu}{\sigma_{\bar{x}}} = \frac{483 - 500}{7.00} = \frac{-17}{7.00} = -2.43$$

Step 8   Apply the decision rule and interpret.
Since the absolute value of the sample test statistic is less than 2.58, accept H_0. The sample result supports the assumption that the population mean $\mu = 500$.

○ EXAMPLE 12.3b

The shape of the population distribution is normal with an assumed mean of 53. A sample of 100 had a sample mean of 54 and a standard deviation of 5. Conduct a right-tail test at the 5% level of significance.

● SOLUTION

$\mu = 53; \bar{x} = 54; n = 100; s = 5; \alpha = 0.05$.

Step 1　$H_0 : \mu = 53; H_a : \mu > 53$

Step 2　Right-tail test is specified.

Step 3　For a one-tail test, $\alpha = 0.05$ represents the area in the right tail.

Step 4　Since $n > 30$ the use of the z statistic is appropriate.

Step 5　Using Table 12.2, for a one-tail test and $\alpha = 0.05$, the critical value of the test statistic $z = 1.64$.

FIGURE 12.5 Critical Regions for Right-Tail Test

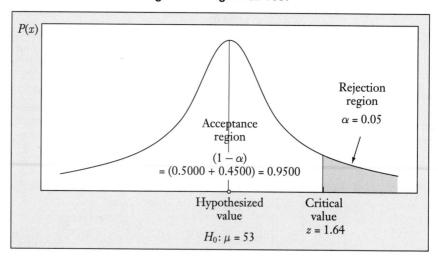

Step 6　Reject H_0 and accept H_a if the absolute value of the sample test statistic > 1.64.

Step 7　Since $s = 5$ and $n = 100$,

$$s_{\bar{x}} = \frac{s}{\sqrt{n}} = \frac{5}{\sqrt{100}} = \frac{5}{10} = 0.50$$

$$z = \frac{\bar{x} - \mu}{s_{\bar{x}}} = \frac{54 - 53}{0.50} = \frac{1}{0.50} = 2.00$$

Step 8　Since the absolute value of the sample test statistic $z > 1.64$, reject H_0 and accept H_a. The sample result does not support the assumption that the population mean is 53.

○ **EXAMPLE 12.3c**

The population mean is assumed to be 6425. A sample of 49 items had a mean of 6320 with a standard deviation of 285. Conduct a left-tail test at the 2% level of significance.

● **SOLUTION**

$\mu = 6425; \overline{x} = 6320; s = 285; n = 49; \alpha = 0.02$.

Step 1 $H_0 : \mu = 6425; H_a : \mu < 6425$

Step 2   Left-tail test is specified.

Step 3 $\alpha = 0.02$

FIGURE 12.6 Critical Regions for Left-Tail Test

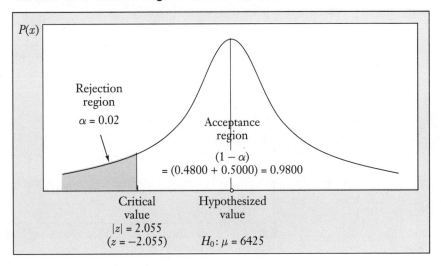

Step 4   Since $n > 30$, the z statistic can be used.

Step 5   Using Table 12.2, for a one-tail test and $\alpha = 0.02$, the critical value of the test statistic $z = 2.055$.

Step 6   Reject H_0 and accept H_a if the absolute value of the sample test statistic > 2.05.

Step 7

$$s_{\overline{x}} = \frac{s}{\sqrt{n}} = \frac{285}{\sqrt{49}} = \frac{285}{7} = 40.71$$

$$z = \frac{\overline{x} - \mu}{s_{\overline{x}}} = \frac{6320 - 6425}{40.71} = \frac{-105}{40.71} = -2.58$$

$$|z| = 2.58$$

Step 8   Since the absolute value of the sample test statistic $z > 2.05$, reject H_0 and accept H_a. The sample result does not support the assumption that $\mu = 6425$.

○ **EXAMPLE 12.3d**

The population mean is assumed to be 400. A sample of 20 had a mean of 381 with a standard deviation of 40. Conduct a two-tail test at the 5% level of significance.

● **SOLUTION**

$\mu = 400$; $\bar{x} = 381$; $s = 40$; $n = 20$; $\alpha = 0.05$.
$H_0 = 400$; $H_a \neq 400$.
Two-tail test is specified.
$\alpha = 0.05$.
Since σ is unknown and $n < 30$, the appropriate test statistic is the t value for $\alpha = 0.05$ and $(n-1) = 19$ degrees of freedom. From the t table (Table 11.4) the critical value of the test statistic $t = 2.093$.

FIGURE 12.7   Critical Regions for Two-Tail Test

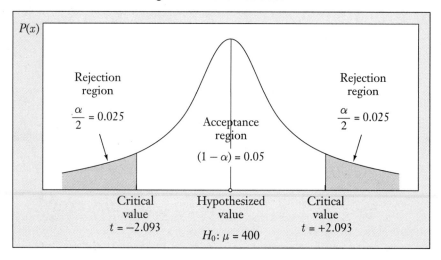

Reject H_0 and accept H_a if the absolute value of the sample test statistic $t > 2.093$.

$$s_{\bar{x}} = \frac{s}{\sqrt{n}} = \frac{40}{\sqrt{20}} = \frac{40}{4.472136} = 8.9443$$

$$t = \frac{\bar{x} - \mu}{s_{\bar{x}}} = \frac{381 - 400}{8.9443} = \frac{-19}{8.9443} = -2.124$$

$$|t| = 2.124$$

Since the absolute value of the sample test statistic $t > 2.093$, reject H_0 and accept H_a. The sample result does not support the assumption that $\mu = 400$.

○ **EXAMPLE 12.3e**

The population mean is assumed to be 4500. A sample of 25 items had a mean of 4620 and a standard deviation of 250. Conduct a right-tail test at the 1% level of significance.

● **SOLUTION**

$\mu = 4500; \bar{x} = 4620; s = 250; n = 25; \alpha = 0.01$.

$H_0 : \mu = 4500;$

$H_a : \mu > 4500$.

Right-tail test is specified.

Since σ is unknown and $n < 30$, the appropriate test statistic is the t value for $(n - 1) = 24$ degrees of freedom. The significance level of 0.01 represents the area in the right tail. The critical value for the test statistic t is obtained from Table 11.4 in the column headed by $2(0.01) = 0.02$; this critical value is $t = 2.492$.

FIGURE 12.8 Critical Regions for Right-Tail Test

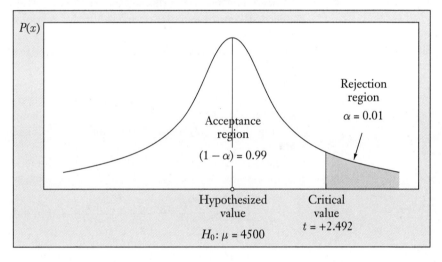

Reject H_0 and accept H_a if the absolute value of the sample test statistic $t > 2.492$.

$$s_{\bar{x}} = \frac{s}{\sqrt{n}} = \frac{250}{\sqrt{25}} = \frac{250}{5} = 50$$

$$t = \frac{\bar{x} - \mu}{s_{\bar{x}}} = \frac{4620 - 4500}{50} = \frac{120}{50} = +2.40$$

Since the absolute value of the sample test statistic $t < 2.492$, accept H_0. The sample results support the assumption that the population mean $\mu = 4500$.

Examples 12.3a to 12.3e illustrate when to use and how to compute the value of the sample test statistic z or t.

In most practical situations the most difficult aspect of hypothesis testing is determining the appropriate alternate hypothesis and the type of test to be performed. The following examples show how to deal with this aspect of testing.

○ EXAMPLE 12.3f

On a university entrance examination, the students' scores are known to be normally distributed with a mean of 500 points and a standard deviation of 75 points. One high-school principal claims that her school's graduates scored higher than the stated mean. To support her claim she asked the examiners to randomly select 50 of her school's students who wrote the examination. She learned that the mean score for this sample was 520 points. Does this result support her claim at the 5% significance level?

● SOLUTION

$\mu = 500; \bar{x} = 520; \sigma = 75; n = 50; \alpha = 0.05$.

Since the claim is that this school's students do better than average, we are concerned that the population mean for this school may be greater than the stated mean μ.

$H_0 : \mu = 500; H_a : \mu > 500$.
The test to be performed is a right-tail test.
$\alpha = 0.05$.

Since the distribution is known to be normal and σ is known, the appropriate test statistic is z.

Using Table 12.2, for $\alpha = 0.05$ and a one-tail test, the critical value of $z = 1.64$.

Reject H_0 and accept H_a if the absolute value of the sample test statistic $z > 1.64$.

$$\sigma_{\bar{x}} = \frac{\sigma}{\sqrt{n}} = \frac{75}{\sqrt{50}} = \frac{75}{7.0711} = 10.6066$$

$$z = \frac{\bar{x} - \mu}{\sigma_{\bar{x}}} = \frac{520 - 500}{10.6066} = \frac{20}{10.6066} = +1.89$$

Since the sample test statistic $z > 1.64$, we reject H_0 and accept H_a. The sample result supports the claim that graduates of this high school score better than the average on the entrance examination.

○ EXAMPLE 12.3g

An industrial engineer determined on the basis of some rough calculations that it should take 150 hours to produce a complex part for a new, advanced spacecraft. After production of the new part had been carried on for some time, the engineer followed the production of nine such parts and found the mean production time to be 147.5 hours with a standard deviation of 2.5 hours. Test at the 5% level of significance whether the original estimate of 150 hours was reasonable.

● SOLUTION

$\mu = 150; \bar{x} = 147.5; s = 2.5; n = 9; \alpha = 0.05$.

Since the engineer's estimate could be either high or low, we are concerned that the true population mean may be either greater than 150 or less than 150.
$H_0 : \mu = 150; H_a : \mu \neq 150$.
The test to be performed is a two-tail test.
$\alpha = 0.05$.
Since the shape of the distribution is not known and $n < 30$, the appropriate test statistic is the t statistic.
Using Table 11.4, for a two-tail test, $\alpha = 0.05$ and $(n-1) = 8$ degrees of freedom, the critical value of the test statistic t is found in the column headed by 0.05. Therefore, $t = 2.306$.
Reject H_0 and accept H_a if the absolute value of the sample test statistic $t > 2.306$.

$$s_{\overline{x}} = \frac{s}{\sqrt{n}} = \frac{2.5}{\sqrt{9}} = \frac{2.5}{3} = 0.8333$$

$$t = \frac{\overline{x} - \mu}{s_{\overline{x}}} = \frac{147.5 - 150.0}{0.8333} = \frac{-2.5}{0.8333} = -3.00$$

$$|t| = 3.00$$

Since the absolute value of the sample test statistic t is greater than 2.306, reject H_0 and accept H_a. The sample test result suggests that the average production time is less than the engineer's original estimate; that is, the original estimate of the time required to produce the new part was too high.

○ EXAMPLE 12.3h

A steel fabrication mill manufactures cotter pins with a mean length of 15 mm. The customer may accept pins that are longer than the mean length specified but not those that are shorter. The customer takes a sample of 10 pins from each batch received and measures their lengths to determine whether the batch should be accepted or rejected at the 1% level of significance.
The sample from a recently arrived batch showed the following lengths (in mm):

14.35	14.65	14.75	14.35	14.15
14.95	14.65	14.85	15.15	14.75

Should the batch be accepted?

● SOLUTION

$\mu = 15.00; n = 10; \alpha = 0.01$.
Since we are willing to accept pins longer than but not shorter than 15 mm, our concern is that the population mean may be less than the specified mean of 15 mm.
$H_0 : \mu = 15.00; H_a : \mu < 15.00$.
The test to be performed is a left-tail test.
$\alpha = 0.01$.

Since $n < 30$ and σ is unknown, the appropriate test statistic is the t statistic.

For a one-tail test, $\alpha = 0.01$ and $(n - 1) = 9$ degrees of freedom, the critical value of the test statistic is found in Table 11.4 in the column headed by $2(0.01) = 0.02$. Therefore, $t = 2.821$.

Reject H_0 and accept H_a if the absolute value of the sample test statistic $t > 2.821$.

To compute the sample test statistic we must first determine the sample mean and the standard error of the sample mean.

x	x^2
14.35	205.9225
14.65	214.6225
14.75	217.5625
14.35	205.9225
14.15	200.2225
14.95	223.5025
14.65	214.6225
14.85	220.5225
15.15	229.5225
14.75	217.5625
$\sum x = 146.60$	$\sum x^2 = 2\,149.9850$

$$\bar{x} = \frac{\sum x}{n} = \frac{146.60}{10} = 14.66$$

$$s^2 = \frac{n(\sum x^2) - (\sum x)^2}{n(n-1)} = \frac{10(2149.9850) - (146.60)^2}{10(9)}$$

$$= \frac{21\,499.850 - 21\,491.560}{90} = \frac{8.29}{90} = 0.09211111$$

$$s = \sqrt{0.09211111} = 0.30349812$$

$$s_{\bar{x}} = \frac{s}{\sqrt{n}} = \frac{0.30349812}{\sqrt{10}} = \frac{0.30349812}{3.16227766} = 0.09597$$

$$t = \frac{\bar{x} - \mu}{s_{\bar{x}}} = \frac{14.66 - 15.00}{0.09597} = \frac{-0.34}{0.09597} = -3.543$$

$$|t| = 3.543$$

Since the absolute value of the sample test statistic $t > 2.821$, reject H_0 and accept H_a. The sample result does not support the assumption that the batch mean is 15.00 mm. The batch should not be accepted.

EXERCISE 12.3

1. Given $H_0 : \mu = 8.0$, $H_a : \mu \neq 8.0$; $\bar{x} = 7.8$; $s = 2.1$; $n = 15$; and $\alpha = 0.05$, should H_0 be accepted or rejected?

2. $H_0 : \mu = 110$; $H_a : \mu < 110$. Given that $\bar{x} = 108$, $s = 18$, and $n = 40$, should the null hypothesis be accepted or rejected at the 98% confidence level?

3. Accept or reject the null hypothesis given the following data:

	Sample test statistic	Critical value of test statistic	Type of test
a)	2.9	3.2	two-tail
b)	−1.5	−1.6	left-tail
c)	1.8	0.8	right-tail
d)	−2.5	−2.3	left-tail

4. Last year, a retailer found that mean credit card sales were $60.00 with a standard deviation of $12.00. Because the cost of accepting credit cards is inversely related to the dollar amount per credit card sale (that is, as the dollar amount per credit card sale goes up, the cost per transaction goes down), it is important for the retailer to know if there has been any change in the average amount per credit card sale.

   To test for any change, the retailer took a sample of 225 credit card slips and determined the mean to be $65.00 and the standard deviation to be $9.00. The retailer wants the chance of error to be no more than 2%.
   a) Set up the decision rule for the test.
   b) Determine the critical value and the computed value of the test statistic.
   c) Should the null hypothesis be rejected?

5. A sample of pressurized tanks received from a supplier are destroyed to determine if the shipment meets minimum pressure requirements of 1000 kPa (kilopascals). From the shipment of 400 tanks, 8 were randomly chosen and found to have an average pressure of 980 kPa and a standard deviation of 8 kPa. The company requires a 99% confidence level for the test.
   a) Why can the shape of the distribution be assumed to be normal?
   b) Should the shipment be accepted?
   c) Why does the company not use a larger sample size to increase the precision of the test?

6. United Way contributions at a Moncton hospital last year averaged $220 per person for 325 employees. The chairperson for this year's campaign hopes to increase the contribution per person by at least 10% through a media blitz. A sample of 40 employees taken following the advertising campaign showed an average contribution of $260 per person with a standard deviation of $18 per person. Test the null hypothesis at $\alpha = 0.05$ to determine if the media blitz was successful.

SECTION 12.4 Hypothesis Testing of a Population Proportion for Large Samples ($n > 30$)

Provided that the sample size is large and the true population proportion π is not too far away from 0.5, we can use the z statistic to perform a hypothesis test of the population proportion. The accepted rule is that the z statistic can be used if $n\pi(1 - \pi) > 3$.

○ **EXAMPLE 12.4a**

The population proportion is assumed to be 80%. Conduct a two-tail test at the 10% level of significance for a sample of 144 items with a proportion of 83%.

● **SOLUTION**

$\pi = 0.80$; $p = 0.83$; $n = 144$; $\alpha = 0.10$.
$H_0 : \pi = 0.80$; $H_a : \pi \neq 0.80$.

Two-tail test is specified.

$\alpha = 0.10$.

$n\pi(1 - \pi) = 144(0.80)(0.20) = 23.04$.

Since $n > 30$ and $n\pi(1 - \pi) > 3$, the z statistic can be used as a test statistic.

Using Table 12.2, for a two-tail test and $\alpha = 0.10$, the critical value of the test statistic $z = 1.64$.

Reject H_0 and accept H_a if the absolute value of the sample test statistic $z > 1.64$.

$$\sigma_p = \sqrt{\frac{\pi(1 - \pi)}{n}} = \sqrt{\frac{(0.80)(0.20)}{144}} = \sqrt{0.00111111}$$

$$= 0.03333333$$

$$z = \frac{p - \pi}{\sigma_p} = \frac{0.83 - 0.80}{0.03333333} = \frac{0.03}{0.03333333} = 0.900$$

Since the absolute value of the sample test statistic $z < 1.64$, accept H_0. The test result supports the null hypothesis.

○ **EXAMPLE 12.4b**

A large retailer considers signing a long-term contract with a supplier of waterproof boots. Before doing so, the retailer wants to be certain that the proportion of defective boots is less than 5%. Specifying a 1% level of significance, the retailer accepted on consignment a shipment of 100 pairs of boots for test marketing and found three pairs to be defective. Should the retailer sign the contract?

● **SOLUTION**

$\pi = 0.05$; $n = 100$; number defective $= 3$; $p = \dfrac{3}{100} = 0.03$; $\alpha = 0.01$.

Since the retailer hopes to establish that the true proportion of defective boots is less than 5%, the concern is that the proportion of defective pairs of boots be less than 5%.

$H_0 : \pi = 0.05$; $H_a : \pi < 0.05$.

Perform a left-tail test.

$\alpha = 0.01$.

Since $n > 30$ and $n\pi(1 - \pi) = 100(0.05)(0.95) = 4.75 > 3$, we can use the z statistic.

For a one-tail test and $\alpha = 0.01$, the critical value of the test statistic $z = 2.33$.

Reject H_0 and accept H_a if the absolute value of the sample test statistic $z > 2.33$.

$$\sigma_p = \sqrt{\frac{\pi(1 - \pi)}{n}} = \sqrt{\frac{(0.05)(0.95)}{100}} = \sqrt{0.000475} = 0.021794$$

$$z = \frac{p - \pi}{\sigma_p} = \frac{0.03 - 0.05}{0.021794} = \frac{-0.02}{0.021794} = -0.92$$

$$|z| = 0.92$$

Since the absolute value of the sample test statistic $z < 2.33$, accept H_0. This means that the sample does not support the alternate assumption that the true proportion is less than 5%. The retailer should not sign the long-term contract.

EXERCISE 12.4

1. A random sample of 120 financial analysts were polled on the issue of company valuation. Ninety-six favoured inclusion of third-party company valuation with any prospectus. Test the claim that more than 75% of all financial analysts are in favour of including third-party valuations with any prospectus at the 1% level of significance.

2. The YUP-PEE retail chain, vendors of fashionable clothing, wish to confirm that 65% of its target market are college and university students. If the true proportion were different from the claimed 65%, a new marketing program would be required. Of 400 potential customers included in a random sample, 160 were not college or university students. Test the claim at the 2% level of significance.

3. In a blindfold taste test, buyers of Brand A coffee were asked to compare their brand with a competing Brand B coffee. Final results of the test showed that 240 preferred Brand A while 200 preferred Brand B. The product manager claims that there is a definite preference for Brand A over Brand B. Is the product manager's claim correct based on the test information? Use $\alpha = 0.05$ to support or contradict the product manager's claim.

4. Lucky Strike Bowling Alley has a Monday night league consisting of 120 bowlers. A vote at the beginning of the season indicated 61% support for a

year-end banquet. Some members of the organizing committee now feel that support for the banquet has dropped. A random sample of 40 bowlers shows support now to be 54%. Test at the 5% level of significance if there has been a drop in support.

Type I and Type II Errors, *p* Values

A. Type I and Type II Errors

As the decision to accept or reject the null hypothesis is based on a sample, we can never be absolutely sure that the decision is correct. There are four possible decisions that can be made about a null hypothesis: two of them correct and two of them incorrect.

The two correct decisions are
1. to accept H_0 when it is true;
2. to reject H_0 when it is not true.

The two incorrect decisions are
1. To reject H_0 when it is true — referred to as **Type I error**.
2. To accept H_0 when it not true — referred to as **Type II error**.

A Type I error occurs when we reject the null hypothesis but should accept it. The probability of making this type of error is the specified chance of error. For this reason, a Type I error is sometimes called an α error.

A Type II error occurs when we accept the null hypothesis but should reject it. The probability of making this type of error is denoted by β (read "beta") and is sometimes called a β error.

The four possible decisions and their associated probabilities are summarized in Table 12.3.

TABLE 12.3 Summary of Type I and Type II Errors

Possible decision	Null hypothesis is true		Null hypothesis is not true	
	Decision	Probability	Decision	Probability
Accept H_0	correct	$1 - \alpha$	incorrect	β Type II error
Reject H_0	incorrect	α Type I error	correct	$1 - \beta$

Table 12.4 summarizes Type I and Type II errors for Example 12.4b.

TABLE 12.4 Type I and Type II Errors for Example 12.4b

Action concerning the null hypothesis H_0	State of nature	
	$H_0 : |z| > 2.33$ (Consignment shipment meets reject specifications)	$H_0 : |z| < 2.33$ (Consignment shipment does not meet reject specifications)
Accept H_0 (Sign the long-term contract)	no error	Type II error (Accept a shipment that does not meet specifications)
Reject H_0 (Do not sign)	Type I error (Reject a shipment that does meet specifications)	no error

A Type I error is also known as the *producer's risk* since it could involve, for example, shutting down a production line on the sample evidence that a machine was incorrectly set when, in fact, it was operating correctly.

Conversely, a Type II error is known as the *consumer's risk* since it could involve, for example, a retailer accepting a consignment of goods on the basis of sample evidence that the consignment met specifications when, in fact, it did not.

Both types of errors can be very costly in terms of lost production time on one hand and customer dissatisfaction on the other hand. As a result, decisions about the levels at which α and β should be set usually involve consideration of economic costs.

The risk of making a Type I error has traditionally been considered the more serious of the two types of errors. For this reason the value of α is usually specified first. In most business situations $\alpha = 0.05$ and $\alpha = 0.01$ are the most common values used.

In very sensitive situations, such as medical research, much smaller values may be chosen. For example, in testing the effectiveness of a new drug, the null hypothesis is that the current practice is the best procedure. The alternate hypothesis is that the new drug is "better." In these situations an extremely serious Type I error could result if H_0 is rejected when, in fact, it represents the better alternative and the new drug proves to have terrible side effects.

B. An Alternative Method of Reporting the Result of a Hypothesis Test — p Values

The hypothesis procedure shown in this chapter follows the traditional approach to testing. In modern practice, however, the results of such tests are frequently given in terms of a *p* **value**.

In Example 12.3g, which involved the testing of the time estimate required to produce a new part for a spacecraft, we determined the critical value of the test statistic $t = 2.306$ while we found the absolute value of the sample test statistic $t = 3.00$.

Since our decision rule was to reject H_0 if the absolute value of the sample test statistic t was greater than 2.306, we rejected H_0 at the 5% level of significance and concluded that the engineer had overestimated the time required to manufacture the new part. Note what happens if we specify a 1% level of significance instead.

For a two-tail test, $\alpha = 0.01$ and $(n-1) = 8$ degrees of freedom, the critical value of the test statistic $t = 3.355$.

Our decision rule would be to accept H_0 if the absolute value of the sample test statistic is less than 3.355. This being the case, we would accept the null hypothesis at the 1% level of significance.

The choice of significance level influences the decision to accept or reject the null hypothesis. Specification of the level of significance before the test is carried out is important since it prevents choosing after the test a level of significance that suits the particular objectives or preconceptions of the decision-maker.

Now consider the values in the t table for 8 degrees of freedom for the selected values of α.

α	0.20	0.10	0.05	0.02	0.01
d.f. = 8	1.397	1.860	2.306	2.896	3.355

In our case the absolute value of the sample test statistic $(t = 3.000)$ is less than the critical value of the test statistic $t = 3.355$. This means that the actual significance level of our result lies somewhere between $\alpha = 0.02$ and $\alpha = 0.01$. However, in the absence of a complete t table, we cannot determine the precise value of the probability of the sample result.

This probability is known as the p value of the test. In our case it can be interpreted as "the probability of obtaining a sample mean as small as 147.5 hours if the true population mean of 150 hours lies somewhere between 2% and 1%."

Stated another way, "a sample mean as low as 147.5 hours would occur by chance only about 1.5 times in 100 if it is true that the population mean is 150 hours." We can therefore be reasonably sure that drawing a sample with a mean of 147.5 hours is not a chance event and that the engineer's estimate was too high.

The p value is the actual value of the level of significance as distinct from a specified value such as $\alpha = 0.05$ or $\alpha = 0.01$. Accordingly, an alternative version of our decision rule is

"Reject H_0 if the p value is less than the specified value of α; otherwise do not reject H_0."

In Example 12.3f, which involved a test of entrance examination results for a high school at the 5% level of significance, the critical value of the test statistic $z = 1.64$, while the absolute value of the sample test statistic

$z = 1.89$. Following the decision rule for a right-tailed test, we rejected the null hypothesis and concluded that the entrance examination test results obtained by the graduates of the high school were above the average at the 5% level of significance.

From the z-table (Table 9.1) we know that the area between μ and $z = 1.89$ is 0.4706. The area above $z = 1.89$ is $(0.5000 - 0.4706) = 0.0294$. The p value = 2.94%.

Note When using the z table, we can determine the p value directly from the table.

Since the p value is less than the specified significance level of 5%, we reject the null hypothesis and interpret the result to indicate that "a mean score as high as 520 points when the population mean score is 500 points would happen by chance only about 3 times in 100." We can therefore be quite confident that it is not a chance event and that these students, in fact, scored higher than average.

EXERCISE 12.5

1. A toothpaste manufacturer tested the following hypothesis:
$$H_0 : \mu = 350 \text{ ml}; \qquad H_a : \mu \neq 350 \text{ ml}$$
   a) Explain the meaning of a Type I error in this example.
   b) Explain the meaning of a Type II error in this example.

2. The credit manager of a credit union set up the following hypotheses:
$$H_0 : \pi = 0.10; \qquad H_a : \pi < 0.10$$
   a) What does a Type I error mean in this case? Explain.
   b) What does a Type II error mean in this case? Explain.

3. Based on return on assets, food, beverage and tobacco stocks were one of the top performers on the Toronto Stock Exchange in 1990. This group of stocks had a return on assets of 6.3% with a standard deviation of 3.9%. A random sample of 38 stocks belonging to the food, beverage and tobacco group taken in 1991 showed an average return on assets of 4.8%.
   a) State a hypothesis for this situation.
   b) Compute the p value for the hypothesis test.

4. A retailer promotes its own brand of tires on the basis of a guaranteed life span of at least 60 000 km. A random sample of 100 tires showed a mean of 62 000 km and a standard deviation of 7800 km.
   a) State the hypotheses for this test.
   b) Compute the p value for the test.

5. A Canada Customs supervisor believes that at least 60% of Canadians returning from the United States are not declaring purchases made in the United States. A survey of 81 car occupants crossing the Canadian border revealed that 55 of them did not declare their purchases.
   a) Compute the p value for a hypothesis test (assume $\alpha = 0.05$).
   b) State your conclusion about the validity of the supervisor's belief.

6. A quarterly survey of 500 business executives measures their perception of the

economy. In January, 80% of the executives believed that the economy was going to go down further. A sample of 42 executives taken in April indicated that 28 thought the economy was going down further.
a) Calculate the p value.
b) At the 10% level of significance would you conclude that there has been a change in attitude?

REVIEW EXERCISE

1. A sailboat charter company in the Virgin Islands claims that rainfall in the month of February is less than 5 mm. State the null and alternate hypotheses.

2. The test weight of a fishing line (e.g., 5 kg) is the minimum weight needed to break the line. State the null and alternate hypotheses.

3. The average number of sick days per month at Hab Corporation was 56. A new fitness program has been started to reduce the number of sick days.
a) State the null and alternate hypotheses.
b) Identify the type of test.

4. Marketing research studies reveal that the average consumer adds two teaspoons of sugar to breakfast cereal.
a) State the null and alternate hypotheses.
b) Identify the type of test.

5. Determine what type of test is required for each of the following:
a) $H_0 : \mu = 30$ ml; $H_a : \mu \neq 30$ ml
b) $H_0 : \pi = 0.05$; $H_a : \pi < 0.05$

6. State the type of test required for the following:
a) $H_0 : \pi = 0.90$; $H_a : \mu > 0.90$
b) $H_0 : \mu = 75$; $H_a : \mu < 75$

7. In a certain year the mean yield on a treasury bill was 9.75% and the standard deviation was 0.20%. One year later a random sample of 100 treasury bills yielded an average of 9.68%. Would you be willing to conclude that the average yield has changed significantly? Assume a 5% risk of making a Type I error.

8. An audit of mortgages held by a local trust company showed the mean mortgage to be $76 500 with a standard deviation of $8900. Six months later the trust company was asked by the regulatory body to supply details regarding its mortgage portfolio. A random sample of 70 mortgages had a mean principal of $80 700. Using a 2% level of significance, determine if the trust company's portfolio has changed.

9. A normal population has a mean of 92 and a standard deviation of 31. Some of the data were found to be erroneous. A random sample of 20 data points had a mean of 98. Does the population mean change significantly ($\alpha = 0.02$) if the erroneous data are removed?

10. Lance Bell is a sales representative for a pharmaceutical company. He averaged 42 sales calls per week with a standard deviation of 3.8 sales calls per week before he was involved in a car accident. A random sample of 10 weeks shows

Lance averaged 37 sales calls per week after the accident. Assume a normal population and 0.05 significance level to test if Lance's performance dropped after the accident.

11. Trylex's corporate objectives require an average inventory value of $850 000 for each of its retail outlets throughout the year (312 days). A random sample of 60 days at a store identified by the code 50712 found the average daily inventory to be $823 000 with a standard deviation of $33 500. Does store 50712 significantly deviate from the corporate objective over a one-year period at the 1% level of significance?

12. Trans-Port-It Trucking uses special corrugated boxes so that the cartons can be piled higher than normal boxes. The special boxes are supposed to withstand 100 kg of weight. The company's records show that 35 damage claims were due to crushed cartons that carried an average weight of 95 kg and a standard deviation of 3.4 kg. Assume $\alpha = 0.02$.
a) Do the crushed cartons differ significantly from the manufacturer's claim?
b) What is suspect about the sample used?

13. Russell Cousins purchased a new rechargeable electric shaver that is supposed to last an average of 14 shaves before requiring recharging. Russell believes his shaver is not living up to the manufacturer's claim. He randomly selected 5 recharging periods and obtained a mean of 12.3 shaves between recharges with a standard deviation of 1.8 shaves. At the 1% level of significance, is Russell's shaver significantly different from the manufacturer's claim?

14. Tally Greenberg is an artist who makes jewellery. Tally has based her prices on the assumption it takes 11 hours to design and make a piece of jewellery for a customer. A random sample of her work showed the following:

Customer	A	B	C	D	E	F	G
Time (hours)	9.8	11.1	10.8	10.1	11.0	10.6	10.0

At the significance level $\alpha = 0.05$, determine if the sample times are significantly different from what Tally expected.

15. R.T. Kolly Advertising was awarded an advertising contract by the government to communicate the need for recycling. To prove the effectiveness of its work the agency surveyed 220 people and found 80 believed in recycling. After the survey, the same 220 people were shown the advertisement. Fifty of them were selected randomly for a follow-up survey. The agency found that 24 of the 50 believed in recycling.
a) State the null and alternate hypotheses.
b) Is there evidence of a change in attitude at the 1% level of significance?
c) Does this measurement of attitude seem valid? Comment.

16. It is believed that 47% of Canadians drink coffee. A consumer study asked 1500 Canadians about their drinking habits and found 41% said they drink coffee. At $\alpha = 0.02$ determine if the study results are significantly different from the assumed proportion.

17. Last year 39% of a high school's graduating class went on to college or university. A sample of 40 graduates from this year's graduating class showed 18 will be going to college or university. The high school graduates 500 students yearly. Has there been a significant change in the proportion of students continuing their education ($\alpha = 0.01$)?

18. One of Toronto's institutional investment companies has a special mailing list of 185 clients who buy their research. Twenty percent of the customers normally act on the research recommendations. A survey of 36 clients after the latest mailing indicated 9 had made a transaction as a result of the information received in the mailing. Allowing for a 5% chance of a Type I error, determine if the latest mailing significantly increased business.

19. Last year Aquamarine Ltd. reported 85% of all sales orders were shipped on or before the required date. A new computerized tracking system was installed to increase the company's compliance rate. A random sample of 120 sales orders shipped after the installation of the new system showed a 93% compliance rate. At the 2% level of significance, has the new system made a positive difference?

20. A marketing research company pays people $5.00 to come to their office, taste new products and answer a questionnaire. Historically, 28% of consumers who said they would participate were no-shows. Because of the time and money lost due to missed bookings, the company decided to increase the payment to $10 per visit. A random sample of 100 bookings taken after the increase in payment showed 15 people did not show up as scheduled. Would you conclude, at the 1% level of significance, that there has been a change in the proportion of missed bookings?

21. A train running between City A and City B averaged 26 minutes per trip with a standard deviation of 7.8 minutes per trip last year. A random sample of 25 trips made this year showed a mean of 30 minutes per trip. The sample seems to indicate that the train takes longer to complete the trip this year.
a) Compute the p value.
b) If the significance level were 0.02, what would you conclude about the time taken to make the trip?

22. Last year Jack's Orchards harvested a mean of 0.65 bushels of apples per tree with a standard deviation of 0.04 bushels. A sample of 200 trees for this year's crop averaged 0.61 bushels per tree. It seems that this year's drier growing season has affected the crop.
a) Compute the p value.
b) What would you conclude about this year's apple harvest at the 1% level of significance?

23. Ellen assembled 240 components per shift when she was paid an hourly rate. Now she is paid a piecework rate. A random sample of 64 of her shifts shows output to average 261 components per shift with a standard deviation of 70 components.
a) Compute the p value.
b) Comment on the effect of changing Ellen's basis of remuneration.

24. A store owner claims the mean amount of delinquent charge accounts is $85. A random sample of eight such accounts showed the following amounts:

$79 $98 $73 $83 $121 $57 $68 $61

a) Determine the p value.
b) Do the sample data support the owner's claim?

25. A job applicant claims a typing speed of 90 words per minute. A random sample

of five letters typed during the probationary period showed the following speeds in words per minute:

  80   75   84   86   90

a) Determine the p value.

b) Do the data support the applicant's typing speed claim?

26. Paul's Diner advertises lunch will be served in 10 minutes or less. A random sample of 50 lunch orders showed a mean serving time of 12.3 minutes per order with a standard deviation of 4.7 minutes. Assume $\alpha = 0.05$.

   a) Determine the p value.

   b) Does the sample substantiate Paul's advertised claim?

27. Suppose you work for a toy company that specializes in making toys for small children. Recently your company has received complaints of small pieces breaking off certain toys, which could be hazardous if swallowed. Before the company accepts a shipment of 8000 kadidles used in the production of the toys, a quality control program is implemented allowing a maximum rejection rate of 2% of the incoming parts. The new program is willing to run a 1% chance of rejecting the shipment when it actually meets the specified rejection rate. In a random sample of 250 kadidles from the shipment, 9 were rejected.

   a) Determine the p value.

   b) Should the shipment be accepted?

28. The director of a city's harbour commission believes that 65% of the sailboats mooring overnight are under 8 m in length. In a sample of 150 moored boats, 45 were over 8 m. Allowing for 5% error, determine the p value to assess the director's estimate.

SELF-TEST

1. The owner of a small shopping mall surveyed the households within a five-block radius and found household income was normally distributed with a mean of $47 000 and a standard deviation of $4800. A potential tenant thinks the income numbers are inflated and undertakes its own survey. The random sample of 25 households shows a mean income of $45 800.

   a) Set up reasonable hypotheses.

   b) Indicate the type of test to be used.

   c) At the 1% level of significance would you accept or reject the null hypothesis?

2. A window manufacturer claims its special tinting will decrease heating costs an average of 10%.

   a) State the null and alternate hypotheses to test this claim.

   b) Identify the type of test to be used.

3. Balline Corporation's typing pool maintains that its people have an average typing speed of 80 words per minute with a standard deviation of 4 words per minute. The corporation's claim is to be tested at the 2% level of significance using a large sample.

a) State the two hypotheses for the test.
b) Identify the test to be used.
c) State the value of α.
d) Identify the test statistic to be used.
e) Determine the critical value of the test statistic.
f) State the decision rule.

4. Given the following data, show all detail to determine whether the null hypothesis should be accepted or rejected.

$H_0 : \mu = 61$; $H_a : \mu < 61$.

$\bar{x} = 60$; $s = 20$; $n = 400$; $\alpha = 0.01$.

5. After the introduction of a new brand of cigarettes, 600 consumers of the new brand were surveyed to determine, among other things, the frequency of purchase. The study found the mean to be 2.7 packs per week. Six months later, a sample of 60 people selected at random from the original group surveyed found the mean frequency of purchase to be 2.5 packs per week with a standard deviation of 0.7 packs per week. The company wonders if the "novelty" has worn off and the frequency of purchase is dropping. At the 5% level of significance, can you conclude that the frequency of purchase has changed?

6. Lauren Thomson, dentist, has determined that her patients wait an average of 28 minutes before seeing her. Instead of booking a person every 10 minutes, the office manager has been instructed to find out the purpose of each patient visit and to make bookings based on the estimate of how long each visit should take. A random sample of 70 patients taken after the change in procedure showed an average waiting time of 18 minutes and a standard deviation of 5 minutes. Assume a 1% chance of a Type I error.
a) State the hypotheses.
b) Has the new procedure significantly decreased the patients' waiting time?

7. The Robotics Industrial Association estimates that 56% of all industrial robots in Canada are used for welding. A random sample of companies with a total of 200 robots showed that 61% of the robots are used for welding. At the 0.05 significance level, would you conclude the sample results support the Association's estimate?

8. Soni Metals just installed a new piece of equipment to form a special fastener. The engineering department supplied the cost accountant with a production standard of 3.8 minutes per fastener. After the first month of operation, the cost accountant identified cost overruns associated with the operation of the new equipment. A random selection of production times (in minutes) was recorded as follows:

    3.9   4.0   3.8   3.9   4.0   3.7   3.7   3.9   4.0   4.1

a) Calculate the mean and standard deviation of the sample.
b) Explain your choice of test statistic for a hypothesis test.
c) If $\alpha = 0.05$, can you conclude that the new machine does not meet the established time standard?

9. Employee satisfaction at Signet Inc. averaged 7.2 points on a 9-point scale. The survey was given to all 2850 employees last spring. Recently a random sample of 49 employees showed an average score of 7.0 points with a standard deviation of 1.19 points. The company is concerned that the employees' level of

satisfaction has changed.
a) State the hypothesis test.
b) Compute the p value.
c) At the 98% confidence level, would you conclude there was a significant change in the level of employee satisfaction?

10. A major credit card company knows from past experience that 38% of its cardholders pay their account balance in full to avoid interest charges. The company also knows that if the proportion of cardholders who pay their account in full drops significantly the number of accounts in arrears increases. A random sample of 121 account statements showed 42 accounts past due. Assuming a 5% level of significance, determine if the company should be concerned about an increase in the number of past-due accounts.

Key Terms

Acceptance region   322
Alternate hypothesis   320
Critical value   322
Null hypothesis   320
p value   339
Rejection region   322
Significance level   321
Type I error   338
Type II error   338

Summary of Formulas

1. **Hypothesis testing of a mean:**

   a) **Population normal:**

   i) When σ is known $z = \dfrac{\overline{x} - \mu}{\sigma_{\overline{x}}} = \dfrac{\overline{x} - \mu}{\dfrac{\sigma}{\sqrt{n}}}$

   ii) When σ is not known $z = \dfrac{\overline{x} - \mu}{s_{\overline{x}}} = \dfrac{\overline{x} - \mu}{\dfrac{s}{\sqrt{n}}}$

   b) **Population shape unknown:**

   I Large sample ($n > 30$):

   i) When σ is known $z = \dfrac{\overline{x} - \mu}{\sigma_{\overline{x}}} = \dfrac{\overline{x} - \mu}{\dfrac{\sigma}{\sqrt{n}}}$

   ii) When σ is not known $z = \dfrac{\overline{x} - \mu}{s_{\overline{x}}} = \dfrac{\overline{x} - \mu}{\dfrac{s}{\sqrt{n}}}$

II Small sample ($n \leq 30$):

i) When σ is known $\qquad t_{n-1} = \dfrac{\overline{x} - \mu}{\sigma_{\overline{x}}} = \dfrac{\overline{x} - \mu}{\dfrac{\sigma}{\sqrt{n}}}$

ii) When σ is not known $\qquad t_{n-1} = \dfrac{\overline{x} - \mu}{s_{\overline{x}}} = \dfrac{\overline{x} - \mu}{\dfrac{s}{\sqrt{n}}}$

2. Hypothesis testing of a proportion:

Large sample ($n > 30$) $\qquad z = \dfrac{p - \pi}{\sigma_p} = \dfrac{p - \pi}{\sqrt{\dfrac{\pi(1 - \pi)}{n}}}$

CHAPTER 13

Simple Linear Regression and Correlation Analyses

Introduction

Simple linear regression and correlation analyses examine a possible relationship between two variables. Simple regression describes the relationship between the variables, while correlation measures the closeness of that relationship. These methods are widely used in business, economic and scientific applications.

Objectives

Upon completion of this chapter you will be able to
1. distinguish between dependent and independent variables;
2. determine the regression equation by the least squares method;
3. plot the regression line on a scatter diagram;
4. interpret the meaning of the regression coefficients;
5. use the regression equation to predict values of the dependent variable for selected values of the independent variable;
6. compute the standard error of estimate;
7. construct forecast intervals;
8. compute the coefficients of determination r^2, non-determination $(1 - r^2)$ and correlation r;
9. interpret the meaning of the coefficients r^2, $(1 - r^2)$, and r.

Simple Linear Regression Analysis

Many situations arise in which two things appear to be related. For example, it is reasonable to speculate that there is a relationship between
a) the number of litres of gasoline sold at service stations and the volume of traffic passing their locations;
b) the number of meals sold in a company cafeteria and the number of company employees;

c) advertising expenditures and sales revenue.

In the analysis of the relationship, the two things (*variables*) are identified by the symbols x and y. The variable denoted by y is called the **dependent variable**. The variable denoted by x is referred to as the **independent variable**. The use of these two labels implies that the value of the variable y depends on the value of the variable x.

The crucial first step in the analysis involves determining which variable is the dependent variable y and which is the independent variable x. A good approach is to ask the following two questions and select the more reasonable one.

1. Does the first variable depend on the second variable? or
2. Does the second variable depend on the first variable?

○ **EXAMPLE 13.1a**

For statements (a), (b) and (c) above, select the dependent and the independent variables.

● **SOLUTION**

For statement (a), the two questions are

1. Does the number of litres of gasoline sold depend on the volume of traffic?
2. Does the volume of traffic depend on the number of litres of gasoline sold?

Question 1 is more reasonable. The number of litres of gasoline sold is the dependent variable and is assigned as variable y. The volume of traffic is the independent variable and assigned as variable x.

For statement (b), the two questions are

1. Does the number of meals sold in a cafeteria depend on the number of employees?
2. Does the number of employees depend on the number of meals served?

Question 1 is more reasonable. The number of meals served is the dependent variable y. The number of employees is the independent variable x.

For statement (c), the two questions are

1. Does advertising expenditure depend on sales revenue?
2. Does sales revenue depend on advertising expenditure?

Question 2 must be deemed to be more reasonable. Sales revenue is the dependent variable y. Advertising expenditure is the independent variable x.

Once it has been established which variable is the dependent variable y and which is the independent variable x, the data can be portrayed graphically in the form of a *scatter diagram*. The diagram may indicate that the relationship can be represented by a straight line. In this case a linear regression model is appropriate for the analysis. If the data graph as a curve, more complex curvilinear models should be used.

The following four scatter diagrams are typical results of plotting data sets involving two variables.

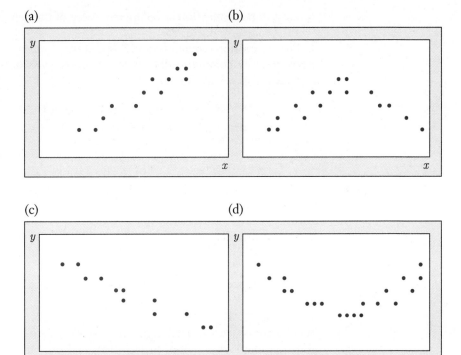

Diagrams (a) and (c) are appropriate for linear analysis, while the data represented by diagrams (b) and (d) require the use of curvilinear models. In this text only the most common straight-line model, referred to as the least squares model, is considered.

SECTION 13.2 The Least Squares Model

The **least squares model** provides a mathematical procedure for determining the equation of a straight line that best fits the data. The equation is referred to as the **regression equation**. The graphical representation of the equation is known as the regression line. This line is called the *line of best fit* since the procedure *minimizes* the sum of the vertical deviations of the data points about the line.

The general form of the least squares equation is given by

$$y_p = a + bx$$

where a is the value of y when $x = 0$; that is, a is the y intercept;

b is the slope of the regression line and indicates the change in the dependent variable y for a change of one unit in the dependent variable x;

x is a selected value of the independent variable;

y_p is the predicted (or computed) value of the dependent variable for a given value of x.

The regression equation for a specific set of data can be uniquely determined by computing the values of the **regression coefficients** a and b from the following formulas:

$$b = \frac{n(\sum xy) - (\sum x)(\sum y)}{n(\sum x^2) - (\sum x)^2}$$

$$a = \frac{\sum y}{n} - b\frac{\sum x}{n}$$

The following procedure is suggested when using the least squares model:
1. Determine the dependent variable.
2. Construct the scatter diagram.
3. If the scatter diagram indicates a reasonable linear relationship between the two variables, use the two formulas to determine a and b.
4. Determine the regression equation.
5. Plot the regression line on the scatter diagram.

○ **EXAMPLE 13.2a**
Many companies involved in assembly operations use aptitude tests on potential employees before hiring them and on current employees before promoting them to more demanding tasks. To obtain a reading on the usefulness of a particular test, the personnel department of Mega Tech, Inc., has administered the aptitude test to a group of employees. The resulting test scores matched to output data are listed in Table 13.1 below.

TABLE 13.1   Test Scores and Output Data

Employee	Output	Test score
A	31	5
B	40	11
C	30	4
D	34	5
E	25	3
F	20	2

a) Determine the dependent variable.
b) Construct the scatter diagram.
c) Determine the regression equation.
d) Plot the regression line.
e) Compute the predicted output for aptitude test scores of 6 and of 10.

● **SOLUTION**
a) To determine the dependent variable, pose the two questions
   1. Does aptitude depend on output?
   2. Does output depend on aptitude?

Question 2 is the more resonable question. Output becomes the dependent variable y; the aptitude test scores become the independent variable x.

b) The scatter diagram is now constructed as shown in Figure 13.1.

FIGURE 13.1   Scatter Diagram of Output versus Aptitude Test Score

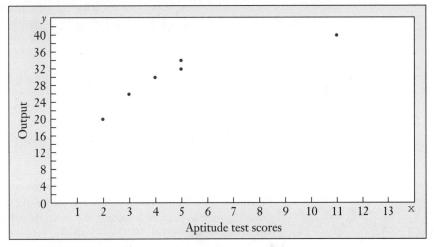

c) The scatter diagram indicates that there is a reasonable linear relationship between the two variables. Use of the least squares method is appropriate.

To determine the specific regression equation we must compute the values of the regression coefficients a and b. To do so we first need to determine $\sum x$, $\sum y$, $\sum xy$ and $\sum x^2$.

In addition, $\sum y^2$ is also computed at the same time since this value is needed in sections 13.3 and 13.5. Details of the calculation of these values are shown in Table 13.2.

TABLE 13.2   Calculations for Finding *a* and *b*

Employee	Output y	Test scores x	xy	x^2	y^2
A	31	5	155	25	961
B	40	11	440	121	1600
C	30	4	120	16	900
D	34	5	170	25	1156
E	25	3	75	9	625
F	20	2	40	4	400
$n = 6$	$\sum y = 180$	$\sum x = 30$	$\sum xy = 1000$	$\sum x^2 = 200$	$\sum y^2 = 5642$

The values of the regression coefficients can now be found by substituting the appropriate values listed in Table 13.2 into the two formulas.

$$b = \frac{n(\sum xy) - (\sum x)(\sum y)}{n(\sum x^2) - (\sum x)^2} = \frac{6(1000) - (30)(180)}{6(200) - (30)^2}$$

$$= \frac{6000 - 5400}{1200 - 900}$$

$$= \frac{600}{300} = 2$$

$$a = \frac{\sum y}{n} - b\frac{\sum x}{n} = \frac{180}{6} - 2(\frac{30}{6}) = 30 - 10 = 20$$

The regression equation is $y_p = 20 + 2x$.

d) To graph the regression line on the scatter diagram we need a minimum of two points on the line. A third point is recommended as a check.

The coordinates of three such points can be obtained by selecting three values of x. These three values are substituted into the regression equation and the corresponding values of y computed.

Any three values of x can be selected but it is preferable to select *convenient* values of x that are as far apart as the scatter diagram allows.

In our case two convenient extreme values are $x = 0$ and $x = 10$. The third value of x should be between the extreme values. In our case $x = 5$ is an appropriate selection.

For $x = 0$, $y_p = 20 + 2(0) = 20$; point (0,20) is on the line.
For $x = 10$, $y_p = 20 + 2(10) = 40$; point (10,40) is on the line.
For $x = 5$, $y_p = 20 + 2(5) = 30$; point (5,30) is on the line.

The three points can now be plotted on the scatter diagram reproduced below from part (b) and the **regression line** drawn by joining the three points (0,20), (5,30) and (10,40).

Note   The three points must lie on a straight line.

FIGURE 13.2   Scatter Diagram with Regression Line

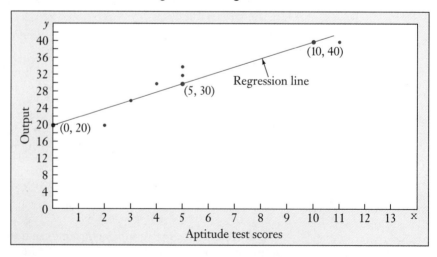

Note that $a = 20$ gives us the y intercept. The value $b = 2$ determines the *slope* of the line and indicates that, for an increase of one unit in x, the value y_p increases by two units.

e) To predict y_p, substitute the chosen values of x in the regression equation and compute y_p.

For $x = 6$, $y_p = 20 + 2(6) = 20 + 12 = 32$.

For $x = 10$, $y_p = 20 + 2(10) = 20 + 20 = 40$.

○ **EXAMPLE 13.2b**

A statistics instructor claims that he knows there is a distinct relationship between a student's statistics mark and the student's class attendance. To prove the point, the instructor has taken a random sample from last year's class and obtained the following information:

Student	A	B	C	D	E	F	G	H
Statistics mark	90	75	55	73	70	85	63	45
Number of classes missed (out of 32)	2	7	13	9	7	4	13	16

a) Determine the dependent variable.
b) Draw the scatter diagram.
c) Obtain the regression equation.
d) Plot the regression line.
e) Interpret the meaning of the regression coefficients a and b.
f) Predict the statistics marks for two students who miss 3 classes and 10 classes respectively.

● **SOLUTION**

a) To identify the dependent variable, pose the two questions
   1. Does class attendance depend on the statistics mark?
   2. Does the statistics mark depend on class attendance?
   Question 2 is more reasonable. The statistics mark data become the dependent variable y.

b) Scatter diagram. See Figure 13.3.

FIGURE 13.3 Scatter Diagram with Regression Line

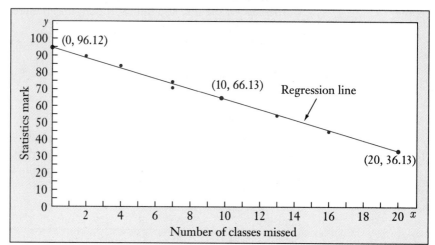

c) The distribution of the points in the scatter diagram indicates that there is a reasonable linear relationship between the statistics mark and the number of classes missed. The use of the least squares method is appropriate.

The values required to obtain the values of the regression coefficients a and b are listed in Table 13.3.

TABLE 13.3 Calculations for Finding a and b

Student	Statistics mark y	Number of classes missed x	xy	x^2	y^2
A	90	2	180	4	8100
B	75	7	525	49	5625
C	55	13	715	169	3025
D	73	9	657	81	5329
E	70	7	490	49	4900
F	85	4	340	16	7225
G	63	13	819	169	3969
H	45	16	720	256	2025
$n = 8$	$\sum y = 556$	$\sum x = 71$	$\sum xy = 4446$	$\sum x^2 = 793$	$\sum y^2 = 40\ 198$

Substituting into the formulas, we obtain

$$b = \frac{n(\sum xy) - (\sum x)(\sum y)}{n(\sum x^2) - (\sum x)^2} = \frac{8(4446) - (71)(556)}{8(793) - (71)^2}$$

$$= \frac{35\ 568 - 39\ 476}{6344 - 5041}$$

$$= \frac{-3908}{1303} = -2.9992$$

$$a = \frac{\sum y}{n} - b\frac{\sum x}{n} = \frac{556}{8} - (-2.9992)\left(\frac{71}{8}\right) = 69.5 + 26.6179 = 96.1179$$

The regression equation is $y_p = 96.1179 - 2.9992\,x$.

d) Graph of the regression line.
Three convenient values of x for plotting the regression line are $x = 0$, $x = 10$, and $x = 20$.
For $x = 0$, $y_p = 96.1179 - 2.9992(0) = 96.1179$.
For $x = 10$, $y_p = 96.1179 - 2.9992(10) = 96.1179 - 29.992 = 66.1259$.
For $x = 20$, $y_p = 96.1179 - 2.9992(20) = 96.1179 - 59.984 = 36.1339$.
The regression line passes through points $(0, 96.1)$, $(10, 66.1)$ and $(20, 36.1)$.

e) The regression coefficient $a = 96.1179$ indicates that a student who misses no classes ($x = 0$) can expect to achieve a statistics mark of 96.
The coefficient $b = -2.9992$ indicates a negative slope of approximately 3; that is, for every class missed the statistics mark can be expected to drop by three marks.

f) The expected statistics mark for a student missing three classes is

$$y_p = 96.1179 - 2.9992(3) = 96.1179 - 8.9976 = 87.1203,$$

that is, 87. The expected mark for a student missing 10 classes is

$$y_p = 96.1179 - 2.9992(10) = 96.1179 - 29.992 = 66.1259,$$

that is, 66.

Note When a regression equation is used for prediction purposes we should theoretically confine our prediction to within the range of the x values of the original data. For Example 13.2a, we should not calculate expected y values for x values below 2 and above 11. Similarly, for Example 13.2b we should restrict our calculation of expected y values to the range 2 to 16. The reason for this restriction is that there is no guarantee that the regression equation is valid beyond these limits. In practice, this restriction is often ignored and predictions are made beyond the limits of the sample data.

EXERCISE 13.2

1. Select the dependent variable that seems most appropriate for each of the following pairs of variables.
   a) The age of manufacturing equipment and the number of rejects produced by the equipment.
   b) Interest rates and the level of foreign investment in Canada.
   c) Sales volume and the number of marketing representatives.

2. Select the independent variable that seems most likely for each of the following pairs of variables.
   a) Number of bankruptcies and gross national product.
   b) Sales commissions and sales volume.
   c) Number of sales transactions and the amount of accounts receivable.

3. State whether the following scatter diagrams look like linear relationships or curvilinear relationships. In the case of a linear relationship, indicate whether the slope of the line is positive or negative.

(a)   (b)   (c)

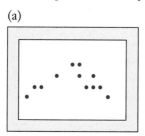   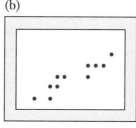   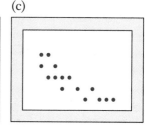

4. State whether the following scatter diagrams indicate linear relationships or curvilinear relationships. For linear relationships state whether the slope of the line is positive or negative.

(a)   (b)   (c)

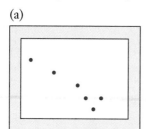   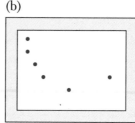   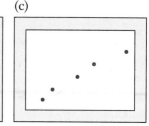

5. The following statistics were calculated from 45 pairs of data points:
   $\sum x = 60; \sum y = 180; \sum xy = 1159; \sum x^2 = 406$.
   a) Determine the least squares equation.
   b) Estimate y_p for $x = 2.5$.

6. Consider the following statistics:
   $n = 4; \sum x = 13.5; \sum y = 47.5; \sum xy = 2600; \sum x^2 = 734$.
   a) Determine the regression equation.
   b) When $x = 10$, what is the estimated value of y?

7. The guidance counsellor of a high school believes a linear relationship exists between English marks and History marks. The counsellor has collected the following data about a group of students taking both courses:

Student	A	B	C	D	E	F	G	H
History mark	45	82	30	80	71	95	67	56
English mark	50	74	37	73	70	91	69	63

a) Determine the dependent variable.

b) Draw a scatter diagram.

c) Determine the regression equation.

d) Interpret the meaning of the regression coefficients.

e) Plot the regression line.

f) Predict the values of y for **(i)** $x = 60$;   **(ii)** $x = 90$.

8. The sales manager of a company wants to gain insight into the relationship between the unit sales price and the number of units sold for a particular product. Using past sales data, the sales department supplied the following information:

Price per unit	$20	$24	$30	$33	$42	$45	$51
Number of units sold (000s)	20	18	16	14	12	10	8

a) Determine the regression equation and interpret the meaning of the regression coefficients.

b) Construct a scatter diagram and plot the regression line.

c) Predict the value of y for $x = 9$ and $x = 17$.

SECTION 13.3 Measuring Variability about the Regression Line

The equations of regression lines fitted to the sample data in examples 13.2a and 13.2b were calculated by the least squares method, which ensures that the lines fit the data with the minimum amount of variation. The associated diagrams (figures 13.2 and 13.3) show that the observed values are unlikely to fall on the fitted regression line but are scattered around it. Some of the data points are located above the line and some below it.

When dealing with a single variable x, we developed a measure of the variability of the observed values around the mean value $\bar{x}$. Called the sample variance, we computed it by using the formula

$$s = \sqrt{\frac{\Sigma(x - \bar{x})^2}{n - 1}}$$

Similarly, when dealing with two variables x and y, a measure of the variability of the observed y values from the predicted y values (y_p), known as the **standard error of estimate**, is given by

$$s_{y \cdot x} = \sqrt{\frac{\Sigma(y - y_p)^2}{n - 2}}$$

where $s_{y \cdot x}$ = standard error of estimate;

y = an observed value of y for a given value of x;

y_p = predicted (expected) value of y for the given value of x;

n = the number of paired observations (x, y).

The computation of the standard error of estimate is facilitated by use of the alternate formula:

$$s_{y \cdot x} = \sqrt{\frac{\sum y^2 - a(\sum y) - b(\sum xy)}{n - 2}}$$

Except for $\sum y^2$, all values in the formula are available from the regression line computations. As done in Table 13.2, $\sum y^2$ is frequently computed at the same time.

○ **EXAMPLE 13.3a**

Compute the standard error of estimate of y based on x for examples 13.2a and 13.2b.

● **SOLUTION**

For Example 13.2a the following values are needed from Table 13.2:
$n = 6$; $\sum y = 180$; $\sum xy = 1000$; $\sum y^2 = 5642$.
Substituting in the formula we obtain

$$s_{y \cdot x} = \sqrt{\frac{5642 - 20(180) - 2(1000)}{6 - 2}}$$

$$= \sqrt{\frac{5642 - 3600 - 2000}{4}} = \sqrt{\frac{42}{4}}$$

$$= \sqrt{10.5} = 3.2404$$

For Example 13.2b the values obtained from Table 13.3 are
$n = 8$; $\sum y = 556$; $\sum xy = 4446$; $\sum y^2 = 40\ 198$.
Substituting in the formula we obtain

$$s_{y \cdot x} = \sqrt{\frac{40\ 198 - 96.1179(556) - (-2.9992)(4446)}{8 - 2}}$$

$$= \sqrt{\frac{40\ 198 - 53\ 441.552 + 13\ 334.443}{6}} = \sqrt{\frac{90.891}{6}}$$

$$= \sqrt{15.148467} = 3.8921$$

The values obtained for $s_{y \cdot x}$ in the above calculations indicate the variation of the data points around the fitted regression lines.

For the fitted regression line $y_p = 20 + 2x$ in Example 13.2a, $s_{y \cdot x} = 3.2404$.

For the fitted regression line $y_p = 96.1179 - 2.9992\,x$ in Example 13.2b, $s_{y \cdot x} = 3.8921$.

The interpretation of the standard error of estimate is similar to that of the sample standard deviation. The sample standard deviation for a single variable x measures the variability of the data about the arithmetic mean $\bar{x}$.

The standard error of estimate for two variables (x, y) measures the variability of the data about the *fitted least square regression line* $y_p = a + bx$.

Furthermore, just as we used the sample standard deviation in developing an interval estimate around the mean $\bar{x}$, we can use the standard error of estimate in developing interval estimates around the expected values y_p.

EXERCISE 13.3

1. Determine the standard error of estimate for Question **7** in Exercise 13.2.
2. Determine the standard error of estimate for Question **8** in Exercise 13.2.

SECTION 13.4 Constructing Forecast Intervals

An important use of the regression line is to forecast (predict) values of the dependent variable y given some *specific* value of the independent variable x;

Two types of forecasts can be developed:
1. a forecast of the *expected mean value* of y for the given (specified) value of x;
2. a forecast of the *expected actual value* of y for the given value of x.

Both types of forecast are sometimes used. For example, Statistics Canada may wish to forecast the average expected level of unemployment in Canada for 1993. A forecast of the expected actual level of unemployment in the first quarter of 1993 might also be required.

The best *point estimate* for both types of forecast is obtained by substituting the given value of x in the regression equation $y_p = a + bx$.

In Example 13.2a the best point estimate of y_p when $x = 5$ is $y_p = 20 + 2(5) = 30$. This means that the best point estimate of the *average output* of all employees who score 5 on the dexterity test is 30 units. Similarly, the best point estimate of the *actual output* of an individual employee who scores 5 on the dexterity test is also 30 units.

While both point estimates equal 30 units, there is a difference when we construct interval estimates for the two types of forecast. An interval estimate is *centred* on a point estimate and the calculation of the upper and lower limits of the interval involves the standard error of estimate.

The *standard error for the mean value* of y for a specified value of x is given by

$$(s_{y \cdot x}) \sqrt{\frac{1}{n} + \frac{n(x - \bar{x})^2}{n \sum x^2 - (\sum x)^2}}$$

The *standard error for the actual value* of y for a specified value of x is given by

$$(s_{y \cdot x}) \sqrt{1 + \frac{1}{n} + \frac{n(x - \bar{x})^2}{n \sum x^2 - (\sum x)^2}}$$

The confidence interval estimate for predicting the *average* value of y is

$$y_p \pm t_{n-2}(s_{y \cdot x}) \sqrt{\frac{1}{n} + \frac{n(x - \bar{x})^2}{n \sum x^2 - (\sum x)^2}}$$

and the confidence interval estimate for predicting the *actual* value of y is

$$y_p \pm t_{n-2}(s_{y \cdot x}) \sqrt{1 + \frac{1}{n} + \frac{n(x - \bar{x})^2}{n \sum x^2 - (\sum x)^2}}$$

where y_p = the point estimate for a specified value of x;
   x = a specified value of the independent variable x;
   $\bar{x}$ = the mean of the observed values of x;
   $s_{y \cdot x}$ = the standard error of estimate of y based on x;
   t_{n-2} = the t table value for $(n - 2)$ degrees of freedom at a specified level of confidence;
   n = the number of paired observations.

○ EXAMPLE 13.4a
With reference to Example 13.2a, determine at the 95% level of confidence
a) the confidence interval for the output of a group of employees with aptitude test scores of 8;
b) the confidence interval for the output of an individual employee with an aptitude test score of 8.

● SOLUTION
From Table 13.2, $n = 6$; $\sum x = 30$; $\sum x^2 = 200$.
The regression equation is $y_p = 20 + 2x$;
for $x = 8$, $y_p = 20 + 2(8) = 20 + 16 = 36$;
$s_{y \cdot x} = 3.2404$ (see solution to Example 13.3a).

$$\bar{x} = \frac{\sum x}{n} = \frac{30}{6} = 5.$$

The number of degrees of freedom $(n - 2) = (6 - 2) = 4$.
t_{n-2} at the 95% level of confidence $= 2.776$.

a) First calculate

$$\sqrt{\frac{1}{n} + \frac{n(x - \bar{x})^2}{n \sum x^2 - (\sum x)^2}}$$

$$= \sqrt{\frac{1}{6} + \frac{6(8 - 5)^2}{6(200) - (30)^2}}$$

$$= \sqrt{0.1667 + \frac{6(9)}{1200 - 900}}$$

$$= \sqrt{0.1667 + 0.18}$$

$$= \sqrt{0.3467}$$

$$= 0.5888$$

The confidence interval for the mean value of y is

$$y_p \pm t_{n-2}(s_{y \cdot x}) \sqrt{\frac{1}{n} + \frac{n(x - \bar{x})^2}{n \sum x^2 - (\sum x)^2}} = 36 \pm 2.776(3.2404)(0.5888) =$$

36 ± 52.965. The confidence limits are 30.7035 and 41.2965.

This means we can be 95% certain that the average output of a group of employees, each of whom scored 8 on the aptitude test, will lie between 31 and 41 units.

b) First calculate $\quad \sqrt{1 + \frac{1}{n} + \frac{n(x - \bar{x})^2}{n \sum x^2 - (\sum x)^2}}$

$$= \sqrt{1 + \frac{1}{6} + \frac{6(8 - 5)^2}{6(200) - (30)^2}}$$

$$= \sqrt{1 + 0.1667 + 0.18} = \sqrt{1 + 0.3467} = \sqrt{1.3467}$$

$$= 1.1605.$$

The confidence interval for the actual value of y is

$$y_p \pm t_{n-2}(s_{y \cdot x}) \sqrt{1 + \frac{1}{n} + \frac{n(x - \bar{x})^2}{n \sum x^2 - (\sum x)^2}} = 36 \pm 2.776(3.2404)(1.1605)$$

$= 36 \pm 10.4391$. The confidence limits are 25.5609 and 46.4391.

This means we can be 95% certain that the output of an individual employee in the group who scored 8 on the aptitude test will lie between 26 and 46 units.

○ **EXAMPLE 13.4b**

For Example 13.2b determine the 90% confidence interval for
a) a group of students who miss 16 classes;
b) an individual student who misses 16 classes.

● **SOLUTION**

From Table 13.3, $n = 8$; $\sum x = 71$; $\sum x^2 = 793$.
The regression equation is $y_p = 96.1179 - 2.9992 x$;
substituting $x = 16$, $y_p = 96.1179 - 2.9992(16) = 48.1307$;
$s_{y \cdot x} = 3.8921$ (see solution to Example 13.3a).

$$\bar{x} = \frac{71}{8} = 8.875.$$

The number of degrees of freedom, $(n - 2) = (8 - 2) = 6$;
t_{n-2} at the 90% level of confidence $= 1.943$.

a) First calculate

$$\sqrt{\frac{1}{8} + \frac{8(16 - 8.875)^2}{8(793) - (71)^2}}$$

$$= \sqrt{\frac{1}{8} + \frac{8(50.7656)}{6344 - 5041}}$$

$$= \sqrt{0.125 + \frac{406.1248}{1303}}$$

$$= \sqrt{0.125 + 0.3117} = \sqrt{0.4367}$$

$$= 0.6608$$

The confidence interval for the mean value of y is

$$y_p \pm t_{n-2}(s_{y \cdot x}) \sqrt{\frac{1}{n} + \frac{n(x - \overline{x})^2}{n\sum x^2 - (\sum x)^2}}$$

$= 48.1307 \pm (1.943)(3.8921)(0.6608) = 48.1307 \pm 4.9971$.
The confidence limits are 43.1336 and 53.1278.

This indicates that we can be 90% certain that the average mark obtained by a group of students in this class who miss 16 classes can be expected to lie between 43 and 53.

b) First calculate

$$\sqrt{1 + \frac{1}{8} + \frac{8(16 - 8.875)^2}{8(793) - (71)^2}} = \sqrt{1 + 0.125 + \frac{406.1248}{1303}}$$

$$= \sqrt{1 + 0.125 + 0.3117} = \sqrt{1.4367}$$

$$= 1.1986$$

The confidence interval for the actual value of y is

$$y_p \pm t_{n-2}(s_{y \cdot x}) \sqrt{1 + \frac{1}{n} + \frac{n(x - \overline{x})^2}{n\sum x^2 - (\sum x)^2}}$$

$= 48.1307 \pm (1.943)(3.8921)(1.1986) = 48.1307 \pm 9.0643$.
The confidence limits are 39.0664 and 57.1950.

This means that we can have 90% confidence that an individual who misses 16 classes can be expected to obtain a mark between 39 and 57.

Shape of the Forecast Intervals

Note that the forecast intervals for the two types of forecast are different because their standard errors are different. Since the standard error for an individual value of y is always larger than the standard error for the mean value of y, the forecast intervals for an individual value of y_p will always be wider than those for the average value of y_p.

For both types of forecast the size of the standard error depends on the specified value of x. As the given value moves farther away from the mean $\overline{x}$, the numerical value of the expression under the square-root sign increases. As a result, the confidence interval becomes wider and the forecast becomes less accurate.

Table 13.4 and Figure 13.4 show the 95% forecast intervals for Example 13.4a for values of x ranging from 0 to 10. Note that the minimum width of both forecast intervals, 7.34 and 19.43, occurs at $x = \overline{x} = 5$. The largest width, 14.69 and 23.23, occurs at $x = 10$, the given value of x that differs most from the mean value $\overline{x} = 5$.

TABLE 13.4   **Forecast Intervals for Example 13.4a**

| x | y_p | Interval forecast for average value of y | | | Interval forecast for individual value of y | | |
		Upper limit	Lower limit	Width of interval	Upper limit	Lower limit	Width of interval
0	20	27.3447	12.6553	14.6893	31.6129	8.3871	23.2259
1	22	28.2753	15.7247	12.5506	32.9679	11.0321	21.9359
2	24	29.2963	18.7037	10.5926	34.4387	13.5613	20.8775
3	26	30.4676	21.5324	8.9352	36.0437	15.9563	20.0874
4	28	31.8864	24.1136	7.7729	37.7990	18.2010	19.5980
5	30	33.6723	26.3277	7.3447	39.7161	20.2839	19.4322
6	32	35.8864	28.1136	7.7729	41.7990	22.2010	19.5980
7	34	38.4676	29.5324	8.9352	44.0437	23.9563	20.0874
8	36	41.2963	30.7037	10.5926	46.4387	25.5613	20.8775
9	38	44.2753	31.7247	12.5506	48.9679	27.0321	21.9359
10	40	47.3447	32.6553	14.6893	51.6129	28.3871	23.2259

FIGURE 13.4   **Confidence Intervals around Regression Line**

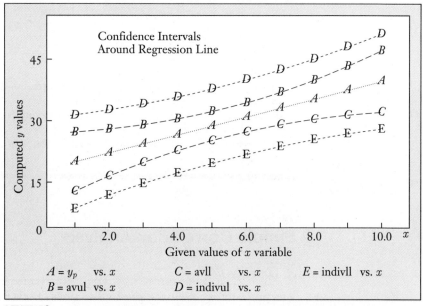

MINITAB®

EXERCISE 13.4

1. For Question **7** in Exercise 13.2, determine the 99% confidence interval in which the history mark will fall
   a) for a group of students who have an English mark of 65;
   b) for Daisy if her English mark is 65.

2. For Question **8** in Exercise 13.2, determine the 98% confidence interval in which
   a) the average number of units sold can be expected to be found if the unit price is $15;
   b) the specific number of units that will be sold at a special sale if the price is set at $15.

3. An engineering department has compiled unit cost data for a series of production runs of different quantities as shown below.

Quantity (000s)	20	13	14	18	13	17	10
Cost per unit ($)	2.10	2.39	2.34	2.19	2.43	2.18	2.50

   a) Obtain the least squares regression equation.
   b) Construct a scatter diagram and plot the regression line.
   c) Compute the standard error of estimate of y based on x.
   d) Calculate the cost per unit associated with a production run of 15 000 units.
   e) Construct the 90% confidence interval for the expected unit cost of production runs of 15 000 units.
   f) Determine the 90% prediction interval for the next run of 15 000 units.

4. A random sample from a class of mathematics students comparing student marks with student class attendance yielded the following data:

Number of classes attended	30	28	16	25	32	32	20	25
Mathematics mark	86	81	41	63	97	90	47	72

   a) Obtain the least squares regression equation.
   b) Construct a scatter diagram and plot the regression line.
   c) Calculate the standard error of estimate.
   d) Compute a point estimate of the mathematics mark for a student attending 18 classes.
   e) Construct the 99% confidence interval for Pat, who will be able to attend just 22 classes.

SECTION 13.5 Simple Correlation Analysis

A. Introduction

The procedures used in the previous sections of this chapter provide the least squares equation, which describes the relationship between two variables. To measure the closeness of association between variables we use **correlation analysis**. The closer the actual data points are to the fitted regression line, the closer is the relationship between the variables.

The three main measures of the strength of the relationship are
1. the coefficient of determination, r^2;
2. the coefficient of non-determination, $(1 - r^2)$;
3. the coefficient of correlation, r.

B. Total Deviation, Explained Deviation and Unexplained Deviation

The three coefficients take into account three related vertical deviations for each value of y.
1. The *total vertical deviation* is the difference between an observed value y and the mean value y.

$$\text{TOTAL DEVIATION} = (y - \bar{y})$$

2. The *explained deviation* is part of the total deviation. It is the difference between y_p, the value of the dependent variable obtained when a particular value of x is substituted into the regression equation, and the average value $\bar{y}$.

$$\text{EXPLAINED DEVIATION} = (y_p - \bar{y})$$

3. The *unexplained deviation* is the remaining part of the total deviation.

$$\text{UNEXPLAINED DEVIATION} = (y - y_p)$$

It follows that

$$\text{TOTAL DEVIATION} = \text{EXPLAINED DEVIATION} + \text{UNEXPLAINED DEVIATION}$$
$$(y - \bar{y}) = (y_p - \bar{y}) + (y - y_p)$$

The concept of the three deviations is illustrated in Figure 13.5, which is based on data from Example 13.2a.

The regression equation for Example 13.2a was $y_p = 20 + 2x$.

Since $n = 6$ and $\sum y = 180$, $\bar{y} = \dfrac{\sum y}{n} = \dfrac{180}{6} = 30$.

FIGURE 13.5   **Illustrations of Vertical Deviations**

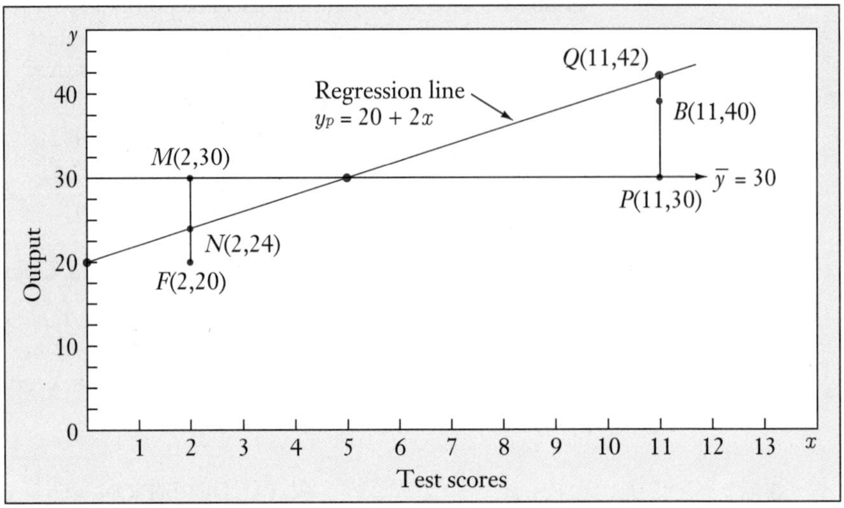

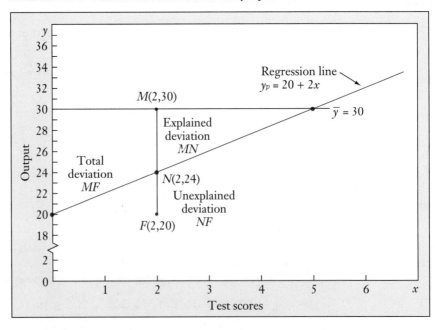

Let us consider the data for Employee F from Example 13.2a.
Output, $y = 20$.
Test score, $x = 2$.
Computed output, $y_p = 20 + 2(2) = 24$.
The data for Employee F are reflected in the excerpt from Figure 13.5 shown as Figure 13.6.

FIGURE 13.6   **Vertical Deviations for Employee F**

The vertical line segment FM joining point $F(2, 20)$ to point $M(2, 30)$ on the $\bar{y}$ line represents the total deviation of the output of Employee F from the average output for the group of employees.

$$\text{TOTAL DEVIATION, } (y - \bar{y}) = (20 - 30) = -10$$

The line segment FM intersects the regression line at N. This point represents the output that is expected for a test score of 2.

$$y_p = 20 + 2(2) = 24$$

The coordinates of N are $(2, 24)$.

The line segment MN represents the difference between expected output $y_p = 24$ for a test score of 2 and the average output $\bar{y}$. Because this deviation is determined by the dependency of y on x, it is referred to as the explained deviation.

$$\text{EXPLAINED DEVIATION, } (y_p - \bar{y}) = (24 - 30) = -6$$

The line segment FN is the remaining part of the total deviation and represents the deviation that cannot be explained by the dependency of y on x. It is referred to as the unexplained deviation.

$$\text{UNEXPLAINED DEVIATION, } (y - y_p) = (20 - 24) = -4$$

From this we obtain

$$\text{EXPLAINED DEVIATION + UNEXPLAINED DEVIATION}$$
$$= (-6) + (-4)$$
$$= -10$$
$$= \text{TOTAL DEVIATION}$$

Now consider the data for Employee B.

Output, $y = 40$.

Test score, $x = 11$.

Computed output, $y_p = 20 + 2(11) = 42$.

As shown in Figure 13.5, the line segment BP represents the total deviation of employee B's output from the average output.

$$\text{TOTAL DEVIATION, } (y - \bar{y}) = (40 - 30) = 10$$

Point $Q(11, 42)$ represents the expected output for a test score of 11. The line segment QP represents the explained deviation for y when $x = 11$.

$$\text{EXPLAINED DEVIATION, } (y_p - \bar{y}) = (42 - 30) = 12$$

The line segment QB represents the unexplained deviation.

$$\text{UNEXPLAINED DEVIATION, } (y - y_p) = (40 - 42) = -2$$

From this we obtain

$$\text{EXPLAINED DEVIATION} + \text{UNEXPLAINED DEVIATION}$$
$$= (12) + (-2)$$
$$= 10$$
$$= \text{TOTAL DEVIATION}$$

C. Total Variation, Explained Variation and Unexplained Variation

The closeness of the association between y and x can be measured by considering the relationship between the explained variation and the total variation for the group of observations.

○ **EXAMPLE 13.5a**

For the data used in Example 13.2a, compute the total variation, explained variation and unexplained variation.

● **SOLUTION**

Step 1   Compute the deviations for the individual observations and then total the deviations as shown in Table 13.5.

TABLE 13.5   Computations of Deviations

Employee	Test score x	Out- put y	Predicted output y_p	Total deviation $y - \bar{y}$	Explained deviation $y_p - \bar{y}$	Unexplained deviation $y - y_p$	Check: EXPLAINED + UNEXPLAINED = TOTAL DEVIATION
A	5	31	30	31 − 30 = 1	30 − 30 = 0	31 − 30 = 1	0 + 1 = 1
B	11	40	42	40 − 30 = 10	42 − 30 = 12	40 − 42 = −2	12 − 2 = 10
C	4	30	28	30 − 30 = 0	28 − 30 = −2	30 − 28 = 2	−2 + 2 = 0
D	5	34	30	34 − 30 = 4	30 − 30 = 0	34 − 30 = 4	0 + 4 = 4
E	3	25	26	25 − 30 = −5	26 − 30 = −4	25 − 26 = −1	−4 − 1 = −5
F	2	20	24	20 − 30 = −10	24 − 30 = −6	20 − 24 = −4	−6 − 4 = −10

Step 2   Use the deviations computed in Table 13.5 to determine the squared deviations as shown in Table 13.6.

TABLE 13.6   Computations of Squared Deviations

Employee	Total squared deviation $(y - \bar{y})^2$	Explained squared deviation $(y_p - \bar{y})^2$	Unexplained squared deviation $(y - y_p)^2$
A	$(1)(1) = 1$	$(0)(0) = 0$	$(1)(1) = 1$
B	$(10)(10) = 100$	$(12)(12) = 144$	$(-2)(-2) = 4$
C	$(0)(0) = 0$	$(2)(-2) = 4$	$(2)(2) = 4$
D	$(4)(4) = 16$	$(0)(0) = 0$	$(4)(4) = 16$
E	$(-5)(-5) = 25$	$(-4)(-4) = 16$	$(-1)(-1) = 1$
F	$(-10)(-10) = 100$	$(-6)(-6) = 36$	$(-4)(-4) = 16$
	$\sum(y - \bar{y})^2 = 242$	$\sum(y_p - \bar{y})^2 = 200$	$\sum(y - y_p)^2 = 42$

The three resulting variations are defined as follows:

1. The sum of the total squared deviations is called the

$$\text{TOTAL VARIATION} = \sum(y - \bar{y})^2 = 242$$

2. The sum of the explained squared deviations is called the

$$\text{EXPLAINED VARIATION} = \sum(y_p - \bar{y})^2 = 200$$

3. The sum of the unexplained squared deviations is called the

$$\text{UNEXPLAINED VARIATION} = \sum(y - y_p)^2 = 42$$

For the three variations,

$$\text{TOTAL VARIATION} = \text{EXPLAINED VARIATION} + \text{UNEXPLAINED VARIATION}$$
$$= 200 + 42 = 242$$

D. The Coefficients of Determination, Non-determination and Correlation

The **coefficient of determination** is the ratio of the explained variation to the total variation.

The **coefficient of non-determination** is the ratio of unexplained variation to total variation.

The **coefficient of correlation** is the square root of the coefficient of determination.

The three coefficients can be determined for Example 13.3a as follows:

$$\text{COEFFICIENT OF DETERMINATION, } r^2 = \frac{\text{EXPLAINED VARIATION}}{\text{TOTAL VARIATION}}$$

$$= \frac{200}{242} = 0.826446$$

$$= 82.6\%$$

The coefficient r^2 indicates that 82.6% of the total variation is explained by the dependency of output on the test scores.

COEFFICIENT OF NON-DETERMINATION, $(1 - r^2)$

$$= \frac{\text{UNEXPLAINED VARIATION}}{\text{TOTAL VARIATION}}$$

$$= \frac{42}{242} = 0.173554 = 17.4\%$$

The coefficient $(1 - r^2)$ indicates that 17.6% of the total variation is caused by factors other than the dependency of output on the test scores.

Note The coefficient of non-determination can be calculated directly from the coefficient of determination as $(1 - r^2) = (1 - 0.826446) = 0.173554 = 17.4\%$.

COEFFICIENT OF CORRELATION, r

$$= \sqrt{\text{COEFFICIENT OF DETERMINATION}}$$

$$= \sqrt{r^2} = \sqrt{0.826446} = 0.9079978 = 90.8\%$$

E. Computation of the Coefficient of Determination r^2 by Formula

Computation of the individual deviations and squared deviations is usually avoided because it is tedious. The value of the coefficient of determination r^2 can be more readily calculated from the values required for determining the regression equation.

$$r^2 = \frac{[n(\sum xy) - (\sum x)(\sum y)]^2}{[n(\sum x^2) - (\sum x)^2][n(\sum y^2) - (\sum y)^2]}$$

○ **EXAMPLE 13.5b**

Compute the coefficients of determination, non-determination and correlation for Example 13.2a.

● **SOLUTION**

From Table 13.2 we know that $n = 6$; $\sum x = 30$; $\sum y = 180$; $\sum xy = 1000$; $\sum x^2 = 200$; $\sum y^2 = 5642$.

Substituting in the formula, the coefficient of determination

$$r^2 = \frac{[6(1000) - 30(180)]^2}{[6(200) - (30)^2][6(5642) - (180)^2]}$$

$$= \frac{(6000 - 5400)^2}{(1200 - 900)(33\ 852 - 32\ 400)}$$

$$= \frac{(600)^2}{(300)(1452)} = \frac{360\ 000}{435\ 600} = 0.825446$$

COEFFICIENT OF NON-DETERMINATION, $(1 - r^2) = (1 - 0.826446)$

$$= 0.173554$$

COEFFICIENT OF CORRELATION, $r = \sqrt{0.826446} = 0.9079978$

For some obvious and some not so obvious reasons the coefficient of correlation r is used much more than the coefficient of determination r^2.

One significant advantage the coefficient of correlation has over the coefficient of determination is that it indicates whether the relationship between the variables is *direct* or *inverse*.

The not-so-obvious reasons for preferring the use of the coefficient of correlation deal with more complex issues related to its arithmetic relationship with the regression equation.

The user of statistical information should be critical of research that suggests a correlation coefficient as low as 0.40 indicates a substantial relationship between two variables. Since $(0.40)^2 = 0.16$, a correlation of this size indicates that only 16% of the variance in one variable is predictable from the other. It takes a coefficient of correlation of 0.707 before 50% of the variance of y is explained by x.

○ **EXAMPLE 13.5c**

Compute the coefficients r^2, $(1 - r^2)$ and r for Example 13.2b.

● **SOLUTION**

From Table 13.3 we know that $n = 8$; $\sum x = 71$; $\sum y = 556$; $\sum xy = 4446$; $\sum x^2 = 793$; $\sum y^2 = 40\ 198$.

Substituting in the formula, the coefficient of determination

$$r^2 = \frac{[8(4446) - 71(556)]^2}{[8(793) - (71)^2][8(40\ 198) - (556)^2]}$$

$$= \frac{(35\ 568 - 39\ 476)^2}{(6344 - 5041)(321\ 584 - 309\ 136)}$$

$$= \frac{(-3908)^2}{(1303)(12\ 448)} = \frac{15\ 272\ 464}{16\ 219\ 744} = 0.9416 = 94.16\%$$

The coefficient of non-determination

$$(1 - r^2) = (1 - 0.9416) = 0.0584 = 5.84\%$$

The coefficient of correlation

$$r = \sqrt{0.9416} = 0.9704 = 97.04\%$$

F. Interpretation of the Coefficients

The coefficient of determination relates the explained variation for a set of data to the total variation. *At most*, the explained variation can *equal* the total variation.

If the explained variation equals the total variation, all the variation in y is explained by the dependency of y on x and $r^2 = 1$.

The smallest value for the explained variation is zero, in which case $r^2 = 0$. In most cases, r^2 has a value somewhere between 0 and 1, that is,

$$0 \leq r^2 \leq 1$$

The coefficient of correlation, as the square root of r^2, can take positive or negative values between +1 and −1, that is,

$$-1 \leq r \leq 1$$

The coefficient of correlation is used as a common index to measure the degree of association between two variables.

Positive values of r indicate a *direct* relationship between the two variables; that is, as x increases y increases. This case is referred to as **positive correlation** and is associated with a regression line that slopes upward to the right. This means that when the value of the regression coefficient b is *positive*, the variables are *positively correlated*.

The closer r is to +1, the stronger the relationship. If $r = 1$, the relationship between x and y is *direct and perfect*. This situation is referred to as *perfect positive correlation*. Representative scatter diagrams indicating positive correlation between two variables are shown below.

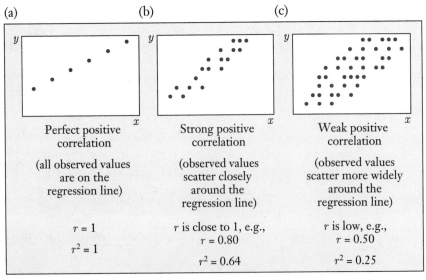

(a)   (b)   (c)

Perfect positive correlation	Strong positive correlation	Weak positive correlation
(all observed values are on the regression line)	(observed values scatter closely around the regression line)	(observed values scatter more widely around the regression line)
$r = 1$	r is close to 1, e.g., $r = 0.80$	r is low, e.g., $r = 0.50$
$r^2 = 1$	$r^2 = 0.64$	$r^2 = 0.25$

Negative values of r indicate an *inverse* relationship between the two variables; that is, as x increases y decreases. This case is referred to as **negative**

correlation and is associated with a regression line that slopes downward to the right. This means that when the regression coefficient b is negative, the variables are negatively correlated.

If $r = -1$ the relationship between x and y is *inverse* and *perfect*. This situation is referred to as *perfect negative correlation*. Representative scatter diagrams indicating negative correlation are shown below.

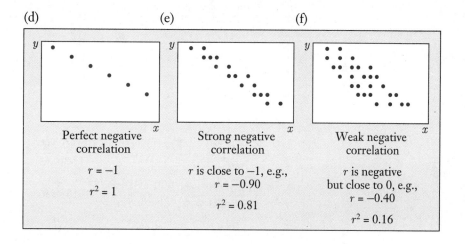

(d)	(e)	(f)
Perfect negative correlation	Strong negative correlation	Weak negative correlation
$r = -1$	r is close to -1, e.g.,	r is negative
$r^2 = 1$	$r = -0.90$	but close to 0, e.g.,
	$r^2 = 0.81$	$r = -0.40$
		$r^2 = 0.16$

○ **EXAMPLE 13.5d**

The following data relate to the results of a test producing an index of digital dexterity and the associated output of a group of assembly-line operators.

Operator	A	B	C	D	E	F	G	H	I	J
Index of dexterity	21	20	22	23	21	22	15	18	18	20
Output	23	19	25	24	25	22	12	16	19	22

a) Obtain the least squares regression equation.
b) Construct the scatter diagram and plot the regression line.
c) Determine the value of y when $x = 16$.
d) Assess the reliability of the test by computing the coefficients of determination and correlation.

● **SOLUTION**

a) Since it is more reasonable to say that output depends on digital dexterity, output is the dependent variable y.

TABLE 13.7 Computations Required

Employee	x	y	xy	x^2	y^2
A	21	23	483	441	529
B	20	19	380	400	361
C	22	25	550	484	625
D	23	24	552	529	576
E	21	25	525	441	625
F	22	22	484	484	484
G	15	12	180	225	144
H	18	16	288	324	256
I	18	19	342	324	361
J	20	22	440	400	484
$n = 10$	200	207	4224	4052	4445

$$b = \frac{n(\sum xy) - (\sum x)(\sum y)}{n(\sum x^2) - (\sum x)^2} = \frac{10(4224) - (200)(207)}{10(4052) - (200)^2}$$

$$= \frac{42\ 240 - 41\ 400}{40\ 520 - 40\ 000} = \frac{840}{520} = 1.6153846$$

$$a = \frac{\sum y}{n} - b\frac{\sum x}{n} = \frac{207}{10} - 1.6153846(\frac{200}{10})$$

$$= 20.7 - 1.6153846(20) = 20.7 - 32.307692 = -11.607692$$

The regression equation is $y_p = -11.6077 + 1.6154\,x$.

FIGURE 13.7 Scatter Diagram with Regression Line

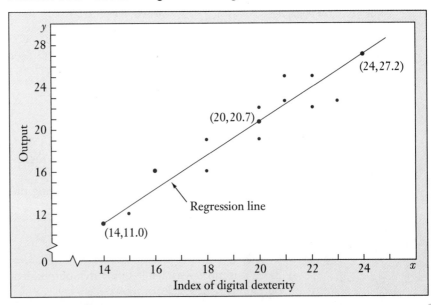

b) For $x = 14$, $y_p = -11.6077 + 1.6154(14)$
$$= -11.6077 + 22.6256 = 11.0079.$$

For $x = 20$, $y_p = -11.6077 + 1.6154(20)$
$$= -11.6077 + 32.3080 = 20.7003.$$

For $x = 24$, $y_p = -11.6077 + 1.6154(24)$
$$= -11.6077 + 38.7696 = 27.1619.$$

To draw the regression line, plot points $(14, 11.0)$, $(20, 20.7)$, and $(24, 27.2)$ and join the three points.

c) The predicted output level for $x = 16$ is

$$y_p = -11.6077 + 1.6154\, x$$
$$= -11.6077 + 1.6154(16)$$
$$= -11.6077 + 25.8464 = 14.2387$$

d) The coefficient of determination is given by

$$r^2 = \frac{[\,n(\sum xy) - (\sum x)(\sum y)\,]^2}{[\,n(\sum x^2) - (\sum x)^2\,][\,n(\sum y^2) - (\sum y)^2\,]}$$

$$= \frac{[\,10(4224) - (200)(207)\,]^2}{[\,10(4052) - (200)^2\,][\,10(4445) - (207)^2\,]}$$

$$= \frac{(42\ 240 - 41\ 400)^2}{(40\ 520 - 40\ 000)(44\ 450 - 42\ 849)}$$

$$= \frac{840^2}{(520)(1601)} = \frac{705\ 600}{832\ 520} = 0.8475472 = 84.75\%$$

This means that 84.75% of the total variation in output y is determined by changes in the index x.

The coefficient of correlation

$$r = \sqrt{0.8475472} = 0.9206 = 92.06\%$$

The coefficient of correlation is positive and high. This indicates that there is a strong direct relationship between output and test results. The index of digital dexterity appears to be a good predictor of output.

EXERCISE 13.5

Refer to graphs A, B and C to answer questions **1** to **6**.

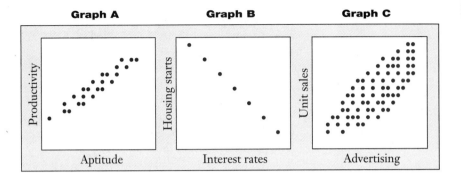

1. **a)** Name the predicted variable in Graph B.
   b) Name the independent variable in Graph C.
2. **a)** What is the dependent variable in Graph A?
   b) What is the independent variable in Graph B?
3. For which graph is the relationship between the two variables weakest?
4. For which graph is the relationship between the two variables strongest?
5. What is the approximate coefficient of correlation for Graph B?
6. Select the most reasonable value as coefficient of correlation for Graph A from the following four values:
   (a) 0.50   **(b)** −0.50   **(c)** 0.80   **(d)** −0.80
7. If the coefficient of determination is 0.81, what is the coefficient of correlation?
8. If the coefficient of correlation is 0.64, what is the coefficient of determination?
9. If the regression coefficient $b = -2.25$ and the coefficient of determination is 0.16, what is the coefficient of correlation?
10. If the regression coefficient $a = -4.00$ and $b = 3.50$ and the coefficient of non-determination is 0.64, what is the coefficient of correlation?
11. A drama instructor claims that he knows from experience that a student's drama grade is related to the student's mathematics mark. He has given you the list of grades of nine students as follows:

Student	J	K	L	M	N	P	Q	R	S
Drama mark	69	65	59	66	78	63	67	68	75
Math mark	62	68	61	75	89	73	86	50	53

   a) Compute the regression equation.
   b) Construct a scatter diagram and plot the regression line.
   c) Compute the coefficient of determination.
   d) Comment on the results.
12. The following are data on mortgage loans and prime rates:

Mortgage loans	80	60	75	40	54	66	47
Prime rate	9%	12%	10%	15%	13%	11%	14%

   a) Compute the regression equation.

b) Construct a scatter diagram and plot the regression line.
c) Compute the coefficient of correlation.
d) Comment on the results.

REVIEW EXERCISE

1. For each of the following sets of data,
   a) determine the regression equation;
   b) construct a scatter diagram and plot the regression line;
   c) compute the coefficient of correlation.

Data set **(i)**

x	10.0	8.0	13.0	9.0	11.0	14.0	6.0	4.0	12.0	7.0	5.0
y	8.04	6.95	7.58	8.81	8.33	9.96	7.24	4.26	10.86	4.82	5.68

Data set **(ii)**

x	10.0	8.0	13.0	9.0	11.0	14.0	6.0	4.0	12.0	7.0	5.0
y	9.14	8.14	8.74	8.77	9.26	8.10	6.13	3.10	9.13	7.26	4.74

Data set **(iii)**

x	10.0	8.0	13.0	9.0	11.0	14.0	6.0	4.0	12.0	7.0	5.0
y	7.46	6.77	12.74	7.11	7.81	8.84	6.08	5.39	8.15	6.42	5.73

Data set **(iv)**

x	8.0	8.0	8.0	8.0	8.0	8.0	8.0	19.0	8.0	8.0	8.0
y	6.58	5.76	7.71	8.84	8.47	7.04	5.25	12.50	5.56	7.91	6.89

2. Comment on the results for Question **1**.
3. Consider the following data:

Student	A	B	C	D	E	F	G	H
Hours of study	5.5	8.2	4.0	8.0	7.1	9.5	6.7	6.6
History mark	50	90	37	73	70	91	73	63

a) Determine the regression equation.
b) Compute the coefficient of correlation.
c) Obtain a point estimate of the history mark for students who studied 7.0 hours.
d) Determine the 90% confidence interval for the students in part **(c)**.
e) Predict the history mark for Kelly, who studied 7.0 hours.

4. A committee studying federal tax reform reviewed the reported revenue and net income statistics of some of Canada's largest corporations. The data for a random sample of seven such firms are listed below.

Company	1989 revenue ($ millions)	1989 net income ($ millions)
Alcan Aluminium Ltd.	10 467	989
Ontario Hydro	6 346	699
Seagram Co. Ltd.	5 339	841
Canadian Wheat Board	4 279	521
IBM Canada Ltd.	4 188	349
Moore Corporation	3 208	239
McDonald's Restaurants	1 378	51

a) Determine the regression equation.
b) Compute the coefficient of correlation.
c) Compute the standard error of estimate.
d) Calculate the net income point estimate when revenue is $2 billion.
e) Determine the 95% confidence interval for corporations reporting net income of $1 500 000 000.

5. The human-resources manager of a real estate company is faced with determining the firm's staffing needs. As a first step the manager considered the overall activity in the housing market and compiled the following information:

One-year mortgage rate	8%	9%	10%	11%	12%	13%	14%
Number of homes sold	1480	1235	890	670	445	250	120

a) Determine the regression equation.
b) Compute the coefficients of determination and correlation.
c) Compute the 98% confidence interval and prediction intervals when the one-year mortgage rate is 9.5%.
d) Comment on the usefulness of the information obtained in parts (a), (b) and (c).

6. The owner of a small auto body shop has compiled quarterly sales data for the last two years and asked his accountant to determine if the data might be useful in operational planning.

Quarter	I	II	III	IV	V	VI	VII	VIII
Economic indicator	13.2	11.6	12.1	15.3	12.2	11.6	9.9	12.5
Sales ($000)	265	290	275	224	285	280	338	265

a) Determine the regression equation.
b) Compute the coefficients of determination, non-determination and correlation.
c) If the forecasted economic indicator for quarter I was 14.4, what sales level should be expected?
d) Compute the 99% confidence and prediction intervals for an economic activity level of 13.8.
e) Interpret the two intervals obtained in part (d).

7. The following data have been collected by a company's controller:

Year	1986	1987	1988	1989	1990	1991
Bad debts ($000)	3	12	10	5	4	16
Index of business activity	210	100	110	140	160	80

a) Explain your choice of dependent variable.
b) Determine the least squares regression equation.
c) Construct a scatter diagram and plot the regression line.
d) Calculate r for the data.
e) Compute $s_{y \cdot x}$.
f) Comment on the usefulness of the information with regard to the company's credit-granting policies.

8. An advertising account representative is hoping to use information gained from the following data to increase business.

Advertising ($000)	2000	200	1000	800	1500	500
Sales ($000 000)	9.0	6.8	7.8	8.0	8.4	7.4

a) Determine the regression equation.
b) Construct a scatter diagram and plot the regression line.
c) Compute r^2 and $s_{y \cdot x}$.
d) What are the predicted sales for an advertising campaign costing $1 800 000?
e) Could the account representative use this information to increase the advertising company's business?

9. The manager of a bakery shop wishes to determine the relationship between the number of loaves of bread per order and the direct labour cost per order. Six orders were randomly selected and the following data were obtained:

Order number	Number of loaves	Direct labour cost per order
5	6	$4.75
8	3	2.45
12	7	5.30
13	10	7.90
18	5	4.25
25	6	4.95

a) Obtain the regression equation.
b) Calculate r^2 and r and comment on the type and closeness of the relationship between the two variables.
c) If the manager receives an order for eight loaves, determine the associated direct labour cost for that order.

10. A labour efficiency consultant compiled the following data for the employees of a production department.

Productivity (parts per hour)	43	52	60	66	59	73	68	65	49	40
Employees stress (scale 0–10)	8	7	5	3	7	0	2	3	9	10

a) Determine the regression equation.
b) Calculate r and comment on the type and the closeness of the relationship between the two variables.
c) Predict with 90% confidence the stress scale score for John who was off work on the day of testing. His production is known to be 62 parts per hour.

SELF-TEST

1. The regression equation $y_p = 19\ 665 + 0.0000027\,x$ shows the relationship between a company's total assets and the number of employees it has.
   a) Is the linear relationship direct or inverse?
   b) How many employees would you estimate to find in a company such as Canadian Pacific with assets of approximately $20 223 500 000?

2. Consider the following data:

Dependent variable	16	10	26	32
Independent variable	4	3	5	6

   a) Construct a scatter diagram.
   b) Determine the least squares equation.
   c) Draw the regression line in the scatter diagram.

3. Consider the following:

x	5	2	6	3
y	11	31	9	24

   a) What type of linear relationship appears to exist between the two variables?
   b) State whether the value of b in the least squares equation is positive or negative.
   c) Compute the coefficients of determination, non-determination and correlation.

4. A research group is interested in predicting the gasoline consumption in a geographic area based on the number of automobiles registered. The sample of randomly selected geographic areas showed the following gas consumption and automobile registrations:

Geographic area	Gasoline consumption (millions of litres)	Automobile registration (000s)
A	16	17
B	8	10
C	3	6
D	9	13
E	18	22
F	6	8
G	15	16
H	4	5
I	9	15
J	10	16

a) Determine the least squares equation.
b) Compute the coefficient of correlation.
c) Estimate gasoline consumption in an area with an automobile registration of 14 000.
d) Determine the 90% confidence interval for a mean value of 14 000.
e) Determine the 90% confidence interval for an actual value of 14 000.

Key Terms

Summary of Formulas

1. Regression line:

$$y_p = a + bx$$

2. Regression coefficients:

$$b = \frac{n(\sum xy) - (\sum x)(\sum y)}{n(\sum x^2) - (\sum x)^2}$$

$$a = \frac{\sum y}{n} - b\frac{\sum x}{n}$$

3. Standard error of estimate:

$$s_{y \cdot x} = \sqrt{\frac{\sum y^2 - a(\sum y) - b(\sum xy)}{n - 2}}$$

4. Confidence interval estimate for predicting the average value of y_p:

$$y_p \pm t_{n-2}(s_{y \cdot x}) \sqrt{\frac{1}{n} + \frac{n(x - \bar{x})^2}{n(\sum x^2) - (\sum x)^2}}$$

5. Confidence interval estimate for predicting the actual value of y_p:

$$y_p \pm t_{n-2}(s_{y \cdot x}) \sqrt{1 + \frac{1}{n} + \frac{n(x - \bar{x})^2}{n\sum x^2 - (\sum x)^2}}$$

6. Coefficient of determination:

$$r^2 = \frac{[n(\sum xy) - (\sum x)(\sum y)]^2}{[n\sum x^2 - (\sum x)^2][n\sum y^2 - (\sum y)^2]}$$

7. Coefficient of correlation $r = \sqrt{r^2}$ and $-1 \leq r \leq 1$

APPENDIX

Statistical Software for Personal Computers

The development of statistical software for computers — first for mainframes, then midis and, in the 1980s, for personal computers — has allowed statisticians to perform more exhaustive, accurate and affordable data analysis. Because the computer can reduce the time it takes to process data, larger data sets can be dealt with more easily.

The first commercial packages for mainframes were developed in the 1960s, although few such packages are still in use today. By contrast, there are now more than 200 commercial statistical packages available for PCs of varying capabilities and usefulness.

Most colleges and universities have at least one statistical package on their mainframe, and your instructor may wish to integrate such a package into this course. In addition, some schools have both a mainframe and a PC version of certain packages (MINITAB, for example) available for student use.

MYSTAT, a student version of SYSTAT, is available with this text. Student versions of MINITAB and SPSS are also available at relatively modest cost for those students who possess their own personal computers. These student versions have most, but not all, of the capabilities of the full versions of these programs and are restricted to handling smaller data sets. Nevertheless, they are still powerful programs and are more than adequate for dealing with all of the material covered in this introductory text.

Although this text does not provide instruction on how to use any particular software package, it is important for those students who will be exposed to the use of such software to understand certain concepts that are common to all programs.

First, we will discuss the following important terms: *case*, *variable*, *value*, and *data set*.

Each person, household, product, or anything else that is the subject of an experiment, is known as a *case*. For example, in a real estate survey the case would be a household. For each household in the survey you might record such things as the age and sex of the householder, his or her marital status, income, and the number of persons living in the house. In other words, *the case is the unit for which measurements are made or responses recorded.*

Each of the attributes that are measured and/or recorded is known as a *variable* — for example, the answer that each householder gives to the question about his or her age is that person's *value* for the *variable* "Age of Householder."

So, if 50 households were sampled and 5 variables recorded for each, there would be 50 cases and 5 variables for a total of 250 values, which then constitute the complete *data set.*

As another example, suppose a production manager were to randomly sample 100 items from a production run and have the quality control laboratory measure 3 attributes on each sample — length, width and weight, for instance. Here we have 100 cases and 3 variables for a total of 300 values making up the data det.

Data Entry and Data Arrangement

Entering raw data into the program is accomplished by typing in the individual values obtained as in the above examples or from a computer file previously prepared on a data disk such as that provided to instructors using this text.

The statistical software arranges the data set in a matrix such that the variables make up the *columns* and the cases make up the *rows* (see Figure 1). In most programs, therefore, it is important to enter the variable data in the same order so that each value falls into its appropriate variable column. Most programs also allow the variable to be referred to either by a column number or by a name assigned to it. Note that, although the matrix arrangement is similar to that of a spreadsheet, it does *not* appear on the screen but remains in the background in the computer's RAM. It can, of course, be displayed on the screen and printed out as a hard copy by a suitable command.

Figure A1

Cases	C1	C2	C3	C4	Variables						C_n
1	–	–	–	–	–	–	–	–	–	–	–
2	–	–	–	–	–	–	–	–	–	–	–
3	–	–	–	–	–	–	–	–	–	–	–
4	–	–	–	Values	–	–	–	–	–	–	–
5	–	–	–	–	–	–	–	–	–	–	–
–	–	–	–	–	–	–	–	–	–	–	–
–	–	–	–	–	–	–	–	–	–	–	–
–	–	–	–	–	–	–	–	–	–	–	–
–	–	–	–	–	–	–	–	–	–	–	–

The complete matrix constitutes the data set.

Batch Mode versus Interactive Mode

Statistical software packages may operate under batch mode or interactive mode, and some packages allow the user to employ either procedure. In both modes of operation all of the values must first be entered into the data matrix and stored in the computer's RAM.

For *batch mode* operation, *all* of the commands that the operator wants the program to perform on the data are then entered. Subsequently, the user enters some form of RUN command, which causes the program to carry out all of the necessary calculations and to provide the answers required by the complete list of commands already entered.

In the *interactive mode*, a command for a *single* operation is entered and acted on by the program immediately. The operator then enters the next command, which again is performed immediately, and so on.

Batch mode operation is especially useful for large statistical analysis projects that need to be done in one session. The disadvantage of batch mode is that great care must be taken to ensure that all of the commands entered are absolutely correct in every detail, otherwise the whole run may be aborted.

Interactive mode is more convenient when only a few procedures need to be done in one session. In addition, it has the advantage that you can see the results of one command before moving on to the next and, if a mistake has been made in entering a command, it can then be re-entered correctly.

Creating New Variables

An important characteristic of the better statistical software packages is that they allow the user to create new variables from the original ones read into the data matrix. For example, a scatter diagram created by the software from original data might show a curvilinear relationship between two variables. Examples are the exponential relationship between the demand for electricity and population growth, or the diminishing-returns relationship between advertising expenditures and sales. Changing one or both of the variables to its square, square root or logarithmic form may result in a better fit of the regression line to the data. The new variables are created by typing appropriate commands to accomplish the mathematical transformations required and to insert the new variables into new columns so that the original data are not lost.

Categorical Selection of Cases

Most packages allow for the selection of specific categories of cases from the total number of cases available in the data matrix. To do so requires the categories that may be of interest to be coded in some unique way so that the software can recognize them and pull them out for analysis. A common example is when you wish to separate the responses or attributes of males versus females from a particular study. In this case, one of the variables would be "sex of respondent" where the response to the question could be coded 1 meaning female and 0 meaning male. Analysis of the data could then be carried out in three ways — *all* respondents, female respondents and male respondents, respectively.

Graphical Analysis

Chapters 1 and 2 of this text describe methods for preparing bar charts, pie charts, histograms, frequency polygons and cumulative frequency graphs by hand. Chapter 13 does the same for scatter diagrams and regression line graphs. All of these, and many other graphical displays of data, can be performed by many of the packages now available. STATGRAPHICS and SYGRAPH (the graphics module of SYSTAT) are particularly strong in this regard and some examples of the output from both are displayed in this text.

MINITAB Executable Files

The following executable files have been developed to allow the student and/or instructor using MINITAB on a personal computer to solve various problems associated with the chapter headings listed. They may be reproduced as MINITAB .MTJ files (or .MTB files in release 8) and used as indicated. It is left as an exercise for the student/instructor to identify the values of the constants (k's) in terms of the mathematical symbols in order to interpret the outputs given in the examples and to relate them back to the worked examples in the text.

Chapter 4 — two executable files as follows:

a) The following is a MINITAB executable file designed to calculate the mean, variance and standard deviation of raw data of a *population* using both the theoretical and the short-cut formulas.

```
no echo
name c1 'x' c2 '(x-Mu)' c3 '(x-Mu)Sq' c4 'xsquared'
let k1 = mean (c1)
let c2 = (c1-k1)
let c3 = c2**2
let k2 = mean (c3)
sqrt k2 k3
let c4 = c1**2
let k4 = mean (c4)
let k5 = k1**2
let k6 = (k4-k5)
sqrt k6 k7
let k8 = sum (c1)
let k9 = sum (c2)
```

```
let k10 = sum (c3)
let k11 = sum (c4)
Print k1–k11 c1–c4
echo
End.
```

To use the file: enter the raw data into column c1 of the MINITAB worksheet followed by

MTB> execute <filename>.

Examples 4.3a and 4.3b can both be calculated using this file.

Enter the test scores in column c1 and execute the file to yield the following output:

```
K1   13.0000
K2   6.75000
K3   2.59808
K4   175.750
K5   169.000
K6   6.75000
K7   2.59808
K8   104.000
K9   0
K10  54.0000
K11  1406.00
```

ROW	x	(x-Mu)	(x-Mu) Sq	xsquared
1	17	4	16	289
2	11	-2	4	121
3	13	0	0	169
4	15	2	4	225
5	9	-4	16	81
6	16	3	9	256
7	12	-1	1	144
8	11	-2	4	121

b) This is a MINITAB executable file designed to calculate the mean, variance and standard deviation of (population) frequency distributions, either grouped or ungrouped.

```
no echo
name c1 'MIDPTx' c2 'f' c3 'fx' c4 'f.xsqrd'
let c3 = c1*c2
let c4 = c3*c1
let k1 = sum (c2)
let k2 = sum (c3)
let k3 = sum (c4)
let k4 = k2/k1
```

```
let k5 = k4**2
let k6 = k3/k1
let k7 = (k6-k5)
sqrt k7 k8
print k1–k8 c1–c4
echo
End.
```

To use the file:

a) for a *grouped* frequency distribution, enter the *mid-point* values of the class intervals into column c1, and the frequencies into column c2, of the MINITAB worksheet.

b) for an *ungrouped* frequency distribution, enter the *individual* data values in column c1 and their corresponding frequencies in column c2.

followed by

MTB>execute <filename> and <enter>.

Example 4.3d can be calculated using this file.

Enter the mid-points of the class intervals in column c1 and their corresponding frequencies into column c2 and execute the file to obtain the following output:

K1 170.000

K2 8330.00

K3 452650

K4 49.0000

K5 2401.00

K6 2662.65

K7 261.647

K8 16.1755

ROW	midpt-x	f	fx	f.xsqrd
1	5	2	10	50
2	15	4	60	900
3	25	12	300	7500
4	35	28	980	34300
5	45	44	1980	89100
6	55	43	2365	130075
7	65	23	1495	97175
8	75	7	525	39375
9	85	5	425	36125
10	95	2	190	18050

Chapter 8

The following is a MINITAB executable file designed to calculate the mean, variance and standard deviation of a probability distribution.

no echo

name c1 'X' c2 'P(x)' c3 'xP(x)' c4 '(x-Mu)' c5 '(x-Mu)Sq'
let c3 = c1*c2
let k1 = Sum(c3)
let c4 = (c1-k1)
let c5 = c4**2
let c6 = c5*c2
let k2 = Sum(c6)
sqrt k2 k3
Print k1-k3 c1-c6
echo
End.

To use the file: enter the values of the variable into column c1 and their corresponding probabilities into column c2 of a MINITAB worksheet, followed by

MTB>execute <filename> and <enter>.

Example 8.3c can be calculated using this file.

Enter the values 2–12 in column c1 and their associated probabilities (from Table 8.5 in text) in column c2 and then execute the file to obtain the following output

K1 7.00001
K2 5.83335
K3 2.41523

ROW	x	P(x)	xP(x)	(x-Mu)	(x-Mu)Sq	C6
1	2	0.027778	0.05556	-5.00001	25.0001	0.694452
2	3	0.055556	0.16667	-4.00001	16.0001	0.888899
3	4	0.083333	0.33333	-3.00001	9.0000	0.750001
4	5	0.111111	0.55555	-2.00001	4.0000	0.444447
5	6	0.138889	0.83333	-1.00001	1.0000	0.138891
6	7	0.166667	1.16667	-0.00001	0.0000	0.000000
7	8	0.138889	1.11111	0.99999	1.0000	0.138887
8	9	0.111111	1.00000	1.99999	4.0000	0.444441
9	10	0.083333	0.83333	2.99999	9.0000	0.749993
10	11	0.055556	0.61112	3.99999	15.9999	0.888893
11	12	0.027778	0.33334	4.99999	24.9999	0.694448

Example 8.3c can be solved using MINITAB as follows:

Enter the numbers 0–8 inclusive in column c1 of the worksheet followed by the following commands:

MTB > pdf c1 c2;
SUBC> binomial 8 0.4.
MTB > cdf c1 c3;
SUBC> binomial 8 0.4.
MTB > print c1–c3

ROW	x	P(x)	CumP(x)
1	0	0.016796	0.01680
2	1	0.089580	0.10638
3	2	0.209019	0.31539
4	3	0.278692	0.59409
5	4	0.232243	0.82633
6	5	0.123863	0.95019
7	6	0.041288	0.99148
8	7	0.007864	0.99934
9	8	0.000655	1.00000

In effect, we have created a binomial probability table for $n=8$, $p=0.4$ that can now be used to answer the parts of Example 8.3c.

a) $P(x=6) = 0.041288$

b) $P(x<3) = P(x=0)+P(x=1)+P(x=2) = CumP(x=2) = 0.31539$

c) $P(x=5 \text{ or } 6 \text{ or } 7) = CumP(x=7)-CumP(x=4)$

$= (0.99934-0.82633)$

$= 0.17301$

d) $P(x>4) = 1.00000-CumP(4) = (1.00000-0.82633) = 0.17367$

Chapter 9

Example 9.3b can be solved using MINITAB as follows:

Enter the z values given in parts (a)–(h) of the example into column c1 followed by the commands

MTB > CDF c1 c2;

SUBC> Normal 0.0 1.0.

MTB > print c1 c2

ROW	z	C2
1	1.50	0.933193
2	-2.20	0.013903
3	2.50	0.993790
4	-1.75	0.040059
5	1.25	0.894350
6	2.96	0.998462
7	-2.07	0.019226
8	-1.03	0.151505
9	-2.33	0.009903
10	1.64	0.949497
11	-2.00	0.022750

Column c2 of the print-out now contains the area from the left tail of the normal distribution to the given z value. Refer to the diagrams in the text when reading the following:

a) Area to the left of $z = 1.50$ is 0.933193.

b) Area to the right of $z = -2.20$ is $(1-0.013903) = 0.986097$.

c) Area above $z = 2.50$ is $(1-0.993790) = 0.00621$.

d) Area below $z = -1.75$ is 0.040059.

e) Area between $z = 1.25$ and $z = 2.96$ is $(0.998462-0.894350) = 0.104112$.

f) Area between $z = -2.07$ and $z = -1.03$ is $(0.151505-0.019226) = 0.0.132279$.

g) Area between $z = -2.33$ and $z = 1.64$ is $(0.949497-0.009903) = 0.939594$.

h) Area below $z = -2.00$ or above $z = 2.00 = 2(0.022750) = 0.0455$

Chapter 11

The following is a MINITAB executable file designed to calculate the mean, variance and standard deviation of a *sample* using both the theoretical and short-cut formulas.

```
no echo
name c1 'x' c2 'devn' c3 'devnsqrd' c4 'xsquared'
let k1 = mean(c1)
let k2 = count(c1)
let k3 = (k2-1)
let c2 = (c1-k1)
let c3 = c2**2
let c4 = c1**2
let k4 = sum(c3)
let k5 = k4/k3
sqrt k5 k6
let k7 = sum(c4)
let k8 = k7*k2
let k9 = sum(c1)**2
let k10 = (k8-k9)
let k11 = k2*k3
let k12 = k10/k11
Print k1-k12 c1-c4
echo
End.
```

To use the file: enter the sample values into column c1 of a MINITAB worksheet, followed by

MTB>execute <filename> and <enter>.

Example 11.3c can be solved using this file to produce the following output:

K1 14.5000

K2 32.0000

K3 31.0000

K4 608.000

K5 19.6129

K6 4.42865

K7 7336.00

K8   234752
K9   215296
K10 19456.0
K11 992.000
K12 19.6129

ROW	x	devn	devnsqrd	xsquared
1	14	-0.5	0.25	196
2	9	-5.5	30.25	81
3	12	-2.5	6.25	144
4	23	8.5	72.25	529
5	11	-3.5	12.25	121
6	15	0.5	0.25	225
7	7	-7.5	56.25	49
8	14	-0.5	0.25	196
9	6	-8.5	72.25	36
10	13	-1.5	2.25	169
11	18	3.5	12.25	324
12	19	4.5	20.25	361
13	21	6.5	42.25	441
14	17	2.5	6.25	289
15	9	-5.5	30.25	81
16	10	-4.5	20.25	100
17	16	1.5	2.25	256
18	18	3.5	12.25	324
19	15	0.5	0.25	225
20	20	5.5	30.25	400
21	8	-6.5	42.25	64
22	13	-1.5	2.25	169
23	17	2.5	6.25	289
24	24	9.5	90.25	576
25	11	-3.5	12.25	121
26	16	1.5	2.25	256
27	14	-0.5	0.25	196
28	13	-1.5	2.25	169
29	18	3.5	12.25	324
30	15	0.5	0.25	225
31	12	-2.5	6.25	144
32	16	1.5	2.25	256

Note A problem like this would normally be solved by using the DESCRIBE command as follows:

MTB > describe c1

	N	MEAN	MEDIAN	TRMEAN	STDEV	SEMEAN	MIN	MAX	Q1	Q3
x	32	14.500	14.500	14.429	4.429	0.783	6.000	24.000	11.250	17.750

Chapter 12

Example 12.3h can be solved using MINITAB — enter the sample data into column c1 followed by the command

MTB > ttest 15.00 c1;
SUBC> ALTERNATIVE = -1.

The output obtained is

TEST OF MU = 15.0000 VS MU L.T. 15.0000

	N	MEAN	STDEV	SE MEAN
x	10	4.6600	0.3035	0.0960

	T	P VALUE
x	-3.54	0.0032

Since the absolute value of the calculated t (3.54) is greater than the critical value (2.821), we reject H_0 and do not accept the batch.

Alternatively, the p value indicates that the probability that a sample mean as low as 14.66 when the population mean is 15.00 would happen by chance only about 3.2 times in 1000. We can therefore be quite confident that it is not a chance event and that the mean length of the pins is less than the specified 15.00 mm.

Chapter 13

This is a MINITAB executable file designed to calculate the values of 'a' and 'b' in a regression line and the values of 'r^2' and 'r' in correlation analysis.

```
no echo
name c1 'x' c2 'y' c3 'xy' c4 'xsquared' c5 'ysquared'
let c3 = c1*c2
let c4 = c1**2
let c5 = c2**2
let k1 = count(c1)
let k2 = sum(c1)
let k3 = sum(c2)
let k4 = sum(c3)
let k5 = sum(c4)
let k6 = sum(c5)
let k7 = k2**2
let k8 = k2*k3
let k9 = k1*k4
let k10 = (k9-k8)
let k11 = k1*k5
let k12 = (k11-k7)
let k13 = k10/k12
let k14 = k3/k1
let k15 = k2/k1
```

```
let k16 = k13*k15
let k17 = (k14-k16)
let k18 = k1*k6
let k19 = k3**2
let k20 = (k18-k19)
let k21 = k10**2
let k22 = k12*k20
let k23 = k21/k22
sqrt k23 k24
Print k1-k24 c1-c5
echo
End.
```

To use the file: enter the values of the independent variable x into column c1, and the corresponding values of the dependent variable y into column c2, of the MINITAB worksheet followed by

MTB>execute <filename> and <enter>.

Example 13.2a can be solved using this file to produce the following output:

```
K1  6.00000
K2  30.0000
K3  180.000
K4  1000.00
K5  200.000
K6  5642.00
K7  900.000
K8  5400.00
K9  6000.00
K10 600.000
K11 1200.00
K12 300.000
K13 2.00000
K14 30.0000
K15 5.00000
K16 10.0000
K17 20.0000
K18 33852.0
K19 32400.0
K20 1452.00
K21 360000
K22 435600
K23 0.826446
K24 0.909091
```

ROW	x	y	xy	xSQUARED
1	5	31	155	25
2	11	40	440	121
3	4	30	120	16
4	5	34	170	25
5	3	25	75	9
6	2	20	40	4

ROW	ySQUARED
1	961
2	1600
3	900
4	1156
5	625
6	400

Note   This type of problem would usually be solved using the REGRESS command and its sub-commands as follows:

MTB > Regress c2 1 c1;

SUBC> Predict 8.

The regression equation is

y = 20.0 + 2.00 x

Predictor	Coef	Stdev	t ratio	p
Constant	20.000	2.646	7.56	0.002
x	2.0000	0.4583	4.36	0.012

s=3.240 R-sq=82.6% R-sq(adj) = 78.3%

Analysis of Variance

SOURCE	DF	SS	MS
Regression	1	200.00	200.00
Error	4	42.00	10.50
Total	5	242.00	

F	p
19.05	0.012

Fit	Stdev.Fit	95% C.I.
36.00	1.91	(30.70, 41.30)

    95% P.I.
(25.56, 46.44)

Note that the output gives the regression equation, the predicted value of y for x = 8 (under the heading Fit), and the 95% confidence intervals for both the average value of y (under 95% C.I.) and an individual value of y (under 95% P.I.). The relevant portion of the output for the predicted value of y when x = 12 is

Fit	Stdev.Fit	95% C.I.
44.00	3.47	(34.36, 53.64)

95% P.I.
(30.81, 57.19) X

X denotes a row with x values away from the centre.

Answers to Selected Problems, Review Exercises, and Self-Tests

CHAPTER 2

EXERCISE 2.1

1. (b) (i) 68 **(ii)** 245 **(iii)** 245 **(d)** less-than cumulative frequencies: 1, 3, 5, 8, 16, 20, 23, 24; less-than cumulative percents: 4.2, 12.5, 20.8, 33.3, 66.7, 83.3, 95.8, 100.0 **(e)** more-than cumulative frequencies: 24, 23, 21, 19, 16, 8, 4, 1; more-than cumulative percents: 100.0, 95.8, 87.5, 79.2, 66.7, 33.3, 16.7, 4.2

3. (b) (i) 28 **(ii)** 30 **(c) (i)** 85% **(ii)** 30%

EXERCISE 2.3

3. $P_{65} = 37.5$, $D_4 = 26.0$, $Q_1 = 20.0$, $P_{50} = 30.0$, $x = 17.5\%$

5. (b) $Q_1 = 157.3$, $Q_3 = 188.15$, $P_{50} = 171.74$, $x = 77.1\%$

REVIEW EXERCISE

1. (b) (i) 61% **(ii)** 35%

3. (b) (i) P.E.I. **(ii)** Ont. **(c) (i)** $3.61 **(ii)** $15.58 **(iii)** $15.83

5. (b) (i) 55 **(ii)** 63 **(iii)** 68 **(d)** less-than cumulative frequencies: 3, 9, 26, 40, 46, 48; less-than cumulative percents: 6.3, 18.8, 54.2, 83.3, 95.8, 100.0 **(e)** more-than cumulative frequencies: 48, 45, 39, 22, 8, 2; more-than cululative percents: 100.0, 93.8, 81.3, 45.8, 16.7, 4.2 **(f) (i)** 66 **(ii)** 77.1 **(g)** 84.6%

9. (b) median **(c)** $756.68 **(d)** $899.46 **(e) (i)** 81.45% **(ii)** 67.45% **(f) (i)** 94.2% **(ii)** 32 **(g)** 143

SELF-TEST

1. (b) (i) 65 **(ii)** 55

2. (c) 91.2% **(d)** 17.13%

3. (a) 16.36 **(b)** 19.77 **(c)** 18.41 **(d)** $26.50 **(e)** 85% **(f)** 6%

CHAPTER 3

EXERCISE 3.2

1. 69.15

3. 40.5

5. 34.51

EXERCISE 3.3

1. 71 units, 41 hours, 31.48

EXERCISE 3.4

1. 63

3. 27.69

EXERCISE 3.5

1. (a) median = 11.5, mode = 8 **(b)** 11.75 **(c)** 13.25 **(d)** positively skewed

5. positively skewed

REVIEW EXERCISE

1. (a) $111.517 billion   (b) $111.60 billion

3. (a) $8.07 billion   (b) $7.05 billion

5. 43 seconds

7. 70 kg

9. $129.95

11. $39.25 per share

13. (a) median = 20, mode = 19   (b) 20.15
(c) 23.5   (d) positively skewed

15. mean = 20, median = 20.67, mode = 22.86
(b) negatively skewed

17. (a) $5.70   (b) positively skewed

19. (a) negatively skewed   (b) positively skewed
(c) symmetrical

SELF-TEST

A. 1. median = 24.5, mode = 24.0

2. 24.75

3. 27.61

4. positively skewed

B. 5. $Q_1 = 5.32$, $Q_3 = 10.38$

6. mean = 12.67, median = 10.18, mode = 6.33

7. positively skewed

C. 9. 1134.25 km/h

CHAPTER 4

EXERCISE 4.1

1. (a) $15 023   (b) business admin., accounting,
marketing, chemistry, mathematics, physics

3. (a) $7.00   (b) $1.75

EXERCISE 4.2

1. 45.33

3. 4.29

EXERCISE 4.3

1. (a) 70.92   (b) 49.07   (c) 7.005

3. (a) 5.46   (b) 2.55   (c) 1.60

5. (a) 29.36   (b) 346.51   (c) 18.615

7. Class A: 0.437, Class B: 0.174; marks for Class
A are much more variable than Class B's.

REVIEW EXERCISE

1. (a) 77   (b) 82.5   (c) 39   (d) 25.75   (e) 10.22
(f) 163.124   (g) 12.772

3. (a) $45 516.67   (b) $44 645.39   (c) $110 000
(d) $32 102.31   (e) $21 710.53   (f) $67 425.74
(g) $257 733.06   (h) $16 054.07

5. (a) 128   (b) (i) 72.5   (ii) 45.25   (iii) 23.94

7. $12 345.71

9. (a) 2.38   (b) variance = 10.259, standard
deviation = 3.203

11. (a) 8128   (b) 392 590   (c) 626.57

13. (a) 72 516.9   (b) 16 771.1   (c) 320 000 000
(d) 18 010

15. (a) 10   (b) 10.56   (c) 10.292   (d) 3.2

17. mean = 2.08, standard deviation = 1.29

19. (a) coefficient of variation for the morning
group = 0.30; afternoon group = 0.39. The
afternoon group's golf scores were about 30% more
variable.   (b) 78.57

21. 6.1%

23. (a) Company A = 0.01, Company B = 0.015,
Company C = 0.05   (b) Choose Company A
because its sales are less variable year to year
compared with its average sales.

SELF-TEST

A. 1. 40

2. $9.94

3. $18.38

4. $18.00

5. variance = $45.946, standard deviation = $6.778

B. 6. $115 000

7. $86.25

8. Mean$_{1990}$ = $321 875.00, mean$_{1991}$ = $265
625.00, median$_{1990}$ = $332 500, median$_{1991}$ = $282
500

9. $54 375.00

10. 1990 = $60 670.09, 1991 = $42 752.01

11. 1990 = 0.188, 1991 = 0.161

12. Average house prices dropped in 1991 but the variability was almost the same as in 1990.

CHAPTER 5

EXERCISE 5.3

1. (a) price: bread = 175.0, milk = 133.3, butter = 133.3; quantity: bread = 180.0, milk = 133.3, butter = 75.0; value: bread = 315.0, milk = 177.8, butter = 100.0 **(b)** 139.6 **(c)** price = 152.0, quantity = 130.7, value = 207.8

EXERCISE 5.4

1. 124.3

EXERCISE 5.5

1. (a) $0.877 **(b)** $41,434.26

EXERCISE 5.6

1. Company A 100 108.8 119.7 130.2 132.4 126.6
Company B 100 104.1 112.0 129.8 138.9 142.5

3. 2.3 7.2 37.4 72.3 100.0 121.0 164.1 249.3 372.3

REVIEW EXERCISE

1. (a) price = 120.7, quantity = 150.0, value = 181.0

3. (a)

	Price	Quantity	Value
Movie pass	120.0	84.0	100.8
Large popcorn	142.9	85.7	122.4
Bus ticket	119.0	75.0	89.3

(b) 123.8
(c) price = 125.1, quantity = 83.7, value 104.8

5. (a) 109.3 **(b)** 190.6 **(c)** 1990 = 144.0, 1991 = 209.9

7. (a) price = 98.4, quantity = 92.6, value = 91.2

9. 100.0 102.8 103.3 106.9 117.9 120.3 123.5

11. (a) Women's: 1990 = 1248, 1992 = 1268; Men's: 1990 = 936, 1992 = 906; Children's: 1990 = 624, 1992 = 664 **(b)** price = 112.6, quantity = 100.6, value = 113.3

13. 80.3

15. 134.1

17. 114.2

19. 1987 = $0.954, 1989 = $0.877

21. 1988 = $136 362 799.30, 1989 = $143 472 807.00

23. (a) 6103.2 6080.2 6099.1 6379.7 6405.6
(b) −0.38% 0.31% 4.60% 0.41%
(c) 4.41% 3.95% 9.80% 5.25%

25. 104.6 119.8 89.9

27. (a) 100.0 114.2 119.2 122.9 125.7 131.5
(b) 81.4 92.9 97.0 100.0 102.3 107.0

29. Canada: 100.0 112.2 125.4 128.2 132.0;
USA: 100.0 97.9 101.8 103.3 103.3

31. 75.5 83.7 88.5 92.4 96.0 100.0 104.4 108.5 114.0 119.5

SELF-TEST

1. (a) 95.2 100.0 105.2 110.2 120.0 125.3
(b) 25.3% **(c)** 13.6%

2. 135.7

3. $31 929.82

4. fewer

5. $0.837

6. Beer = 135.7, CPI = 119.5

7. Country A

8. (a) price = 135.2, quantity = 121.2, value 164.1

9. (a) 92.0

10. (a) 83.8 92.3 100.0 108.8 110.6 105.7
(b) Company A: 100.0 110.1 119.3 129.8 131.9 126.1; Company B: 100.0 104.4 110.1 116.4 125.3 126.8

CHAPTER 6

EXERCISE 6.2

1. (b) $y_p = 10.4 + 0.7x$ **(c) (i)** 15.3 **(ii)** 16.7

EXERCISE 6.3

1. $y_p = 10.4 + 0.7x$

3. (b) $y_p = 15.0 - 0.8x$ **(d) (i)** 6.1   **(ii)** 2.9

EXERCISE 6.4

1. 101.6   99.8   91.8   107.5   105.7   87.5   109.6   96.4

EXERCISE 6.5

1. 10.2   10.6   11.4   12.2   13.4   14.4   13.8   13.2

3. 8.5   9.3   10.3   11.3   12.0   13.0   13.5   14.8   15.3   14.5

EXERCISE 6.6

1. (a) $Q_1 = 57.5, Q_2 = 108.0, Q_3 = 74.2,$ $Q_4 = 160.3$
(b) 5.4   5.0   5.2   5.6   6.7   6.2   7.0   7.4   8.1   8.1   10.4   9.3   9.4   10.0   8.7   8.3   5.4   7.5   7.0   8.3

3. (a) Jan = 86.4, Feb = 76.2,   Mar = 86.8, Apr = 91.6, May = 93.8, Jun = 94.9, Jul = 80.5, Aug = 89.7, Sep = 85.9, Oct = 105.4, Nov 120.3, Dec = 188.7
(b) 1988: 13.9   13.1   13.8   15.3   13.9   13.7   13.7   13.4   14.0   14.2   14.1   14.3
1989:   13.9   14.4   17.3   14.2   14.9   14.8   14.9   12.3   12.8   15.2   16.6   17.0
1990:   16.2   15.8   16.1   16.4   16.0   15.8   16.2   16.7   16.3   15.2   15.8   16.4
1991:   17.4   17.1   17.3   17.5   18.1   19.0   19.9   21.2   19.8   19.0   19.1   19.1
1992:   18.5   18.4   18.4   19.7   19.2   20.0   21.1   23.4   21.0   19.9   21.6   21.2

REVIEW EXERCISE

1. (b) $y_p = 3690.2 + 147.0x$ **(c)** 147 million per year   **(d) (i)** 5012.7 million   **(ii)** 5306.6 million

3. (b) $y_p = 5.4 - 1.0x$ **(c)** $1.40 per share

5. (b) $y_p = 2854.7 - 217.3x$ (1985 = 0)
(c) (i) 899.0   **(ii)** 464.4

7. (a) $550   **(b)** $30   **(c)** $880

9. (c) 64.6   72.6   80.6   88.6   96.6
(d) 102.2   100.6   96.8   98.2   102.5

11. (a) $y_p = 612.8 + 42.1x$ (1984 = 0)

(c) $1033.8 million   **(d)** 104.3   99.2   98.9   96.7   97.1   100.4   103.7

13. 62.2   68.6   72.2   69.2   63.6   60.8   49.6   43.2   38.0   34.0   32.0   31.8

15. (a) Computed trend: 129.1   138.8   148.5   158.2   167.9   177.7   187.4   197.1   206.8   216.6   226.3   236.0   245.7   255.4   265.2   274.9   284.6   294.3   304.1   313.8   323.5   333.2   342.9
(b) 95.3   103.0   103.0   99.2   90.5   105.3   104.6   101.0   91.9   104.4   107.8   105.1   90.3   100.6   98.8   99.7   98.4   97.8   106.6   98.5   94.3   98.1   105.6

17. (a) 23 100   23 125   23 150   23 125   22 875   23 125   23   150   23 350   23 375   23 125   23   050   23 100   22 950   23 150   23 350   23   175   23 425   23 650   23 525   23 825   24 025   24   275   24 700   24   850   25 125   25 400   25 775   25 875   26 150   26 350   26 650   25   800   25   583.3
(b) Jan = 143.5, Feb = 178.5, Mar = 116.7, Apr = 75.2, May = 85.4,   Jun = 89.2, Jul = 130.8, Aug = 135.5, Sep 98.3, Oct = 66.2, Nov = 36.8, Dec = 43.9
(c) 22 164.3   22 522.9   24 941.0   25 930.1   18 968.1   24 547.0   21   322.4   23 244.5   25 944.4   23 106.4   22 015.6   23 244.5   22 373.4   22   691.0   24 683.8   21 940.9   22 480.8   24 883.2   23 156.6   23 465.9   22   892.1   21 747.2   23   646.3   19 142.5   24 046.2   24 035.6   22 884.0   25   930.1   25 642.1   23 201.9   25 907.9   25   236.9   25 944.4   29 449.3   28   538.7   26 662.8   26 346.3   26 556.8   23 912.5   30 318.3   28   452.2   27   237.1   18 112.6   23 318.3

19. (a) 661.50   645.00   597.25   547.75   553.25   544.50   543.25   522.25   517.50   502.00   497.00   527.25   549.75
(c) $Q_1 = 114.0, Q_2 = 91.4, Q_3 = 90.4, Q_4 = 104.2$
(d) 628.9   758.0   766.4   521.4   571.0   549.1   547.5   542.5   540.3   543.6   454.6   524.2   485.9   521.7   588.4   610.6

SELF-TEST

1. (a) $y_p = 11.4 + 1.1x$ (1984 = 0)   **(c) (i)** $22.10   **(ii)** $24.30

2. (a) $y_p = 562.1 - 12.5x$ (1987 = 0)
(b) −$12 500 per year   **(c)** $462 500
(d) 99.6   100.4   100.1   99.5   100.9   99.4

3. (a) $110\ 000$ per year   **(b)** $1\ 380\ 000$
(c) $2\ 150\ 000$

4. (a) 33 875.3   31 828.0   30 312.3   28 989.8
28 100.0   30 585.0   32 929.5   36 673.5
38 968.0

5. (a) $Q_1 = 98.7$, $Q_2 = 100.0$, $Q_3 = 99.8$,
$Q_4 = 101.5$
(b) 8 789.4   8 631.5   8 805.5   9 080.8   9 275.9
9 508.2   9 739.0   9 994.8   10 208.3   10 636.7
10 964.9   11 042.7   11 635.4   11 797.2   12
234.8   12 169.4

CHAPTER 7

EXERCISE 7.1

1. (a) 6   **(b)** 36
3. a, b, d
5. (a) 2   **(b)** 3   **(c)** 4
7. 30.8%
9. 95%
11. (a) 95%   **(b)** 29 days

EXERCISE 7.2

1. (a) 2   **(b)** 4   **(c)** 8
3. c, d
5. (a) 8   **(b)** 2

EXERCISE 7.3

1. 0.80
3. (a) 0.077   **(b)** 1   **(c)** 0.539   **(d)** 0.385
5. 0.667
7. 0.125
9. 0.25
11. (a) 0.053   **(b)** 0.553   **(c)** 0.197   **(d)** 0.395

EXERCISE 7.5

1. 120
3. 220
5. 720
7. 125, 970
9. 72

REVIEW EXERCISE

1. 5
3. 9
5. 3
7. 0.75
9. 0.65
11. (a) 0.26   **(b)** 0.33
13. (a) 0.75   **(b)** 0.85
15. (a) 2   **(b)** 3
19. 0.139
21. 0.75
23. 0.20
25. (a) 0.60   **(b)** 0.40   **(c)** 0.125
27. (a) 0.0045   **(b)** 0.006   **(c)** 0.012
29. (a) 0.479   **(b)** 0.011

SELF-TEST

1. (a) (i) 0.475   **(ii)** 0.400   **(iii)** 0.075   **(iv)** 0.70
(b) (i) 0.095   **(ii)** 0.158   **(iii)** 0.125
2. (b) 16   **(c)** 7   **(d) (i)** 0.125   **(ii)** 0.438
(iii) 0.188
3. (b) 40   **(c)** 50
4. (a) 56   **(b)** 28
5. (a) 0.01   **(b)** 0.03   **(c)** 0.088

CHAPTER 8

EXERCISE 8.2

1. $1,310,000
3. (a) 2.2   **(b)** variance = 1.56, standard deviation
= 1.249

EXERCISE 8.3

1. 0.1563
3. (a) 0.0055   **(b)** 0.9526
5. mean = 1.6, standard deviation = 0.9798

REVIEW EXERCISE

1. $209 800
3. expected average sales increase = 16.75%,
standard deviation = 5.31%

5. (a) 0.7164 **(b)** 0.2836

7. (a) 0.0312 **(b)** 5 **(c)** 2.1794

9. (a) 0.0105 **(b)** 0.0676 **(c)** 0.0

SELF-TEST

1. (a) 8.35 mm **(b)** 3.087 mm

1. (a) 0.0183 **(b)** 0.2047 **(c)** 0.2284

3. mean = 7.08, standard deviation = 1.7038

CHAPTER 9

EXERCISE 9.2

1. (a) 0.6826 **(b)** 0.1574 **(c)** 0.8185 **(d)** 0.1587
(e) 0.0013

EXERCISE 9.3

1. (a) equal **(b)** mean **(c)** standard deviation
(d) 0.500

3. (a) −1.0 **(b)** −1.5 **(c)** 2.0 **(d)** 0.1 **(e)** 2.83
(f) −2.40

5. (a) 0.4332 **(b)** 0.9987 **(c)** 0.1574 **(d)** 0.0456
(e) 0.8400 **(f)** 0.9903

EXERCISE 9.4

1. (a) $36 **(b)** $72 **(c)** 99.74%, 2394, $36, $72
(d) $66, 97.72%, 2345, $66

3. (a) 4.076 **(b)** 3.830 **(c)** 4.028 **(d)** 3.406
(e) 3.216 **(f)** 3.1 and 3.9

EXERCISE 9.5

1. (a) 0.73% **(b)** 5.16% **(c)** 39.74%

3. (a) 0.62% **(b)** 15.87% **(c)** 27.81%

REVIEW EXERCISE

1. (a) 0.2422 **(b)** 0.4656 **(c)** 0.4988 **(d)** 0.4901

3. (a) 0.0569 **(b)** 0.0516 **(c)** 0.3497 **(d)** 0.0096
(e) 0.8990

5. (a) 2.33 **(b)** −1.645 **(c)** −2.05 **(d)** 1.28
(e) ±1.96 **(f)** ±2.33

7. (a) 0.50 **(b)** −0.17 **(c)** −1.67 **(d)** 2.50
(e) 3.33 **(f)** −3.00

9. $52 560

11. 8164 shoes

13. 93.94%

SELF-TEST

1. Company ST

2. 192

3. 477

4. 304 and 496

5. 465 and 515

6. 529 invoices

7. 0.15%

8. (a) 0.89 **(b)** 9.18% **(c)** 191 orders

9. (a) 0.19% **(b)** 98.46%

CHAPTER 10

EXERCISE 10.2

1. (a) mean = 6, standard deviation = 1.7889
(b) $\overline{ABC}$ = 4.6667 $\quad \overline{ABD}$ = 5.3333 $\quad \overline{ABE}$ = 6.0000
$\overline{BCD}$ = 5.6667 $\quad \overline{BCE}$ = 6.3333 $\quad \overline{CDE}$ = 7.0000
$\overline{ACD}$ = 5.3333 $\quad \overline{ACE}$ = 6.0000 $\quad \overline{ADE}$ = 6.6667
$\overline{BDE}$ = 7.0000
(c) $\mu_{\overline{x}}$ = 6 **(d)** $\sigma_{\overline{x}}$ = 0.7303 **(e)** 0.7303

EXERCISE 10.3

1. (a) 6.75 **(b)** 820.25 **(c)** 779.75

3. (a) 400 **(b)** 6.944 **(c)** normal, population
normal

5. (a) 7.5 **(b)** 2.8

7. (a) $200 **(b)** $2.5 **(c)** n = 64 **(d)** $195 and
$205 **(e)** 95.44%

EXERCISE 10.4

1. 90.1%

3. (a) 350 **(b)** 40 **(c)** 0.85

5. (a) 0.0816 **(b)** 0.0657 **(c)** 0.0453 **(d)** 0.02

7. 92.16%

REVIEW EXERCISE

1. (a) 0.71 and 0.79 **(b)** 27.2602 and 28.7398

3. (a) 32.38   **(b)** 349.84

5. 0.19%

7. (a) 0.44%   **(b)** 99.68%

9. (a) 638.73 hours   **(b)** 13 949.3 and 16 050.7 hours   **(c)** 2.07%

11. (a) $21 146.03   **(b)** 2   **(c)** 0.0535   **(d)** 0.91%
(e) 14.17%

13. 12.3%

15. (a) 69.85   **(b)** 11.12%

17. 36.32%

19. 0.68 and 0.88

21. 1.16%

SELF-TEST

1. (a) $\mu_{\bar{x}}$ **(b)** σ_p **(c)** μ_p **(d)** $\sigma_{\bar{x}}$

2. 97.59%

3. (a) 1.5513
(b) 60.1 and 64.9 bpm

4. (a) $54 804.55 and higher   **(b)** $46 616.91

5. (a) 2702   **(b)** 52.68%   **(c)** 95.46%   **(d)** Part
b references households whereas part c references
sample means.

6. 66.87%

7. 99.81%

8. 70

CHAPTER 11

EXERCISE 11.2

1. point estimate, interval estimation, estimator,
estimate

1. (a) $163.72   **(b)** normal, population is normal

5. 51.60%

7. 99.58%

EXERCISE 11.3

1. $207 056.23 and $212 943.68

3. (a) 2000   **(b)** 36   **(c)** 41   **(d)** 2   **(e)** 0.3304
(f) 98%

5. 121.2 and 128.8 L

7. 414.0 and 424.7 calls per day

EXERCISE 11.4

1. 0.211 and 0.429

3. 0.514 and 0.686

5. (a) 0.72   **(b)** 0.0224   **(c)** 0.243 and 0.317

EXERCISE 11.5

1. (a) 1521   **(b)** 256   **(c)** 145   **(d)** 553   **(e)** 560

3. 877

5. (a) 3394   **(b)** 2851

EXERCISE 11.6

1. (a) 2.797   **(b)** 1.328   **(c)** 2.132   **(d)** 2.764
(e) 2.045

3. 7.98 and 8.02 mm

5. 16.5 and 19.3 kg

REVIEW EXERCISE

1. (a) 294.1 and 305.9   **(b)** 30.34 and 41.66
(c) 1299.6 and 1300.4   **(d)** 15.57 and 16.43

3. (a) 64   **(b)** 98%   **(c)** 67.20   **(d)** 8.4
(e) 19.57

5. 374.3 and 375.7 g

7. 20.16 and 21.54 months

9. (a) 0.54 and 0.71 kg   **(b)** accept shipment
(c) 0.59 and 0.66 kg

11. $15.06 and $17.78

13. (a) 1.729   **(b)** 1.345   **(c)** 2.306   **(d)** 2.787

15. 3.19 and 6.01

17. (a) sample mean = 9.9, unbiased standard
deviation = 0.4452   **(b)** 9.5 and 10.3 minutes

19. 150.8 and 165.2 km/h

21. 0.276 and 0.404

23. 0.4454 and 0.7146

25. 403

27. 327

29. (a) 212   **(b)** standard error increases 18.96%,
cost savings of $10,120

31. 241

SELF-TEST

1. (a) 8.6 and 11.4　**(b)** 8.7 and 11.3

2. (a) $32.00 and $33.70　**(b)** 0.2339 and 0.4661

3. 5.9 and 7.7 pieces

4. 0.0349 and 0.1611

5. (a) $\bar{x} = 5.2333$, $s = 0.1506$　**(b)** 5.03 and 5.44

6. 1537

7. 35

CHAPTER 12

EXERCISE 12.2

1. (a) (ii) $H_0 : \mu \leq 10\%$, $H_a : \mu > 10\%$
(iii) one-tail　**(b) (ii)** $H_0 : \mu = 1\,000\,000$
characters, $H_a : \mu \neq 1\,000\,000$ characters　**(iii)**
two-tail　**(c) (ii)** $H_0 : \mu >$ current level, $H_a : \mu <$
current level　**(iii)** one-tail

EXERCISE 12.3

1. $(|t_{\bar{x}}| = 0.3689) < (t_c = 2.145)$, accept H_0

3. (a) accept H_0　**(b)** accept H_0　**(c)** reject H_0
(d) reject H_0

5. $(|t_{\bar{x}}| = 7.134) > (t_c = 3.499)$, reject H_0

EXERCISE 12.4

1. $(|z_p| = 1.369) < (z_c = 2.33)$, accept H_0

3. $(|z_p| = 2.098) > (z_c = 1.645)$, reject H_0

EXERCISE 12.5

3. (a) $H_0 : \mu = 6.3\%$, $H_a : \mu \neq 6.3\%$
(b) p value $= 1.8\%$

5. (a) p value $= 14.2\%$　**(b)** p value $> \alpha$, accept
H_0

REVIEW EXERCISE

1. $H_0 : \mu \geq 5\,mm$, $H_a : \mu < 5$ mm

3. (a) $H_0 : \mu = 56$, $H_a : \mu < 56$　**(b)** one-tail

5. (a) two-tail　**(b)** one-tail

7. $(|z_{\bar{x}}| = 3.5) > (z_c = 1.96)$, reject H_0

9. $(|z_{\bar{x}}| = 0.866) < (z_c = 2.33)$, accept H_0

11. $(|z_{\bar{x}}| = 6.936) > (z_c = 2.33)$, reject H_0

13. $(|t_{\bar{x}}| = 2.118) < (t_c = 3.747)$, accept H_0

15. (a) $H_o : \pi = 0.364$, $Ha : \pi > 0.364$
(b) $(|z_p| = 1.933) < (z_c = 2.33)$, accept H_0

17. $(|z_p| = 0.81) < (z_c = 2.575)$, accept H_0

19. $(|z_p| = 2.454) > (z_c = 2.05)$, reject H_0

21. (a) p value $< 1\%$, one-tail　**(b)** p value $< \alpha$,
reject H_0

23. (a) p value $= 0.82\%$, one tail

25. (a) $2.5\% \leq p$ value $\leq 5\%$, one-tail

27. (a) p value $= 3.3\%$, one-tail　**(b)** p value $> \alpha$,
accept the shipment

SELF-TEST

1. $H_0 : \mu = \$47\,000$, $H_a : \mu < \$47\,000$
(b) one-tail
(c) $(|z_{\bar{x}}| = 1.25) < (z_c = 2.33)$, accept H_0

2. $H_0 : \mu = 10\%1$, $Ha = \mu < 10\%$

3. (a) $H_0 : \mu = 80$, $H_a : \mu < 80$
(b) one-tail　**(c)** $\alpha = 0.02$　**(d)** normal distribution
(e) $z_c = 2.05$
(f) accept H_0 if $|z_{\bar{x}}| < 2.05$

4. $(|z_{\bar{x}}| = 1) < (z_c = 2.33)$, accept H_0

5. $(|z_{\bar{x}}|2.33) > (z_c = 1.645)$, reject H_0

6. (a) $H_0 : \mu = 28$, $H_a : \mu < 28$
(b) $(|z_{\bar{x}}| = 16.73) > (z_c = 2.33)$, reject H_0

7. $(z_p = 1.425) < (z_c = 1.645)$, accept H_0

8. (a) mean $= 3.9$, standard deviation $= 0.1333$
(b) t distribution, small sample
(c) $(|t_{\bar{x}}| = 2.372) > (tc = 1.833)$, reject H_0

9. (a) $H_0 : \mu = 7.2$, $H_a : \mu < 7.2$
(b) p value $= 11.7\%$
(c) p value $> \alpha$, accept H_0

10. $(|z_p| = 0.746) < (z_c = 1.645)$, accept H_0

CHAPTER 13

EXERCISE 13.2

5. (a) $y_p = 0.2413 + 2.819\,x$ (b) 7.29

7. (c) $y_p = -18.2736 + 1.2755\,x$ (f) (i) 58   (ii) 97

EXERCISE 13.3

1. 4.3789

EXERCISE 13.4

1. (a) 64 to 66   (b) 61 to 69

3. (a) $y_p = 2.940 - 0.042\,x$ (c) 0.0274   (d) \$2.30
(e) \$2.28 and \$2.33   (f) \$2.23 and \$2.38

EXERCISE 13.5

1. (a) Housing starts   (b) Advertising

3. Graph C

5. −1

7. 0.9

9. −0.4

11. (a) $y_p = 62.830 + 0.0722\,x$ (c) 0.0284

REVIEW EXERCISE

1. (a) (i) $y_p = 2.997 + 0.5006\,x$
(ii) $y_p = 3.001 + 0.500\,x$ (iii) $y_p = 3.0024 + 0.4997\,x$
(iv) $y_p = 3.0017 + 0.4999\,x$ (c) (i) 0.816   (ii) 0.816
(iii) 0.816   (iv) 0.816

3. (a) $y_p = 3.863 + 10.394\,x$ (b) 0.9579   (c) 69
(d) 65 to 73   (e) 57 to 81

5. (a) $y_p = 3278.4 - 23196\,x$ (b) $r^2 = 0.985$,
$r = 0.9922$ (c) 966 to 1185; 819 to 1332

7. (b) $y_p = 21.706 - 0.1\,x$ (d) −0.916   (e) 2.3093

9. (a) $y_p = 0.2527 + 0.7590\,x$ (b) $r^2 = 0.9907$,
$r = 0.9953$ (c) \$6.32

SELF-TEST

1. (a) direct   (b) 74 269 employees

2. (a) $y_p = -13.2 + 7.6\,x$

3. (a) inverse   (b) negative   (c) $r^2 = 0.9764$,
$1 - r^2 = 0.0236$, $r = 0.988$

4. (a) $y_p = -1.361 + 0.8719\,x$ (b) 0.9338
(c) 10.846 million litres   (d) 10.242 and 11.450
(e) 8.891 and 12.801

Index